Musical Bodies, Musical N

Musical Bodies, Musical Minds

Enactive Cognitive Science and the Meaning of Human Musicality

Dylan van der Schyff, Andrea Schiavio, and David J. Elliott

The MIT Press
Cambridge, Massachusetts
London, England

© 2022 Massachusetts Institute of Technology

This work is subject to a Creative Commons CC-BY-NC-ND license. Subject to such license, all rights are reserved.

The MIT Press would like to thank the anonymous peer reviewers who provided comments on drafts of this book. The generous work of academic experts is essential for establishing the authority and quality of our publications. We acknowledge with gratitude the contributions of these otherwise uncredited readers.

This book was set in Stone Serif and Stone Sans by Westchester Publishing Services. Printed and bound in the United States of America.

Library of Congress Cataloging-in-Publication Data

Names: Van der Schyff, Dylan, 1970– author. | Schiavio, Andrea, author. | Elliott, David J. (David James), 1948–
Title: Musical bodies, musical minds : enactive cognitive science and the meaning of human musicality / Dylan van der Schyff, Andrea Schiavio, and David J. Elliott.
Description: Cambridge : The MIT Press, 2022. | Includes bibliographical references and index.
Identifiers: LCCN 2021045986 | ISBN 9780262045223 (paperback)
Subjects: LCSH: Music—Psychological aspects. | Cognitive psychology. | Cognition.
Classification: LCC ML3830 .V36 2022 | DDC 781.1/1—dc23
LC record available at https://lccn.loc.gov/2021045986

10 9 8 7 6 5 4 3 2 1

Contents

Acknowledgments vii

1 Getting Situated 1
2 Basic Principles of Enactive Cognitive Science 25
3 Music and Consciousness 51
4 Phenomenology and the Musical Body 69
5 Music and Emotion 91
6 The Empathic Connection 109
7 The Evolution of the Musical Mind 131
8 Teleomusicality 155
9 Creative Musical Bodies 169
10 Praxis 189

Notes 213
References 237
Index 301

Acknowledgments

The collection of ideas, arguments, and insights contained in *Musical Bodies, Musical Minds* is the result of almost a decade's collaboration between its three authors. The book reflects our backgrounds as musical performers and music educators, as well as our interests as scholars. As such, *Musical Bodies, Musical Minds* develops a perspective on human musicality that integrates knowledge from across a range of domains, including the cognitive and biological sciences, developmental studies, pedagogical theory, affective science, philosophical traditions, various branches of music research, and more. In line with this, we hope that *Musical Bodies, Musical Minds* will contribute to the interdisciplinary orientation that characterizes current musicology and especially to scholarship that explores the "embodied" and "ecological" dimensions of musical perception, cognition, and practice. This research area has produced a number of inspiring books, including Eric Clarke's *Ways of Listening* (2005), David Borgo's *Sync or Swarm* (2005), Marc Leman's *Embodied Music Cognition* (2007), Arnie Cox's *Music and Embodied Cognition* (2016), Jonathan De Souza's *Music at Hand* (2017), Simon Høffding's *A Phenomenology of Musical Absorption* (2018), Mariuz Kozak's *Enacting Musical Time* (2019), and Mark Reybrouck's *Musical Sense-Making* (2020). These texts connect in various ways with the account we offer in *Musical Bodies, Musical Minds*. However, to our knowledge, *Musical Bodies, Musical Minds* is the first monograph fully dedicated to developing a comprehensive enactive/4E view of human musicality.

In addition to the authors just mentioned, we would also like to acknowledge that many of the chapters that comprise *Musical Bodies, Musical Minds* began as research articles, some of which involved additional collaborators who contributed in important ways to the ideas presented in this book.

Chapter 5, "Music and Emotion," is based on an article authored by Schiavio and van der Schyff in collaboration with Julian Cespedes-Guevara and Mark Reybrouck (Schiavio et al., 2017). We thank Julian and Mark for their contribution to that article and for their suggestions on the present chapter. Thanks also to Mark for his comments on chapter 1. Chapter 6, "The Empathic Connection," is based on an article by van der Schyff and Joel Krueger (2019). Thanks to Joel for reading and commenting on this chapter. Chapter 8, "Teleomusicality," develops sections that originally appeared in an article coauthored by Schiavio, van der Schyff, Silke Kruse-Weber, and Renee Timmers (2017). Thanks to Renee and Silke for reading and commenting on this revised version. Chapter 9, "Creative Musical Bodies," is based on an article written with Valerio Velardo, Ashley Walton, and Anthony Chemero (van der Schyff et al., 2018). We thank Valerio, Ashley, and Anthony for their contributions to the ideas presented in this chapter.

In the first endnote of each chapter, we indicate our previously published articles that contribute to the discussion in *Musical Bodies, Musical Minds*. We thank the journal editors and the many anonymous reviewers from whom we received important critical feedback that helped us sharpen our arguments. We would also like to express our appreciation to Ezequiel Di Paolo, who commented on chapter 2; to Tiger Roholt, who commented on chapters 1 and 3; to Luca Barlassina, who commented on an early version of chapter 3; to Tom Froese, who commented on a previous draft of chapter 7; and to Mathias Benedek, who commented on chapter 9.

A huge thank you must also go to Eric Clarke, who facilitated and mentored van der Schyff's 2017–2019 postdoctoral fellowship hosted by the University of Oxford. Professor Clarke read and made detailed comments on drafts of the opening chapters. He also engaged in numerous informal discussions that helped to clarify many of the themes developed throughout the book. van der Schyff would like to acknowledge the Social Science and Humanities Research Council of Canada (SSHRC) for their financial support of this fellowship. He would also like to thank Professors Nikki Dibben and Susan O'Neill for their generous supervision and support. Schiavio would like to thank Professor Richard Parncutt and his colleagues at the Centre for Systematic Musicology of the University of Graz for extended conversations about and constructive feedback on many of the themes developed in the book. He is also grateful to many friends and colleagues with whom he has recently collaborated, including Mathias Benedek, Michele Biasutti,

Acknowledgments

Nikki Moran, Kevin Ryan, Jan Stupacher, Renee Timmers, Jonna Vuoskoski, and many others. We thank Leigh van der Schyff for her lovely drawings in chapter 2. We would also like to thank Philip Laughlin and the team at MIT Press for their patience and expert assistance. And we are grateful to the anonymous reviewers who read and made useful comments on the final drafts of the manuscript. Last, we express our gratitude and love to our families and friends who supported us during the writing of this book.

1 Getting Situated

Every known human society, past and present, engages in activities that can be described as "musical." These activities involve singing, drumming, dancing, listening, and a range of other expressive behaviors and coordinated actions that take place in lived contexts associated with work, play, worship, ritual, entertainment, and more (Cross, 2012; Elliott, 1995; Small, 1998; Tomlinson, 2015). Songs, melodies, rhythms, and musical styles are used to reinforce feelings of belonging and identity in social groups, cultures, and communities (Blacking, 1995). Our day-to-day experiences of the world also involve all kinds of sounds, movements, and relationships that we can perceive as musical—from the noises and rhythms of the city to the sounds and movements of nature (Schafer, 1994). Music affords forms of emotional regulation and nonverbal communication, and is often employed as a therapeutic aid in both clinical and everyday contexts (DeNora, 2000; Sacks, 2007). Since the earliest days of our species, we have used music to enhance and give meaning to the various places we inhabit, from singing and dancing around a campfire or in a place of worship to the use of personal listening devices while exercising, or when traveling on buses, trains, and planes (Bull, 2000; Fritz et al., 2013; Mithen, 2005; Skånland, 2013). Music plays an important role in how we develop cultural understandings; it drives the complex and powerfully transformative experiences associated with different (e.g., religious) ceremonies, symphonic performances, or rock concerts. But it also involves more mundane phenomena we may hardly notice—like tapping one's foot to the beat of a tune on the radio and the subtle shift in mood that may occur as one does so.

In all, the activities, perceptions, and meanings associated with the word "music" span an impressive range of what humans experience. Whatever else we are, we are also musical. Human minds are therefore also musical minds. But why and how is this so? What is meant by "musical mind"? Why did musicality[1] evolve in the human phenotype? What is the relevance of music for human survival and well-being? How does music bring meaning to our social and cultural environments? And how might our responses to such questions inform how we think about, use, and experience music in contexts such as education, therapy, performance, composition, musicological research, and everyday life?

Over the last few decades, researchers have shown a growing interest in exploring these kinds of questions. As a result, music scholarship has developed into a fascinating interdisciplinary field that looks beyond its traditional interests in historical and compositional analysis, drawing on knowledge across the sciences and humanities to examine the psychological, biological, developmental, social, and cultural dimensions of musical experience (see Parncutt, 2007). Additionally, current music research is offering new insights into the nature of the human mind more generally and is therefore becoming an important area of interest for researchers working in the cognitive sciences. This book aims to contribute to this interdisciplinary orientation by offering a novel approach to the musical mind that is based on the enactive approach to cognition (Thompson, 2007; Varela et al., 1991). Where traditional models of mind often equate cognition with information processing confined to the brain (the "mind-as-computer" model), the enactive perspective traces a deep continuity between biological, corporeal, and mental processes, highlighting the central role of the active, situated, living body for cognition. As such, it closely integrates body, brain, and the environment as different aspects of the same evolving system, offering a holistic and nonreductive view of what cognition entails.

In the chapters that follow, we explore questions like those posed previously by extending the enactive perspective into areas such as musical consciousness, musical emotion and empathy, music and human evolution, musical development in infancy, musical creativity, and music pedagogy. In doing so, we develop connections between a diverse array of ideas and knowledge drawn from neuroscience, theoretical biology, psychology, affective science, archeology, developmental studies, social cognition, education, different philosophical traditions, and more. We should note at the outset that some of the enactivist concepts we develop may seem radical to some readers—for example, the ideas that cognition is not grounded in representational processes or that minds extend beyond the brains and bodies of individuals. While we will support such claims in various ways, we should make it clear that the chief aim of this book is not to defend the enactive orientation from its detractors. Prominent philosophers and cognitive scientists have already provided eloquent responses to critics, and we see no need to rehearse these arguments again in detail.[2] Instead, our focus will be on showing how an enactive perspective can provide ways of thinking about the nature and meaning of human musicality that have useful applications for theory, research, and musical practice and for how we conceive of the meaning(s) of music in everyday life.

In a nutshell, our take is that in being *enactive*, (musical) cognition is necessarily *embodied*, *embedded*, and *extended*. By this light, musical minds are explored as active musical bodies that are embedded within, and that extend into, the social, material, and cultural ecologies they inhabit and actively shape or "enact." This orientation—often referred to as the "4E" approach—is currently being developed, discussed, and applied across a range of research areas (Newen et al., 2018). We should note that, because of its interdisciplinary reach, the 4E approach is not always understood in precisely the same way. There are different varieties of extended (and externalist) proposals (Hurley, 2010; Menary, 2010a); competing approaches to embodiment and embeddedness have been offered (Gallagher, 2011); and differing conceptions of enactivism itself populate the discourse (see, e.g., Hutto & Myin, 2012). Nevertheless, there are good reasons to consider an approach to musical cognition based on enactivist and 4E principles. Indeed, because this orientation considers the mind as something that goes beyond the brain, it provides levels of description that are not available from other perspectives (Barrett, 2011). In chapter 2, we outline the main principles of enactive cognitive science and discuss its connections to ecological psychology, dynamical systems theory, and theoretical biology. Here, we explore in more detail the "embodied," "embedded," and "extended" dimensions that are fundamental to enactivist thinking. In doing so, we will attempt to alleviate some of the concerns just mentioned.

In this opening chapter, we provide a few historical perspectives on musical cognition in conjunction with concurrent trends in philosophy of mind and cognitive science. This will help to situate subsequent chapters by highlighting the possible origins of certain assumptions that have tended to guide our understanding of what mind and music entail. It will also aid in tracing some antecedents to the embodied and "biological" perspective we develop further on in association with the enactivist approach. Before we do this, we should note that the discussion that follows is not intended as a history of philosophy and music. This would go well beyond the scope of the present book. Instead, we offer a selective view of what is in fact a wide swath of writings that originate in antiquity; we do this from a certain perspective and with specific theoretical goals in mind. The aim here is simply to introduce some of the ideas, themes, and critical concerns that motivate the approach we develop in subsequent chapters.

Perspectives on the Musical Mind

Humans have been fascinated with music's relationship to thought, feeling, and action since antiquity and probably well into prehistoric times. The archaeological record shows evidence of human musical activity dating back to the Upper Paleolithic period—in the form of bone flutes that were fabricated 40,000 years ago (figure 1.1). But it seems very likely that music would have been present in human life long before this. Indeed, researchers have speculated that music-like behaviors played an important role in the lives of our prehuman ancestors—for example, for social bonding, child rearing, and more generally because it allowed for embodied and

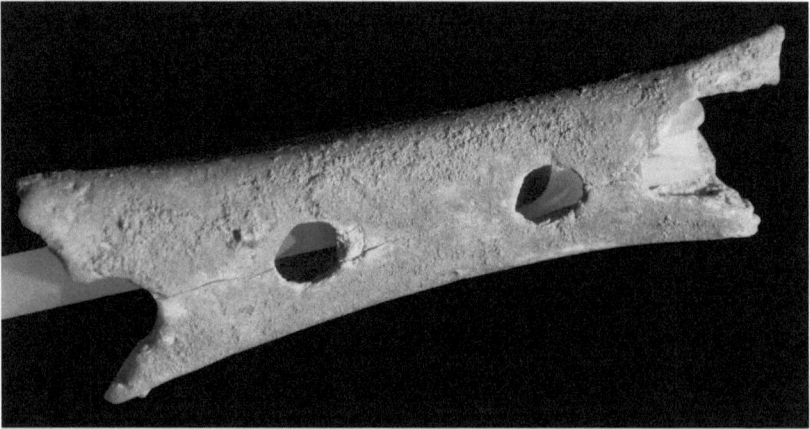

Figure 1.1
A surviving fragment of the Divje Babe Flute (sometimes referred to as the "Neanderthal Flute"). It was discovered in 1995 at the Divje Babe archeological park in northwestern Slovenia. It is made from the femur of a juvenile cave bear and is dated at 43,000 (± 700) years old. (Courtesy of the National Museum of Slovenia, Ljubljana. Photo license: CC BY 2.0.)

emotional forms of communication in prelinguistic societies (Dissanayake, 1995; Mithen, 2005; Tomlinson, 2015).

Some of the earliest writings involving music are found in Ancient Greek texts (Barker, 1989). Here, music is inspired by a divine Olympian source—the Muses, the daughters of Zeus. The influence of music, however, spans the world of plants and animals, the domain of humans and heroes, as well as the eternal realm of the gods. Consider, for example, the Homeric Hymn to Hermes, in which the young god fashions the first lyre (the *chelys*) using a tortoise shell as a resonator and the entrails of cattle as strings (figure 1.2).

Figure 1.2
This image comes from a fifth-century BCE drinking cup found in the tomb of a Delphic priest. It depicts Apollo pouring a libation with his right hand while holding the tortoise shell chelys lyre in his left hand. (Courtesy of the Archaeological Museum of Delphi, Inv. 8140, room XII. Photo license: CC BY-SA 2.0 DE.)

Hermes uses the instrument to soothe the anger of his half-bother Apollo, who becomes so enamored with it that he gives a golden lyre to his son, Orpheus. Orpheus is trained by the Muses and becomes the greatest musician and poet, capable of charming gods, humans, animals, plants, and even rocks with his beautiful singing and playing.

Notably, these ancient stories reveal a conception of music that traverses an *ontologically extended psyche*—one that includes the biological, the material-technological (the fashioning of tools and instruments), and the spiritual or divine. Orpheus himself was a quasi-sacred figure for the Greeks, conceived of variably as a god, a human, or a demigod hero; and the Orphic cults that emerged in the Archaic period (700–480 BCE) exerted considerable influence over Greek society until well into the Hellenistic era (323–31 BCE). In all, the Greeks saw music (along with dance and poetry) as a driving force in the emergence of their people and culture, and as essential to their relationship with the gods who guided their fates. Music was therefore omnipresent in their everyday lives—in their athletic events, religious ceremonies, medicine, theatre, and leisure activities (Anderson, 1994; Bundrick, 2005; Comotti, 1989).

In the sixth century BCE, Pythagoras (570–495 BCE), a reformer of the Orphic tradition, initiated the study of harmonics based on his calculations of vibrating strings and the overtones they produce (Barker, 2007). This led him to discover musical-acoustical laws based on mathematical principles and to instantiate a musical ethos for the use of intervals and modes that would align the human mind with eternal truth (the "Music of the Spheres"). This Pythagorean rational mysticism had a profound influence on the thought of Plato—a pivotal figure in the history of philosophy. Most famously, Plato abstracted the realm of "mind" from the changing world of biological and material reality. He aligned it instead with the eternal and unchanging realm of the "forms"—a transcendental, quasi-mathematical reality that is (partially) revealed in the material world by geometry. Importantly, Plato also recognized the power of music, extolling both its virtues and dangers for the soul and society.[3] He thus claimed that our musical lives should not be guided by our senses—by what appeals to the ear—but by the mathematical principles that govern the rules of musical relationships. It is an understanding of these laws, and not the bodily sense of what is pleasurable, that guides virtuous music making (Anderson, 1966).

Plato's abstract conceptions of mind and ethics were rejected by his most famous student, Aristotle. As a proto-biologist, Aristotle argued that mind and knowledge do not originate in some ideal non-material realm, but here in the world as we experience it—most notably in the transformations of matter and form as things and beings come into and go out of existence. This view is worth expanding upon briefly because it prefigures certain key insights associated with the enactive orientation.

One way to understand Aristotle's perspective is to first consider the distinction he makes between the ontological status of natural and artificial entities, or the "living" and the "made" (in *Physics II* lines 192b8–32; see McKeon, 2001). For him, living beings are not created as such. Instead, their coming-into-existence involves an innate principle of change special to the living organism itself—living creatures are essentially *self-moving* and *self-making* entities that actively "reach out" to the world as they strive to maintain a flourishing life.[4] For example, we see the egg become a tadpole, and the tadpole become a frog; we observe the seed become a sapling, and the sapling become a tree. Each has within it its own self-making principle (*archê*), and each moves seamlessly toward its actuality or purpose (*télos*) (i.e., biological flourishing). A made entity, by contrast, carries no such principle of change or movement within it. A table, a house, a ship, or a computer has its principle of motion and being outside of itself—humans bring these objects into existence through craft (*technê*) and give them the attributes that make them what they are, both in the physical and abstract sense. The purpose, form, and coming-into-being of artificial objects are not self-actualizing; they are wholly dependent on human need, desire, and *technê*. Put another way, artificial entities have their ontological footing outside of themselves: trees beget trees, but tables do not beget tables. In line with this, Aristotle adopts a tripartite theory of mind that is continuous with processes of organic movement, which include growth, bodily action, reproduction, and thought. In various stages of biological complexity, then, the manifestations of mind (*psuchê*) emerge respectively as nutritive (e.g., plants), locomotive (insects and animals), and, indeed, "rational" (human). Accordingly, Aristotle not only asserts that natural entities—not just humans—are intrinsically purposeful but also that all living things have minds.[5]

As we will see, there is a sense in which these contrasting notions of mind—on one hand as a discrete, abstract, and rational category (Plato) or, on the other, as a biologically continuous phenomenon (Aristotle)—are echoed by current debates between traditional information-processing or "computational" approaches to cognition and the embodied, biological, or "life-based" orientation associated with the enactive perspective. For now, it is instructive to consider the view of musical perception provided by Aristotle's pupil, Aristoxenus, who departs from Pythagorean-Platonic musical theory in several ways. Most importantly, Aristoxenus argues that musical understandings emerge first from the senses: "The nature of melody is best discovered by the perception of sense . . . there is no other way of arriving at the knowledge of music . . . for just as it is not necessary for him who writes an Iambic to attend to the arithmetical proportions of the feet of which it is composed, so it is not necessary for him who writes a Phrygian song to attend to the ratios of the sounds proper thereto" (see Hawkins, 1868, p. 67). He also notes: "we must bear in mind that musical cognition implies the simultaneous cognition of a permanent and of a changeable element, and that this applies without limitation or qualification to every branch of music. We shall be sure to miss the truth unless we place the supreme and ultimate [(the ruling principle or *archê*)], not in the thing determined, but in the *activity that determines*" (see Stunk 1950, p. 31; also Bamberger, 2006). Aristoxenus appears to claim that, although aspects of music can be described in terms of abstract mathematical ratios, music cannot be reduced to them, nor are these aspects required for musical phenomena to be perceived, understood, and judged. Rather, it is the situated, active, and sensory aspects of music-making that form the basis of musical experience and knowledge.

It was, however, the abstract Pythagorean-Platonic model of music that remained most influential throughout the early Christian and the medieval periods, where it was adopted to support a devotional and increasingly rule-driven conception of music that mediated between the earthy domain and a heavenly God. It was also during this time that musical notation began to emerge as a central aspect of Western music. Among other things, notation provided a way of fixing musical practices within the liturgy, ensuring that they adhered to the ideals of the church. By the eleventh century, three important developments had emerged that paved the way for the period of Common Practice (roughly 1650–1900), which encompasses the evolution of Baroque, Classical, and Romantic styles. These include the growing use of polyphony, the development of principles of musical form, and the establishment of the composition and the composer as the central components in the production of Western art music.

In the Renaissance, Western thinkers began to return to the original Greek philosophical and musicological sources. Accounts from this period become less concerned with adapting ancient thought to meet religious ideals and show more interest in understanding music's ability to transform biological and psychological states—reflecting a growing fascination with the discoveries of science. As musicologist Bernardo Fantini (2014) notes, interesting examples of this can be found in musicological work of the late Renaissance and early Baroque periods, which demonstrate commonalities in "forms of thought between music and science" (p. 264). More specifically, this entails the development of two models of life that played influential roles in the music theory of the seventeenth and eighteenth centuries. The first model involves the Baroque "theory of fibers," which relates the power of music not to abstract mathematical relationships, but to vibration, movement, muscular (fibral) actions, and bodily resonances. This is paralleled aesthetically in the autonomous yet intertwining contrapuntal lines of a Bach fugue. The second model entails the eighteenth-century interest in elementary organisms or "cells." This is paralleled in how music was understood and constructed in the Classical period, where the focus shifted to distinct formal components (e.g., sonata form). These developments reveal an interesting alignment between how musical and biological processes were understood in these periods—the appositional, interchangeable, and sinuous approach of the Baroque biological-musical view came to be replaced with a cellular, hierarchical, and autonomous perspective in the Classical era.

This orientation resonates with the enactive perspective on the musical mind, which draws on current trends in theoretical biology to examine the deep relationship between music, mind, and the fundamental processes of life. We should add, however, that although embodied-biological approaches are currently (re)gaining influence in Western musicology and the sciences of mind, they were marginalized in both domains for most of the nineteenth and twentieth centuries. This is due, in part, to prevailing philosophical notions of knowledge, aesthetics, and mind received from Enlightenment thinking. An important influence on these ways of thinking is found in the mind-body "dualism" associated with the seventeenth-century philosopher René Descartes, who (like Plato) argued that the mind and body (i.e., the living body and the "extended" material world of things and objects more generally) were in fact two distinct ontological categories or "substances." He claimed that the former, nonmaterial domain must be the proper locus of knowledge and thought.

This notion of a detached rational mind had a significant effect on emerging aesthetic theories in the eighteenth and nineteenth centuries. Notably, the philosopher Immanuel Kant rejected the value of emotion and sensuous qualities for aesthetic experience, emphasizing instead the purely formal properties of music and art more generally. To be clear, the Kantian claim is not that musical experiences do not involve bodily engagements, emotional episodes, and experiences of movement and space; such an argument would be in direct violation of experience. But, by this light, bodily and emotional episodes—or feelings—are subjective, nonrational dimensions that have no bearing on an understanding of "beauty" from a philosophical point of view. This perspective is expressed by nineteenth-century music critic and theorist Eduard Hanslick (1891), who writes that "theorists who ground the beautiful in music on the feelings it excites build upon a most uncertain foundation, scientifically speaking . . . and can therefore, at best, only indulge in speculations and flights of fancy. An interpretation of music based on the feelings cannot be acceptable either to art or science" (p. 120). Following Kant, Hanslick sees the aesthetic experience of music as an entirely intellectual phenomenon: musical beauty emerges from formal characteristics of the musical work and exists independently of the listener; any emotional response is merely a by-product of the listener's subjective experience. It should be noted that several writers in the first half of the twentieth century became interested in addressing the question of emotional expression for musical aesthetics. But here too, the locus of music's (emotionally) expressive content was generally understood to be found in the formal qualities of the composed work itself—whatever music expresses is "in the music" a priori (Kivy, 1990, 2002; see also Meyer, 1956).

We will consider these perspectives in more detail in chapters 5 and 6, where we explore musical emotions and empathy. For now, we can say that, although the strong versions of "aesthetic formalism" associated with Kant and Hanslick are now discussed mostly for their historical relevance, they nevertheless played an important role in driving the more general and long-lasting assumption that a proper aesthetic understanding of a musical work has little to do with the actual lives of individual listeners. What does matter is the possession of the appropriate mental apparatus (and training) to correctly perceive, understand, and reproduce the supposedly objective structural relationships encoded in the score by the composer.[6]

While the composer and the score do continue to hold privileged places in the Western musical discourse, this perspective has been seriously questioned by a new generation of critical musicologists who have traced such attitudes to cultural developments in the late nineteenth and early twentieth centuries (see also Cook, 2006, 2013). These developments include the establishment of the Classical canon and the concert hall as a cultural and entrepreneurial strategy, the "cult" of the elite (white, male, and often dead) composer genius, the emergence of a bureaucratic "culturally administered" bourgeois society, and the mechanical reproduction of printed and recorded music (De Nora, 1986, 2000; Goehr, 1992; Nettl, 1974). These factors are now understood to have played important roles in promoting the notion of the transcendent musical object or "work," as well as the growing commodification of music more generally.[7] In brief, the assumption that the locus of musical expressivity and meaning (whether emotional or intellectual in nature) should be "possessed by" the formal structural relationships of the "music itself" is now increasingly problematized as a historical construction (Bohlman, 1999; Elliott, 1989, 1995; Small, 1998), and many current musicologists offer much broader and more inclusive perspectives on musical ontology and value.

Later, we examine the implications these changing attitudes have had for musical aesthetics and pedagogy—including the so-called "praxial" approach to music education that draws on enactivist theory (Elliott & Silverman, 2015; Silverman, 2012). For the moment, we address an important point raised by musicologist Ian Cross (2010), who reminds us that what we know of music in philosophical and scientific terms "is constrained by a conception of music that is narrowly shaped by historical and cultural notions of what constitutes 'music'" (p. 2). Indeed, the same aesthetic and cultural assumptions just considered also influenced early work in music cognition, where research tended to be focused on measuring psychophysical and cognitive responses to various musical stimuli, with the goal of articulating the rules that govern how the mind extracts and processes the information present in the score and performance thereof (Seashore, 1938; Sloboda, 1985).[8] Here, the basic assumption was that musical cognition is realized by a kind of input-output or stimulus-response schema, whereby objective properties intrinsic to the formal structure of the music (intervallic relationships, harmonic cadences, and so on) cause relevant reactions in listeners. Researchers naturally drew on the accepted scientific theories of day to explain such processes, and by the mid-twentieth century studies were increasingly framed and interpreted largely in terms of post-behaviorist, information-processing models of cognition.[9]

Music and the Computational Mind

Very generally, the information-processing approach conceives of cognition as confined to the individual, and it involves the assumption that all mental processes are essentially computational and representational.[10] The senses receive stimuli from the external world. This information is then transduced into symbols—instantiated by patterns of neural firing—that are manipulated as representations of the sensory data according to the rules (the "language" or "syntax") of the system. This proceeds in a hierarchical manner leading to the production of increasingly complex internal constructs (representations) that generate appropriate behavioral outputs (e.g., experiences, emotions, actions). The main tenets of this orientation are captured in the following passage by Bechtel and colleagues (1998): "To be a cognizer is to possess a system of syntactically structured symbols-in-the-head (mind/brain), which undergo processing that is sensitive to that structure. Cognition, in all its forms, from the simplest perception of a stimulus to the most complex judgment concerning the grammaticality of an utterance consists of manipulating symbols in the head in accord with that syntax. The system of primitive, innate symbols-in-the-head and their syntactic, sentence-like structures is sometimes called 'mentalese'" (pp. 63–64).

To account for the incredible computing capabilities of the human mind, it was also posited that such processing takes place in discrete "modules," each adapted by natural selection to perform a specific cognitive task.[11] Moreover, the processing in these modules is understood to take place at the preconscious or "subpersonal" level, meaning that we have no direct access to the world: all of our experience is located in the brain, which creates virtual, internal representations of relevant aspects the environment.

A consequence of this approach is that the body plays no substantial role in cognition as such: it merely provides the biological scaffolding to support a properly cognizing brain. In line with this, it has been argued that, although the classic substance dualism associated with Descartes is (in most cases) a nonstarter in current debates (as it raises seemingly intractable problems associated with how a nonmaterial mind might communicate with a material brain),[12] traces of it nevertheless remain at the core of the computational approach to mind. As the neuroscientist Antonio Damasio (1994) points out, the dominant idea is that "mind and brain are related but only in the sense that the mind is the software run in a piece of computer hardware called the brain; or that the brain and body are related but only in the sense that the former cannot survive without the life support of the latter" (pp. 247–248). Importantly, this information-processing, or "mind-as-computer," approach relies upon two general assumptions. The first is that cognition—conceived of as a hierarchical and quasi-linguistic (or rule-based) computational process—is necessarily an abstract and "rational" phenomenon. In line with this, the second assumption involves a disembodied, "internalist" view that sees the mind as contained strictly within the skull. The first of these assumptions, as we saw, has philosophical antecedents going back at least to Plato—it has been weakened considerably in recent years in music cognition studies, philosophy of mind, and cognitive (neuro) science. The second assumption, however, remains remarkably resilient.

We consider problems associated with this perspective in more detail in later chapters. For now, it is sufficient to say that researchers in music cognition have traditionally endorsed an information-processing orientation, albeit with varying degrees of specificity. Consider, for example, the following passage on pitch perception by the music psychologist Diana Deutsch (1999): "We shall examine the ways in which pitch combinations are abstracted by the perceptual system. First, we shall inquire into the types of abstraction [(representation)] that give rise to the perception of local features, such as intervals, chords, and pitch classes. . . . We shall then examine how higher-level abstractions [(representations)] are themselves combined according to various rules" (p. 349).

Additionally, the idea that musical cognition should be most fundamentally a hierarchical and rule-based process has also prompted much research into the relationship between music and language as cognitive systems. As we have just seen, the computational approach to the mind understands cognition as proceeding according to a kind of unconscious "language of thought" or "mentalese" (Fodor, 1983). This process is thought to have a kind of conscious correlate in spoken and written language, which functions via the organization of symbolic representations into hierarchical structures according to syntactic rules (Chomsky, 1975, 1980; Pinker, 1997). Furthermore, because language is understood to be the crowning adaptation of the human species, the evolutionary origin and functioning of the musical mind has traditionally been examined largely in terms of comparisons to language—with the assumption often being that musicality is not a proper adaptation; it is dependent on information-processing mechanisms (modules) that evolved to support language and other related functions (Patel, 2008; Pinker, 1997) (more on this in chapter 7). Accordingly, music is discussed in terms of its relationship to language at structural, perceptual, neurobiological, and evolutionary levels (Patel, 2008; Rebuschat et al., 2012), and a "music-as-language" analogy often pervades common understandings of musical experience (see Johnson, 2007).

Language and the Rules of Musical Perception

Perhaps the most well-known attempt to understand the relationship between music and language involves the apparent similarities between the Schenkerian approach[13] to musical analysis and Chomsky's universal grammar theory (Chomsky, 1980).[14] This was developed most famously by Fred Lerdahl and Ray Jackendoff (1983), who posited a "generative theory" for tonal music, in which a given musical work is parsed into hierarchies of pitch, grouping, and meter according to a quasi-syntactic schema. This led some prominent researchers to posit that musical cognition depends on linguistic capacities. As the cognitive psychologist and linguist Steven Pinker (1997) puts it, "Music may borrow some of the mental software for language. And just as the world's languages conform to an abstract Universal Grammar, the world's musical idioms conform to an abstract Universal Musical Grammar" (p. 529; see also Sloboda, 1985, 1988).

Research into music's relationship to language has also been greatly expanded by recent technological developments, most notably in the areas of computer modeling (Wiggins, 2012) and neural imaging (Grahn, 2012). Among other things, studies in the latter area have suggested an overlap between brain areas associated with linguistic syntax and those thought to be involved with the processing of tonal music (Koelsch, 2005, 2012; Koelsch et al., 2002; Patel, 2003, 2008). Interestingly, the results of such studies appear to contrast with research in neuropsychology involving patients diagnosed with *amusia* and *aphasia* (Peretz, 1993, 2006, 2012). Amusia refers to the inability to recognize or produce musical tones—it can be congenital or acquired through injury. Aphasia is a symptom of brain damage that involves impairments in expressing oneself through speech or writing (expressive aphasia) and/or difficulties in understanding written and spoken language (receptive aphasia). Amusia and aphasia can take different forms and exhibit degrees of severity depending on the extent of damage to the brain areas involved. However, studies have shown that the loss of ability in one domain does not necessarily lead to losses in the other. This research suggests, in other words, that there may in fact be dissociations between brain areas thought to process pitch and those related to language (see Van Orden et al., 2001). This apparent discrepancy has been countered by theorists who posit a cognitive "resource-sharing framework" for tonal music and language based on the idea that linguistic and musical cognition employ domain-specific representations that may be shared when necessary (Koelsch, 2012; Patel, 2012). Put simply, this theory argues that the cognitive processing of both music and language requires the ability to compute mental representations of structural hierarchies between sequential elements (Koelsch, 2012; Krumhansl & Kessler, 1982). As such, in music cognition, the representational outputs from the domain of pitch processing are thought to be shared with those from language at higher levels of processing. In this way, the perception of musical sounds (e.g., sequences of pitches, simultaneously occurring pitches) is understood to be transferred into a "cognitive representation of the location of tones and chords within the tonal hierarchy of a key" (Koelsch, 2012, p. 226; see also Krumhansl & Toiviainen, 2001).

Along similar lines, substantial attention has also been placed on theorizing about how structural variations built into a composition may set up and break musical rules and how this creates tension-resolution patterns that allow for emotions to be perceived as a property of the music and to be felt by listeners themselves (Koelsch et al., 2008; Steinbeis & Koelsch, 2008; Steinbeis et al., 2006). This research has resulted in several models in which musical emotions are explained in terms of the computational outputs of neural components (e.g., those adapted to process statistical responses associated with the satisfaction and violation of expectation) (Huron, 2006; Meyer, 1956; Scherer & Zentner, 2001).

Embodied Music Cognition

The research just discussed has produced a wealth of fascinating data and some compelling theories. However, it also views the perception and cognition of music (and language) almost wholly in terms of abstract representational processes that play out in the brain, which, again, paints a rather disembodied, skull-bound picture of the musical mind. The field of "embodied music cognition" has recently made considerable steps toward addressing this concern. As the name suggests, this orientation aims to illuminate the central role played by the body in constituting musical phenomena (Iyer, 2002, 2004; Leman, 2007; Leman & Maes, 2014; Leman et al., 2018; Maes et al., 2014; Reybrouck, 2005a, 2006a, 2017a). Important early work in this area can be found in the studies by music psychologist Jane Davidson and colleagues (e.g., 1993, 2005, 2012; Clarke & Davidson, 1998; Davidson & Correia, 2001), who examined the effects of bodily movement on the perception of emotional expression and meaning in musical performance.[15] More recently, scholars have explored how, and to what degree, the motor system influences the perception and experience of musical events, suggesting that musical stimuli are processed in the brain as representations that are action relevant and corporeally based (see D'Ausilio, 2007, 2009; Novembre et al., 2012, 2013).

While this approach appears to maintain the "internalist" perspective discussed previously (i.e., that cognition is restricted the brain, although shaped by various resources distributed across the rest of the body[16]), theorists in this field have traded the traditional modular hierarchy for a more plastic perspective on the mind-brain relationship that explains how neural connections arise, strengthen, and rewire when necessary through experience. It should be noted that this so-called "connectionist" approach is well received beyond the musical research. For many cognitive neuroscientists (see Pollack, 1989), it provides a model of the brain that better highlights its adaptive and creative capacities, offering a revised and arguably more parsimonious conception of what mental representation entails (more on this later).

While the implications of "connectionism" have been considered in a variety of ways, perhaps the most compelling suggestion is that the mental content that guides most behavior need not be understood first in terms of a language of thought and the complex unconscious mental gymnastics that this would entail.[17] Instead, the formation of neural networks and

Getting Situated 17

the representations they produce are thought to be guided largely by the agent's developmental history within a sociomaterial environment. This implies a conception of cognition and representation that, while still essentially brain-bound, is far less abstract and more grounded in the enactment of (action-relevant) patterns of behavior between neural and bodily components. For example, instead of framing musical cognition first in terms of the adherence to or violation of quasi-syntactic rules, embodied music cognition posits that it can be approached in terms of body-based predictive processing, whereby musically relevant "movements" (visual-sonic-tactile) are anticipated in the brain and where unexpected variations are adapted to in real time (see, e.g., Herholz et al., 2008; Zatorre & Salimpoor, 2013; Zatorre et al., 2007). These adaptive processes involve the negotiation of dynamical patterns of interference and coherence, of entropy and stability, between various neural networks through the constant feedback/feedforward interactions of "bottom-up" and "top-down" processing. In all, embodied music cognition offers a welcome new perspective as it places more emphasis on examining how the body mediates musical interactions within specific environments. As we discuss in chapter 6, this approach is also beginning to offer accounts of the social dimensions of musical experience in terms of how agents internally simulate the musical sounds, movements, and intentions of others via the mirror neuron mechanism.[18]

Interdisciplinary Musicology and Enactive Music Cognition

In addition to the growing interest in the embodied nature of music cognition, a range of new sociocultural, philosophical, and scientific perspectives are being integrated in musicological research. This interdisciplinary orientation is revealing a much wider range of possibilities for understanding what human musicality involves. Consider, for example, the work of sociologist Tia DeNora (2000, 2011), who draws on studies in musical development, music therapy, and the uses of music in everyday life. Notably, DeNora develops an approach to musical meaning as a process that plays out in various ways within the evolving sociocultural and personally enacted contexts of lived experience—music as action, as a therapeutic "force for bio-cognitive organization," and as part of an enacted aesthetic environment through which cultural and individual identities may be constructed and deconstructed. Music is thus a "resource for meaning-making."[19]

Relatedly, Elliott (1995) has argued that music should be conceptualized as a verb: "Fundamentally, music is something that people do" (p. 39). He uses the term "musicing" in the collective sense to mean the actions and the personal, social, and cultural values that emerge from performing, listening, improvising, composing, arranging, and conducting, as well as from "musicing" and dancing, music and worshipping, music and celebrating, and so forth, in their specific sociocultural contexts (p. 129). He adds that because music is an "open" global reality (p. 128), not a unitary entity, it is appropriate to replace the conventional term "music" with "Musics" (p. 43).

The cultural theorist and ethnomusicologist Christopher Small (1998) agrees that music is best understood as a verb rather than a noun. His theory of "musicking" considers human musicality as a multifaceted activity: "The fundamental nature and meaning of music lies not in objects, not in musical works at all, but in action, in what people say and do. . . . To music is to take part, in any capacity in a musical performance, whether by performing, by listening, by rehearsing or practicing, by providing material for the performance (what is called composing), or by dancing" (p. 9). Likewise, the inclusion of non-Western and other traditionally marginalized perspectives has decentered the traditional focus on the relationship between composed form and expression (Elliott, 1989, 1995; Lewis, 2007, 2009; Nettl, 1974; Nettl & Russell, 1988). This research highlights the active, interpersonal-communal, and creative nature of human musicality and the unique meanings that are generated by music-making in specific contexts. Additionally, comparative studies of musical (or music-like) activity in nonhuman animals (Bannan, 2016; Fitch, 2006; Merchant et al., 2015; Patel & Iversen, 2014) have revealed interesting possibilities for understanding the relevance of musicality for our prehuman ancestors and, more generally, for the development and flourishing of the living systems that engage in such forms of behavior.

This expanding view of what musicality involves poses new challenges to traditional thinking and research in music cognition—which, as we have seen, has tended to restrict itself to explaining the stimulus-response mechanisms and the abstract forms of information processing that musical perception is thought to entail. As we also considered, research and theory in embodied music cognition have contributed to these developments by offering an important new perspective that bases musical cognition in the outputs of motor-based processing in the brain, highlighting the relevance of bodily engagement for musical perception (whether actual or simulated internally). However, another "embodied" approach is emerging that is based in recent developments in cognitive science and theoretical biology associated with the increasingly influential enactive approach to cognition. "Enactive music cognition" offers a view of the musical mind that, although consistent with certain aspects of embodied music cognition, integrates body, brain, and environment in a more direct way. It does so by drawing out fundamental continuities between mental and biological processes, tracing the origins of mind to the primordial manifestations of life. Put simply, where embodied music cognition may understand the musical mind in terms of representations of musically relevant movements—or corporally based simulations—that play out in a cognizing brain, the enactive approach explores the environmentally situated body as a cognitive domain in its own right.

The enactive approach takes the interactive, creative, or "world-making" nature of living embodied minds as foundational for cognition. In doing so, it casts the musical mind in a more ecological and "radically embodied" (Chemero, 2009) light as it seeks to overcome the lingering internalist assumptions that often permeate the embodied cognition orientation. From an enactive perspective, musicality is explored as a manifestation of human sense-making—as continuous with (but not simply reducible to) the forms of adaptive interactivity that characterize how even very simple life forms enact survival-relevant relationships within the sociomaterial environments they inhabit (Reybrouck, 2005b, 2006b). In all, the enactive approach aims to form an understanding of the mind that trades an abstract computational analogy for a biological reality. Accordingly, enactive music cognition takes the living, sensing, moving, affective-emotional, and environmentally situated body as the starting point for understanding the musical mind. As is evident in the chapters to come, this perspective has important implications in many areas of interest to current musicology. Indeed, enactivist approaches are currently being applied across a range of domains in the sciences and humanities. As a theoretical framework, it is well suited to engage with the interdisciplinary orientation that characterizes current musicological research.

We outline the main principles of the enactive perspective in detail in the next chapter. Before we do this, however, we would like to note that an enactive approach to the music mind has some important antecedents. This includes the musicologist Eric Clarke's (2005) influential monograph,

Ways of Listening: An Ecological Approach to the Perception of Musical Meaning. Clarke argues that the (often-tacit) acceptance of the information-processing approach has tended to reduce musical experience to a kind of abstract "reasoning or problem-solving process," in which "perception is treated as a kind of disinterested contemplation with no connection to action—which bears little relationship to the essentially exploratory function of perception in the life of an organism" (p. 15). He further questions the validity of this approach as it appears to contradict direct experience. As he notes, we tend to understand music first in terms of its meaning in our lives and how it makes us feel, and only subsequently in terms of its constituent structural elements—an activity that often requires difficult (and conscious) analysis, as well as sustained training. In response to these concerns, Clarke offers a perspective based in connectionist neuroscience, ecological psychology, and first-person experience. In the process, he develops an approach that highlights the active, creative, embodied, and environmentally situated nature of the musical mind. And indeed, the final section of *Ways of Listening* is entitled "The Affordances of Music and the Enactment of Musical Meaning." On the penultimate page he writes, "This book should be no more than a part of a larger project on the *enactment* of musical meaning" (p. 205, italics in original). In recent years Clarke has offered a number of articles that develop enactivist and 4E models (e.g., Linson & Clarke, 2017).

Similarly, the ethnomusicologist David Borgo's (2005) *Sync or Swarm: Improvising Music in a Complex Age* draws on enactivist theory, ecological psychology, and complex systems theory to analyze the performances of collaborating improvisers. In doing so, he explores a practice (free improvisation) that has traditionally been marginalized in Western musicology, revealing the dynamics of musical interaction as it unfolds in real time.

A third important precedent can be found in the work of musicologist Mark Reybrouck (2001, 2005a,b, 2006, 2010, 2012, 2015a,b, 2016, 2017a,b, 2020), who draws on enactivist principles, ecological psychology, and biosemiotics (among other areas) to explore musical sense-making as continuous with processes of niche construction and adaptive behavior in the context of a real-time lived experience. Philosopher Joel Krueger (2009, 2011a,b,c, 2013, 2014, 2015, 2018a,b) has also contributed a number of important papers and chapters that develop phenomenological, enactivist, and 4E perspectives, with a special emphasis on the social aspects of musical experience. Additionally, theorists such as Marissa Silverman (2012, 2020; Elliott

& Silverman, 2015) and Wayne Bowman (2004) have drawn on enactivist and related phenomenological models to develop ethical frameworks for music education. We consider pedagogical perspectives in more detail toward the end of the book.

Looking Ahead

Let us now outline what's to come in the chapters of this book. In chapter 2, we introduce the main features of the enactive approach to mind. Here we consider enactivism's relationship with ecological psychology and distinguish it from other embodied approaches to cognition. Following the seminal work by Francisco Varela, Evan Thompson, and Eleanor Rosch (1991), as well as more recent contributions (e.g., Di Paolo et al., 2017; Thompson, 2007), we then identify three main principles central to the enactive position: autopoiesis, autonomy, and sense-making. In doing so, we explore the enactivist claim that cognition is not best understood first in terms of computations and representational content confined to brains; and we begin to consider the implications of the enactivist conception of mind as primarily rooted in the processes of sense-making that arise between living bodies and the sociomaterial environments in which they are embedded. To conclude, we draw connections with supporting fields such as dynamical systems theory, and we outline the 4E framework (embodied, embedded, extended, and enactive) that will help to guide our musical discussion in subsequent chapters.

In chapter 3, we tackle the tricky subject of music and consciousness. We first consider the ways the issue of consciousness has traditionally been approached from an information-processing orientation. Here we place a special focus on two important perspectives—those of philosophers Daniel Dennett (1988, 1991) and Diana Raffman (1993), respectively. We choose these examples because, for us, they represent two of the most compelling computational accounts of the experiencing mind. Moreover, Raffman offers a fascinating critical extension of Dennett's position, using musical nuance as a paradigmatic example of where their approaches diverge. Drawing on thinkers associated with the phenomenological tradition (Merleau-Ponty, 1945; Roholt, 2014), we argue that Raffman and Dennett maintain disembodied and internalist assumptions that prevent them from engaging fully with what the experience of music entails. We then begin to develop a perspective that highlights the primacy of the situated living body for musical consciousness.

Chapter 4 develops this embodied approach further, outlining the importance of the "phenomenological attitude" for enactive music cognition. Here, we introduce some examples intended to engage readers in phenomenological inquiry. These involve the exploration of multi-stable visual and musical phenomena (i.e., Ghanaian polyrhythm). As we show, these experience-based examples reveal observations that align closely with the enactive approach and the 4E framework—most notably that musical cognition requires an active embodied, embedded, and extended engagement with a social and material environment. Following this, we outline an embodied perspective on musical experience. Here we draw on complementary accounts from phenomenological philosophy (Johnson, 2007) and neuroscience (Ramachandran, 2011) to examine how our "metaphorical" ability to enact cross-modal, embodied, affective-emotional relationships forms the basis for what it means to be an experiencing (musical) being.

In chapters 5 and 6, we explore the emotional and empathic aspects of musical experience, respectively. We review some prominent philosophical and psychological approaches, as well as relevant work in embodied music cognition. We then offer an alternative perspective based in the emerging work on emotion and social cognition associated with enactivism (Colombetti, 2014; Gallagher, 2020). This approach trades the focus on neurally instantiated affect programs, internal representations, and simulations for more dynamical interpretations that encompass the contingent moment-to-moment engagements of living systems across multiple time scales. As we argue, this perspective appears to better address the experience of music as a relational phenomenon that involves direct, embodied forms of interactivity between musical agents and their environments. To conclude, we begin to develop a 4E approach to musical empathy and emotion by exploring the ecological concepts of "musical scaffolding" and "empathic space" (Krueger, 2011a,b; Krueger & Szanto, 2016).

In chapter 7, we consider what an enactive approach may reveal about the origins of musicality in the human phenotype. The field of evolutionary musicology has tended to divide into two main positions: those who argue that music should be understood as a naturally selected adaptation and those who claim that music is a product of culture with little or no relevance for the survival of the species. In light of this, we consider a recent "biocultural" proposal that appears to offer a way though this apparent dichotomy as it posits a more integrated model that sees biological and cultural dimensions as aspects of the same evolving system (Tomlinson, 2015). Here we make connections between the biocultural approach and the enactive perspective, exploring possibilities for how a 4E framework can help us think about the ways (proto)musical behavior could have emerged within the material environments and social spaces that were inhabited and shaped by our prehuman ancestors.

In chapter 8, we discuss musical development in infancy and childhood. In doing so, we examine the central role played by the active sense-making body—highlighting the ways in which infants and children actively pursue developing relationships between manipulated objects and the forms of sound making they afford and how this develops into patterns of behavior that become meaningful over time. In chapter 9, these insights are developed toward an enactive approach to musical creativity. Here we consider how dynamical processes of adaptive body-brain-environment interactivity drive musical creativity across a range of contexts, leading to a 4E perspective that extends the idea of "musical creativity" beyond the personal (inner) domain of musical agents and out into the material and social worlds they inhabit and influence.

Chapter 10 concludes with an exploration of the practical and ethical implications of an enactive approach to the musical mind. Here we explore how musicality, understood as human manifestation of the biological processes of autopoiesis, autonomy, and sense-making, can reveal important aspects of human being and knowing that are often lost or obscured in the modern, technologically driven world. This, we argue, may have profound implications for thought and action in the areas of music education and community music, music therapy, and for the ways we engage with music in everyday life.

In all, we hope that this book will help spark conversations and inspire new approaches to research in academic environments. But equally we hope that the ideas offered here are useful to those who wish to reflect on their own relationship with music and what it means to them in the context of their daily lives. Indeed, our intention is not to provide the final word on the musical mind, nor to debunk or supplant the other perspectives we discuss. Rather, we wish to introduce a compelling new way of thinking about the nature and meaning of human musicality that aligns closely with the actual experience of music in human life. We hope, then, that the ideas and possibilities offered in this book will be interesting and provocative; that they will inspire critical feedback and future refinements; and that this will lead to richer understandings of the complex range of action, thought, and experience we associate with the words "music" and "mind." We should also note that many of the chapters that make up this volume began as academic journal articles. So, in addition to our collaborative efforts, the discussion that follows owes a great deal to the many peer reviewers who helped us sharpen our arguments, to the colleagues and friends who read and commented on chapter drafts, and to other coauthors who provided ideas that influenced our thinking. These contributions are indicated in the first end note of each chapter (and in the acknowledgment). We can now move on to explore the basic concepts associated with the enactive approach to mind.

2 Basic Principles of Enactive Cognitive Science

In the previous chapter, we began to consider how the enactive perspective contrasts with the enduring assumption that mental life should be a fundamentally representational and brain-based (or "internal") phenomenon. The implications of this will come into sharper focus shortly when we examine the basic principles of the enactive approach in more detail. Before we do that, however, it is important to outline the key principles of ecological psychology as it is one of the main sources of inspiration for general theorizing in 4E cognition. This will also help us better distinguish the differences and similarities between the perspective present in much embodied music cognition research—in which cognitive processes may be distributed across the body but not necessarily coupled with extrabodily and nonbiological resources—and the more radically embodied and ecologically extended view of cognition associated with enactive music cognition and the 4E framework.[1]

The Ecological Stance

Ecological psychology is most closely associated with the work of psychologist James Gibson (1966, 1979). Gibson's chief interest was in vision, regarding which he argued persuasively that there is sufficient information in the ambient light to allow organisms to "pick up" or "attend to" invariances in the ecological topology and that the properties of the environment are therefore specified directly. Accordingly, he claimed that symbolic-representational processes are not necessarily required to explain perception. As the cognitive scientist and philosopher Anthony Chemero (2009) explains, Gibson's approach can be described in terms of three main tenets that contrast with an information-processing orientation. The first posits that perception is direct; it is not mediated by representational mental content. The second argues that perception is not first and foremost for information gathering but rather for the guidance of action, that is, for actively engaging with the world. Following from the first two, the third tenet holds that perception is of "affordances." This refers to the possibilities for action offered by the environment to a given organism in relation to its corporeal structure and motor abilities (e.g., a chair affords sitting for a child or an adult, but not for an infant nor a fish). Processes such as growth, skill acquisition, learning, and evolution can alter the set of affordances available to a given animal or species. While Gibson's own account of affordances is perhaps somewhat ambiguous, it has nevertheless proven to be an influential and theoretically useful concept, albeit one that has been interpreted and applied in various ways (Chemero & Turvey, 2007; Reed, 1996; Stoffregen, 2003; Turvey, 1992). Indeed, theorists associated with embodied cognitive science and enactivism have developed revised versions of Gibson's ideas. Understanding the differences between these orientations is important for our discussion later. Thus, we require a little more history.

Although it may seem reasonable to assume that Gibson's approach emerged as a reaction to the dominant computational orientation, this is not the case. In fact, ecological psychology traces its origins further back to the pragmatist philosophical psychology (or the so-called "American naturalism") of Charles Sanders Peirce, William James, and John Dewey (see Heft, 2001), and to the phenomenology of Maurice Merleau-Ponty (1945). More broadly, Chemero argues that ecological psychology is a branch of an "eliminativist" approach to mind that emerged in the late nineteenth century in contrast to the "representational" conceptions associated with Wundt and Titchener (Chemero, 2009, pp. 20–33). In brief, ecological psychology is best viewed as continuous with an orientation that developed in parallel with representational-computational approaches—one that was marginalized for much of the twentieth century and that has enjoyed a strong resurgence in recent decades.

The field of embodied cognitive science began to take shape in the mid-1980s with George Lakoff and Mark Johnson's work in cognitive linguistics (1980, 1999), Rodney Brooks' (1986, 1989) research in behavior-based robotics, and the move toward a dynamical approach that incorporates bodily, environmental, and neural factors to describe cognitive systems (see Kelso, 1995; Thelen & Smith, 1994). By the late 1980s and early 1990s this work began to be integrated with theory and research in ecological psychology. The insights produced by the merging of these fields led some researchers to posit nonrepresentational accounts of cognition—that is, as driven primarily by the bodily enactment of affordances (see Chemero, 2009). However, others argued that cognition as such should still be understood as a representational phenomenon, but that the conception of what this entails should be revised to include inputs from the body and from motor-processing areas in the brain (e.g., Clark, 1997, 2001; Shapiro, 2011). In other words, these researchers continued to develop the Gibsonian insights that perception is *for* relevant action and *of* affordances (environmental possibilities for action). But here cognitive processes do not involve direct perception—they are mediated by "action-oriented" representations that are generally understood in terms of the connectionist framework we began to describe in the previous chapter (see Churchland, 2002; Grush, 1997, 2004; Millikan, 1995). Put very simply, then, this approach to embodied cognition adopts a Gibsonian perspective that is essentially stripped of its eliminativist or nonrepresentational core.

While it is this representational orientation that is generally referred to with the term "embodied cognitive science," other researchers have continued to develop ecological perspectives in ways that embrace direct perception (Chemero, 2009; Thompson, 2007). A key insight that drives this work involves the bidirectional (or "co-arising") nature of organism and environment, of mind and world. Gibson suggests this relationship in passages of his writing that describe affordances as guided equally by environmental features and the behavior of the animal (see Chemero, 2009, pp. 135–162): "An affordance is neither an objective property nor a subjective property; or it is both if you like. An affordance cuts across the dichotomy of subjective-objective and helps us to understand its inadequacy. It is equally a fact of the environment and a fact of behavior. It is both physical and psychical, yet neither. An affordance points both ways, to the environment and to the observer" (Gibson, 1979, p. 129). However, Gibson arguably does not discuss this relational conception of affordances with much precision (see Chemero, 2003). In response to this, researchers associated with the enactive approach have developed revised interpretations of embodiment and affordances that provide more detailed accounts of the recursive relationship between organism and environment (Thompson, 2007).[2]

Autopoiesis and Autonomy

The enactive perspective emerged in the late twentieth century as a counter to the dominant computational model of mind (Varela et al., 1991). And while enactivism should not be understood as a type of ecological psychology, as such, it is nevertheless partly inspired by the latter's key tenets: that perception is direct, that perception and action are closely integrated, and that perception is of affordances. It also draws on research in neuroscience, biology, and dynamical systems theory, as well as current trends in developmental science, affective science, and evolutionary theory. Generally speaking, the enactive approach may be considered as a form of what Chemero (2009) refers to as "radical embodied cognitive science," as it does not restrict cognition to in-the-head representational content, even in a revised connectionist sense. More specifically, the defining aspect of enactive cognitive science is its assertion that there exists a deep continuity between life and mind, between biological processes and mental processes (Thompson, 2007). It thus offers several new possibilities for thinking about what mental life entails.

From the enactive perspective, cognition is understood to be based, first and foremost, in the perceptually guided action of self-organizing living systems that exhibit "autonomy" (Maturana & Varela, 1980; Varela, 1979, 1988). To begin to clarify this idea, consider how the input-output functions of a computer depend on externally imposed designs (hardware and software), information processing rules (system language), and interpretations of outputs by independent agents (i.e., humans). A computer cannot function meaningfully in a fully autonomous fashion: its operations are dependent on the external entities that impose significance, functionality, and form upon it. It is therefore described as an *operationally open* system. Computing machines, in other words, have their ontological footing (how they come into being, what they are, and what they mean) outside of themselves.[3]

Living systems are different. As Varela, Thompson, and Rosch (1991) write, "Under very restricted circumstances, we can speak as if we could specify the operation of a cell or an organism through input/output relations. In general, though, the meaning of this or that interaction for a living system is not prescribed from outside but is the result of the organization and history of the system itself" (p. 157).[4] Living systems bring themselves into existence and actively participate in their own continuation. Such

Basic Principles of Enactive Cognitive Science

systems are understood to be autonomous because the world of significance that is brought forth through such processes is not externally imposed but arises from the system's own self-organizing activity. A living system is not "made," nor is it best understood as a "thing." It is, rather, "a process with the particular property of *engendering itself* indefinitely" (Stewart et al., 2010, p. 2; emphasis in the original). This process involves cycles of reproduction, whereby the components of the living system are constantly self-generated so that the organism can maintain itself within the demands of a contingent environment. Such processes are referred to as "autopoiesis" (literally, "self-making"), the most fundamental example of which can be found in the living cell (figure 2.1).

Membrane
selective permeability
resists-adapts to
perturbations

Bounded metabolism
self-generating
self-regulating

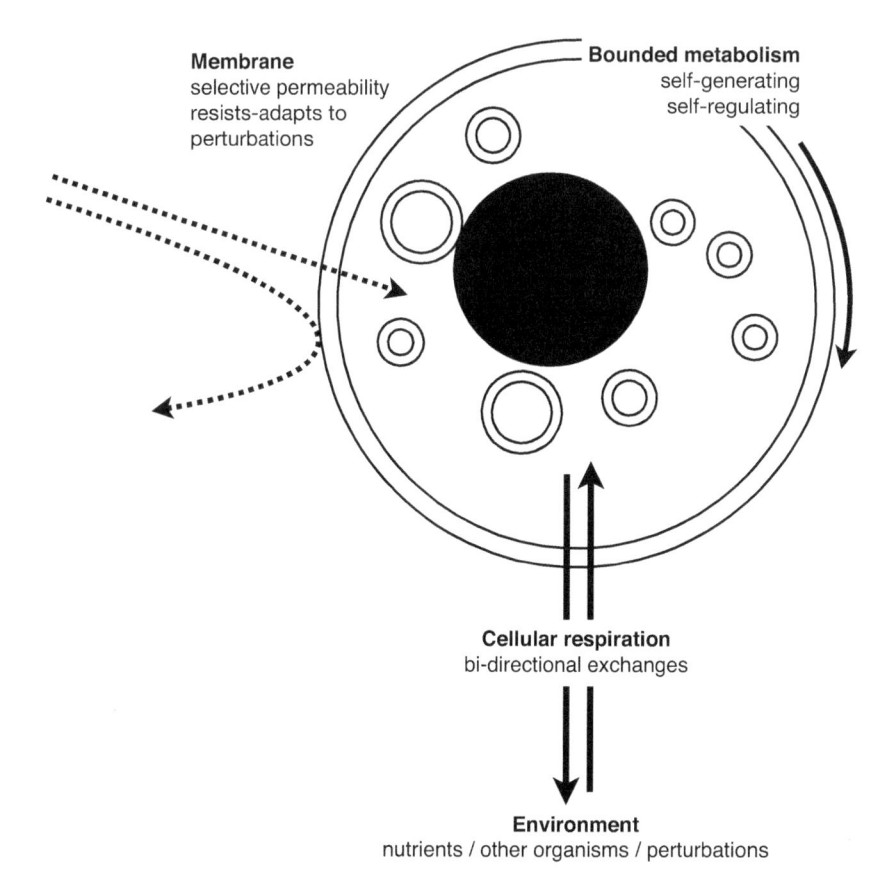

Cellular respiration
bi-directional exchanges

Environment
nutrients / other organisms / perturbations

Figure 2.1
A simple autopoietic entity: the living cell.

The notion of autopoiesis was originally developed by Humberto Maturana and Francisco Varela (1980, 1984) to describe the self-organizing nature of living systems. Most fundamentally, this involves the development and maintenance of a bounded metabolism (Jonas, 1966; Thompson, 2007). As Varela (1979) observes, processes of self-regulation maintain the cellular organization as a unity by keeping metabolic dynamics within ranges that are conducive to survival. Importantly, what lies outside of this biological system does not provide "input" for these processes to unfold: the living cell self-produces everything that is needed for its maintenance within its material boundary, the membrane (Weber, 2001; Weber & Varela, 2002). This means that, unlike the manufactured computing machine, the living organism is *operationally closed*: its material existence and how it makes sense of the world (what it is and what it means) are driven by its own history of action-as-perception within the contingent environment it inhabits and shapes. Operational closure, then, applies to both the "biological" and the "cognitive" aspects of a living system, revealing that these are not, in fact, separate ontological domains. They are, instead, continuous with each other, entailing different ways of looking at the same integrated process (i.e., life).

It is important to note here that the cell's autonomous status does not leave it in isolation—that is, without a unique interactive relationship with its niche. On the contrary, to survive it must actively develop viable relationships with its environment: it must be situated (Di Paolo, 2005). And indeed, important processes—for example, exchange of energy (nutrients)—are realized through the regulatory functions of the membrane itself, which can modify its permeability when needed to adapt to the shifting contingencies of the organism's environment. Therefore, autonomous systems also remain *thermodynamically open* to the environment (Thompson, 2007).

So, while living systems may be characterized, most primordially, as a bounded metabolic unity, they also exhibit the necessary "meta-metabolic" (Barrett, 2018) ability to interact with the environment in ways that are relevant to their continued well-being (figure 2.2).[5] Accordingly, autopoiesis alone is insufficient for cognition—*adaptivity* is also a crucial component:

> Mere autopoiesis—the operationally closed self-production of a chemically bounded network—provides only the all-or-nothing conservation of identity through material turnover and external perturbations to the system, but not the active regulation of interactions with the outside world. Sense-making is normative, but the only

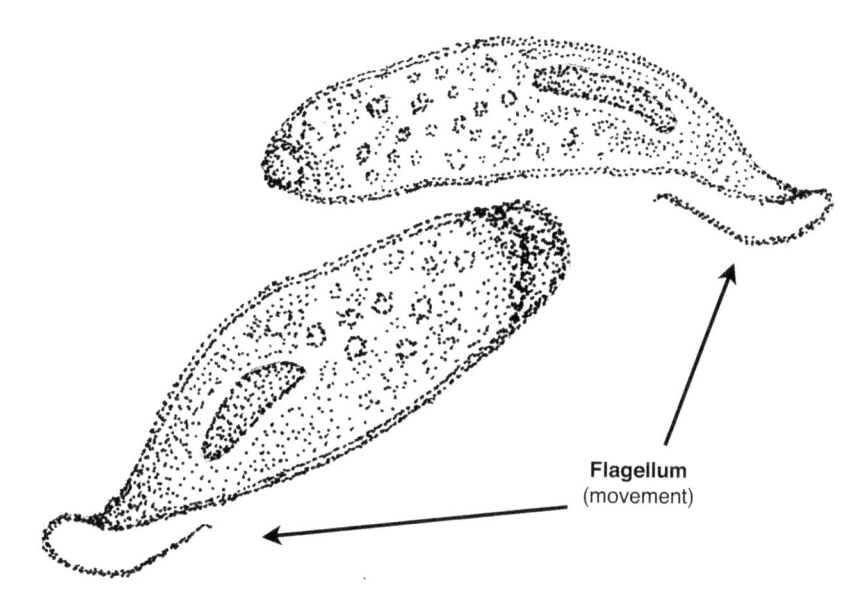

Basic Principles of Enactive Cognitive Science

Flagellum
(movement)

Figure 2.2
Two nonphotosynthetic single-celled eukaryotes measuring 1–150 μm with structures of a multicellular organism. Like more complex creatures, these organisms can sense and move in relation to their environments—a primordial example of sense-making or (proto)cognition as perceptually guided action. (Drawing by Leigh van der Schyff [used with permission])

> norm that autopoiesis can provide is the all-or-nothing norm of self-continuance, not the graded norms of vitality (health, sickness, stress, fatigue) implied by an organism's regulating its activity in ways that improve its conditions for autonomy (as when a bacterium swims up a sucrose gradient or swims away from a noxious substance). An adaptive autopoietic system, however, is one that can regulate its states with respect to its conditions of viability in its environment and thereby modify its milieu according to the internal norms of its activity. (Thompson and Stapleton 2009, p. 3)

In brief, it is a living being's inherent adaptivity that allows it to maintain a primordial "point of view" (a "proto-self") in *relation* to its environment—this is essential for sustaining itself as an autonomous entity in the world.

In being adaptive, the organism necessarily brings forth a world of viable (significant-meaningful) relationships within the environment.[6] Taking this a step further, we can consider how creatures with nervous systems and brains "embody more complex forms of adaptive sensorimotor autonomy"

(Thompson & Stapleton, 2009, p. 3). Indeed, a nervous system does not passively receive information from the environment; it actively "creates a world by defining which configurations of the milieu are stimuli" (Weber, 2001, p. 15). Because of this, insects, reptiles, and mammals are able to enact increasingly richer repertoires of action and perception, including the ways some animals reach out to and interact with "social" worlds through various forms of embodied action.[7]

Sense-Making

This interdependence of autopoiesis, autonomy, and adaptivity leads to an intriguing bio-ontological relationship between creature and world. As we described previously, an autonomous living system must be thermodynamically open to the environment to maintain the operational closure "that actively generates and sustains its identity under precarious conditions" (Thompson & Stapleton, 2009, p. 3).[8] This means that while the boundary between organism and environment may be clearly defined, the two are nevertheless inseparable. The biologist Richard Lewontin (1983) explains, "Just as there is no organism without an environment, so there is no environment without an organism. The organism and environment are not actually separately determined. The environment is not a structure imposed on living beings from outside but is in fact a creation of those beings. The environment is not an autonomous process but a reflection of the biology of the species" (p. 99).

Accordingly, the cognitive scientist Ezequiel Di Paolo (2005) uses the term "interactional asymmetry" to describe the ongoing history of adaptive structural coupling between organism and environment, which contributes to maintaining and restoring the organism's internal balance and its dynamical network of relations with the world. He writes, "The self-regulating processes to keep the agent's conservation as auto-sufficient establish the dialectic between agent and environment: whence the intriguing paradoxicality proper to an autonomous identity: the living system must distinguish itself from its environment, while at the same time maintaining its coupling; this linkage cannot be detached since it is against this very environment from which the organism arises, comes forth" (p. 85). Put simply, the enactive approach considers organism and environment not as a pregiven duality but as dependently "co-arising" through the embodied, perceptually guided activity of the organism as it actively brings forth a world. Therefore, "Living is a process of sense-making, of bringing forth significance and value" (Thompson, 2007, p. 158). This insight is further unpacked by Thompson (2004) in five key points:

1. Life=autopoiesis. By this I mean the thesis that the three criteria of autopoiesis—(i) a boundary, containing (ii) a molecular reaction network, that (iii) produces and regenerates itself and the boundary—are necessary and sufficient for the organization of minimal life.
2. Autopoiesis entails emergence of a self. A physical autopoietic system, by virtue of its operational closure, gives rise to an individual or self in the form of a living body, an organism.
3. Emergence of a self entails the emergence of a world. The emergence of a self is also by necessity the emergence of a correlative domain of interactions proper to that self, an Umwelt.
4. Emergence of self and world=sense-making. The organism's world is the sense it makes of the environment. This world is a place of significance and valence, as a result of the global action of the organism.
5. Sense-making=cognition (perception/action). Sense-making is tantamount to cognition, in the minimal sense of viable sensorimotor conduct. Such conduct is oriented toward and subject to signification and valence. Signification and valence do not pre-exist "out there" but are enacted or constituted by the living being. Living entails sense-making, which equals cognition. (pp. 386–387)

To some, it might seem as if the notion of sense-making could risk a slide into nondialectical constructivism, or that cognition involves making sense out of an implicitly nonsensical environment (that cognition-as-sense-making implies a one-directional schema whereby an agent imposes meaning on the world). This, of course, would be antithetical to the *dialectical* organism-environment relationship that lies at the heart of the enactive approach—that is, where the world is "brought forth" by the relationality of what we can artificially separate as organism and environment but are actually mutually defining and therefore meaningless as separate "components." We can alleviate this concern by again reinforcing the continuity between biological and cognitive processes, this time in terms of the idea of "interactional asymmetry" introduced earlier.

Recall Di Paolo's comment. As we saw, maintaining this asymmetrical relationship (autonomy) requires constant adaptivity. This means that cognition-as-sense-making is based on the ways living beings move and sense *within their environments*, in their fundamental proclivity to seek out regularities and adapt to changes (which are sometimes initiated by the activity of the organism itself). Again, significance and meaning are not "imposed" by the organism, nor are they "pregiven" aspects of the environment; rather, they arise and evolve in a bidirectional way. The idea of sense-making, then, casts Gibson's comment on affordances as "equally a fact of the environment and a fact of behavior" in a richer biocognitive light: it reveals affordances as emergent properties of the adaptive history of structural coupling between organism and environment (where potentials and constraints exist on both sides of the system).

This deep coupling of organism and environment also raises the question of what cognition as sense-making means for the idea of agency (an important aspect of musical cognition). We can begin to answer this by noting that *adaptivity*—understood as an operational concept that plays out within the specific life-world of an organism—and *agency* are closely aligned. By an enactive light, agency begins with the range of sensorimotor processes involved in the self-organization living systems, and, notably, where such systems are understood to be *acting on their own behalf* (see Di Paolo et al., 2017, p. 111). This ability to *act* (as differentiated by other physical exchanges the system undergoes) is guided by and shapes the cycles of sense-making activity that give rise to self-individuation (interactional asymmetry) and normativity (organism-environment regulation). Such "epistemic" cycles extend from *minimal agency*—which describes the fundamental biological operations (e.g., cellular interactions) by which an autonomous organism maintains itself (e.g., sensing, nutrition)—to the *intersubjective operations* that define the social body and the forms of sense-making involved in social organization. For humans, the latter includes musical and linguistic acts (Di Paolo et al., 2017, 2018). Crucially, this perspective trades the traditional idea of agency as an essentially abstract and solitary phenomenon for one that highlights the primary role of relational sensorimotor activity for sense-making, selfhood, communication, and social life.[9]

In all, then, the equation of cognition and (action-based) sense-making allows for the overcoming of disembodied views about mental life and its properties. This insight emphasizes the key role of the environmentally situated body-in-action for the realization of cognition and, at the same time, transforms the traditional focus on computations and hierarchical cognitive structures into a complex network of ongoing relational dynamics between the system's autonomous agency, its domain of sensorimotor interaction with the world, and the emergence of "significance" and "meaning" as properties central to this network (figure 2.3). By this view, sense-making (cognition) cannot be only a matter of representational recoveries of an external world, nor can it be fully described in terms of mental states such as beliefs or thoughts. As useful as those theoretical constructs may be in explaining certain aspects of a cognizer's high-level mental properties, focusing on them as the defining aspects of cognition downplays a number of other important factors: how the brain-body system is dynamically integrated with the environment, how meaningful organism-environment relationships emerge through active sensorimotor processes, and how interaction with other agents and objects in the sociomaterial ecology shape mental life more generally.[10] Here it is also important to note that because such embodied sense-making processes occur under precarious conditions (i.e., survival within a contingent environment), they cannot be understood as indifferent (Kyselo, 2014). In other words, basic cognitive activity is always

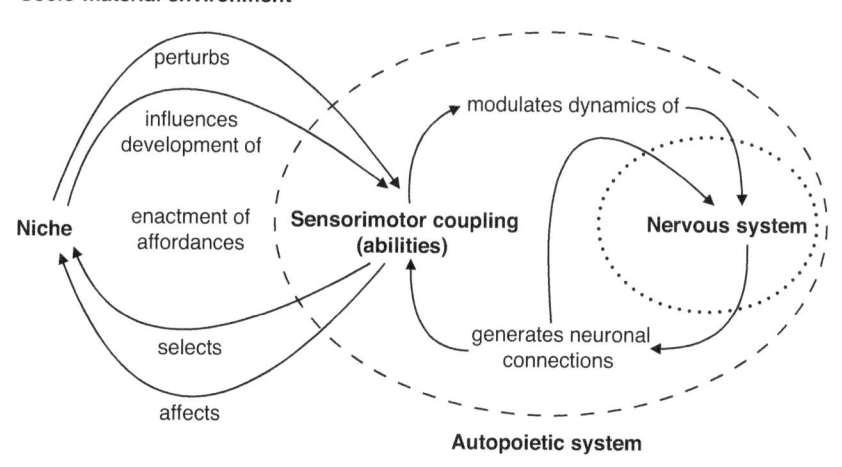

Figure 2.3
The structure of organic sense-making through the coupling of the animal-environment system. Notice the recursive or "nonlinear" dynamics that characterize each part of the system: aspects of the niche influence the development of the organism, which in turn influences the niche, and so on. This synergistic relationality involves the enactment of possibilities for action (affordances) and with it a world of significance and value. (Adapted from Chemero, 2009, p. 153)

characterized by a "primordial affectivity" that motivates relevant action (Colombetti, 2014; Thompson, 2007).

Most fundamentally, then, a living agent makes sense of its world through affectively motivated forms of action-as-perception. In doing so, it constructs a viable niche, transforming the world into "a place of salience, meaning, and value—into an environment (Umwelt) in the proper biological sense of the term" (Thompson & Stapleton, 2009, p. 3). This involves the *enactment of affordances*, which are seen here as emergent properties associated with the (evolutionary and ontogenetic) history of the structural coupling between organisms and their environments.

For example, we might think of the symbiotic and coemergent relationship between honeybees and flowers (figure 2.4). Here autonomous agents exist as environments in relation to each other: the development of their phenotypes and the affordances they enact for themselves and each other are inextricably enmeshed day to day and over evolutionary time (Hutto & Myin, 2012). Likewise, more complex forms of interactivity can be discerned

Figure 2.4
Flowers and honeybees: a symbiotic relationship between two autonomous organisms. (Drawing by Leigh van der Schyff [used with permission])

between other living systems. As mentioned previously, this includes the forms of social cognition found in human beings and other animals, what enactivists refer to as *participatory sense-making* (De Jaegher & Di Paolo, 2007). Consider, for example, the shared worlds of meaning brought forth in the interactions between infants and primary caregivers. In this case, participatory sense-making involves the coenactment of a repertoire of corporeal and emotional gestures and utterances that are meaningful to the personal domain of each agent—their feelings and needs—but that are also shared between both (Fantasia et al., 2014).

Later, we consider how the idea of participatory sense-making can help us think about the kinds of embodied-affective forms of meaning making that characterize collective musical environments. For the moment, we should note that participatory sense-making also reflects the relational conceptions of autonomy (and agency) discussed earlier, this time in an explicitly social context. Moreover, interpersonal forms of cognition cannot be explained only in terms of internal mental content: they also need to be examined in terms of the direct interactivity between embodied minds and the sociomaterial ecologies they inhabit and collectively shape. With this in mind, let us briefly consider some arguments and implications associated with a nonrepresentational approach to cognition.

Basic Cognition without Mental Representation?

In contrast to the dominant trends in cognitive science, enactivists claim that basic sense-making processes do not depend first on mental representations. One reason for this involves the observation that although simple single-celled creatures do not possess the appropriate hardware (e.g., a brain) for such forms of abstract processing, they can nevertheless move purposefully and develop viable relationships within the changing environments that they inhabit (Thompson, 2007).[11] It should be noted that some theorists have argued that complex neural structures may not in fact be required for systems to exhibit representational properties (see Bechtel, 1990, 2008). Such systems include simple cybernetic devices, in which the state of one part of the system may represent and control aspects of another part.[12] Other researchers have countered this position by showing that the representational descriptions of cybernetic devices are more a matter of perspective than necessity and that the functioning of such devices can in fact be explained more parsimoniously in nonrepresentational terms. Indeed, the mathematical equations employed to model these devices are neutral with regard to representations; thus it is argued that evoking representations introduces unnecessary complications (for a discussion see Chemero, 2009, pp. 67–83, 105–134).

Drawing on these insights, the psychologist Louise Barrett (2011) offers numerous examples of how creatures (insects and spiders) with relatively simple neural structures are nevertheless able to perform surprisingly complex cognitive feats by including their bodies and environments as integrated aspects of their cognitive domain, highlighting the corporeally and environmentally extended nature of living cognition. Likewise, researchers in robotics (Beer, 2003; Brooks, 1991) have shown how simple forms of intelligent behavior can emerge organically as artificial systems interact directly with an environment, as opposed to having to rely on the symbolic or representational models that artificial intelligence researchers have traditionally assumed had to be programmed into them.[13]

Related arguments have recently been playing out in cognitive neuroscience over the discovery of two classes of sensorimotor neurons: mirror neurons and canonical neurons. Mirror neurons fire when performing a goal-directed action and when seeing another individual doing the same action (di Pellegrino et al., 1992; Gallese et al., 1996). As neuroscientist Vittorio Gallese writes, "When we observe goal-related behaviors ... specific sectors of our pre-motor cortex become active. These cortical sectors are those same sectors that are active when we actually perform the same actions. In other words, when we observe actions performed by other individuals our motor system "resonates" along with that of the observed agent" (2001, p. 38).

Mirror neurons were originally discovered in the ventral premotor cortex of macaque monkeys and were then observed in the brains of (human) surgical patients (Mukamel et al., 2010). Additionally, a subclass of mirror neurons becomes active not only during the observation or performance a given action but also when hearing the sound produced by the action itself (e.g., the sound of a peanut being broken activates the same motor neurons involved in the action of breaking the peanut) (see Kohler et al., 2002). Therefore, it may be that the goal-directed action of another agent can be perceived independently of the modality of the sensory information, as long as one possesses the motor knowledge necessary to perform the act being witnessed (seen or heard).[14] Canonical neurons, instead, are involved in how agents perceive possibilities (i.e., affordances) for manipulating objects according to their corporeal structure and abilities. These neurons fire both when an individual observes a graspable object in his or her peripersonal space and when he or she actually grasps it (see Rizzolatti & Sinigaglia, 2008).[15] This highlights the tight coupling of perception and action in the brain (see also Jeannerod et al., 1995), showing how the visual properties of an object may be understood to afford possibilities for action.

In embodied cognitive science, the discovery of mirror and canonical neurons may appear to confirm that our capacity to understand the actions of others and to perceive affordances in the environment is grounded in representations produced by the brain, that is, by patterns of neural firing that produce "simulations" or "bodily formatted representations" (for a discussion, see Gallagher, 2017). However, the role of these neurons can also be explained in nonrepresentational ways (i.e., through the types of representation-neutral dynamical models just mentioned). Moreover, and as we will see in greater detail later, it is posited that understanding the development and functioning of these sensorimotor neurons involves examining the brain, body, and environment as an integrated and evolving complex system.

Indeed, an enactive perspective on these neurons (and neural activity more generally) will be useful for understanding important aspects of musical experience, which necessarily entails the development of various forms of bodily, neural, emotional-affective, and empathic relationships with the people and things that make up the extended environments we inhabit.[16] For now, we can note that, although the enactive approach does not deny that human minds are capable of representational forms of thinking and cognizing, it argues that mental life cannot be reduced to this. Neurons and brains do not produce representations (or think); people do (Hutto & Myin, 2012). And because people do not live in a vacuum, the ways they do this always involves a history of bodily interaction within a sociomaterial and cultural environment (Cuffari et al., 2014; Hurley, 1999, 2001; Hutto & Myin, 2017; Malafouris, 2013, 2015). In other words, while certain forms of cognition may well involve representational content of some kind (e.g., imagination, declarative memory), the enactive position does not see this as foundational for cognitive processes; rather, representation is understood as a secondary or derivative aspect. And while some enactivists are willing to entertain the possibility that it may be useful to understand aspects of living systems in terms of representations in certain contexts, they are also wary of suggesting that this should imply an input-output structure—where cognition involves the internal recovery of a pregiven external reality or where the generation of meaning is externally imposed (e.g., computing devices)—as this would obscure the (relationally) autonomous status of living cognitive agents (Colombetti, 2014, p. 57).

The Dynamical Perspective

As we have seen, enactive theorists think of the "co-arising" organism-environment relationship as an ongoing "circular" phenomenon, where sense-making involves all parts of the body-brain-world system influencing each other in a nonlinear, or recursive, way. As such, the enactive principles of autopoiesis, autonomy, and adaptivity resonate with the phenomenon of "self-organization," which characterizes complex dynamical systems more generally (including nonbiological varieties). In line with this, "dynamical systems theory" (DST) has helped to guide enactivist thinking since its inception as it offers tools for describing how self-making (i.e., autopoietic) systems emerge and develop (Varela, Thompson, & Rosch, 1991).

DST is a branch of mathematics that explores how complex systems—from weather and climate patterns to insect colonies and more (Strogatz, 1994, 2001)—self-organize, maintain structural coherence, generate recurrent patterns of behavior, and evolve over time through networks of mutually influencing processes (Beer, 1995a,b; Thelen & Smith, 1994). Because such systems are inherently interactive, they cannot be properly described using linear mathematical modeling (e.g., static point slope equations). This is because complex systems are not best examined in terms of discrete events, fixed properties, or linear types of causality, involving, for example, a strict input-output structure. Rather, such systems develop in a recursive way and must be examined in terms of the continuous interaction of temporal trajectories. These trajectories can be expressed mathematically in terms of nonlinear differential equations, which allow the functions of a dynamical system to be described in relation to its derivatives (or its rates of change). This approach allows one to map a much wider range of relationships between temporal variables and to make distinctions between local and global features in ways that are not possible with linear modeling.

Put another way, DST can help to reveal how the trajectories of a complex system converge and diverge as patterns of relative stability and instability—constraint and entropy—and how this results in the development of various relationships and patterns that characterize the state of the system over various time scales. These areas of stability and instability are referred to as "attractors" and "repellors," respectively, and are often represented as a topographical space, or a "phase portrait," which describes how the possible states of a given system evolve. Areas where the system's state tends to evolve toward an attractor are shown as "basins of attraction." Over time, perturbations to constraints of the system can lead to phase transitions that produce qualitative shifts in the global state of the system, which is described by a new topology. These perturbations are understood to occur at two levels of description. To see how this is so we can think about how changes in heat added to an oil-filled pan perturb the local interactions of the oil molecules (first-order constraints), which, in turn, affect the global behavior of the oil in its totality (second-order constraints). Such macrolevel patterns, which are observable as changes in the amplitude of convection rolls of the oil, then impose further reciprocal constraints on the movement of the molecules (Haken, 1977). The term "emergence" is used to refer to distinct properties or patterns of self-organizing behavior that arise from the recursive interactions of such systems (Friston, 2009; Thompson, 2007).

Additionally, nonlinear equations can also model how the trajectories of two or more systems dynamically interact with each other. A relatively simple example of this can be observed when wall-mounted pendulums mutually constrain one another, resulting in synchronization or "entrainment" over time (see Clark, 2001). Similar insights can be applied when thinking about the relationship between living systems and their environments. Consider again the model of the animal-environment system outlined in figure 2.3. The recursive relationship between the two can be expressed with the following equation provided by Randall Beer (1995a):

$$X_A = A[X_A; S(X_E)]$$
$$X_E = E[X_E; M(X_A)]$$

Here A and E describe animal and environment as dynamical systems. Note that each equation (system) contains the other within it as a variable, so a change in the state of one will necessarily entail a change in the other. These systems are coupled *nonlinearly*, where the coupling parameters are described by the functions $S(X_E)$ and $M(X_A)$. The first function concerns environmental variables on organism parameters, while the second concerns the organismic variables on environmental parameters. Importantly, this dynamical stance shows that organism and environment are not best understood as separate domains, but as aspects of "one non-decomposable system" that evolves over time (Chemero, 2009, p. 26).

DST also can help model how multiple living systems co-organize patterns of behavior that are relevant to their continued survival and well-being in contingent and sometimes rapidly changing environments. Such processes of participatory sense-making occur over various time scales, spanning the level of the individual and the collective (or even an entire ecosystem), where interacting agents may be understood to influence and help sustain each other's behavioral dynamics (to act as environments to each other). This results in what is sometimes referred to as "higher-order" self-organizing systems. Here cognitive agents act adaptively as "constraints" on each other that keep the shared system from dissolving into nondifferentiation (decay, death). That is, they "work" to transform energy into meaningful activity: movement, nutrition growth, reproduction, socialization, and the development of shared actions and signals that express the state of the system. Importantly, such processes are understood as inherently goal directed, and thus living cognitive systems may be understood as "teleodynamic" systems (Deacon, 2012; Walton et al., 2014).

This dynamical perspective will help us explore a range of musical phenomena associated with musical emotions and empathy, musical development in infancy, music and human evolution, and musical creativity. Indeed, listening, improvising, and coordinating musical actions more generally all require reaching out to, and transforming, musical environments in interactive ways. As we discuss, such engagements involve (collaborative) moment-to-moment adaptations to changes in the shared musical ecology so as to maintain stable relationships. However, they also entail initiating perturbations that influence the state of the system, which result in the self-organization of new relationships, perceptions, and shared possibilities.

Life, Mind, and Culture

Enactivists trace a continuity between the basic, affectively motivated sense-making of simpler organisms and the richer manifestations of mind found in more complex biological forms (Di Paolo et al., 2017; Froese & Di Paolo, 2011). Where the meaningful actions of single-celled and other simple creatures are associated with basic factors related to nutrition and reproduction, more complicated creatures will engage in ever richer forms of sense-making activity and thus exhibit a wider range of cognitive-emotional behaviors (Froese & Di Paolo, 2011). For social animals, this includes *participatory* forms of sense-making that involve the enactment of emotional-affective and empathic modes of communication between agents and social groups, which coincide with the development of shared repertoires of coordinated action (De Jaegher & Di Paolo, 2007; Di Paolo, 2009). It follows, then, that for highly complex organisms such as human beings, the life-mind continuity also involves a lived developmental history of social embodiment that is embedded within a domain of "consensual action and cultural history" (Varela et al., 1991, p. 149).

In line with these insights, the subsequent chapters explore human musicality as a fundamental human sense-making capacity that reflects the adaptive, creative, emotional, improvisational, sociocultural, and technological facets of the embodied human mind. To do this, we draw on the DST perspective discussed here as well as the recently developed 4E (embodied, embedded, extended, and enactive) framework to help describe the self-organizing and participatory sense-making processes associated with an enactive music cognition approach. Let us then conclude this chapter by outlining this 4E perspective.

The 4E Framework

We should note first that there are several ways of thinking about "4E" cognition. That is, the embodied, embedded, extended, and enactive dimensions that constitute the 4Es can each be understood in terms of their own programs of research and theory (Newen et al., 2018). Earlier we began to consider the differing conceptions of what "embodied cognition" entails, which remains a topic of congoing debate in cognitive science and philosophy of mind (Chemero, 2009; Rowlands, 1999, 2010; Shapiro, 2011; Wheeler, 2005). Likewise, the "embedded" dimension has connections to ecological psychology, and both the embedded and extended dimensions have precedents in so-called "distributed cognition" (Kirsh, 1995; Sutton, 2006, 2010; Sutton et al., 2010). Some approaches in these areas maintain a reliance on representational content and are therefore not wholly compatible with the enactive perspective (see Chemero, 2009; regarding the "extended mind thesis" see Di Paolo, 2009; Gallagher & Crisafi, 2009; Wheeler, 2010). Moreover, while some theorists agree that the body plays a fundamental role in cognitive processes (e.g., in the generation of corporeal and action-based representations), they nevertheless deny the possibility that cognition can be in any sense "extended" beyond the body and brain (Adams & Aizawa, 2009). While these debates are fascinating, it is not our intention to rehearse them here as they are well documented in other books and articles (e.g., Menary, 2010a,d). We are aware, however, that the apparent disparities between these schools of thought may result in a misconception that a 4E approach to (music) cognition risks being theoretically disjointed. To alleviate this issue, we should point out that although insights from the various perspectives associated with "E" cognition inform our approach, our conception of a 4E framework for music cognition is guided more by how the embodied, embedded, and extended dimensions emerge and interact within the broader context of the life-mind continuity central to enactivism. Indeed, there is a strong sense in which these dimensions have been present, or at least strongly implied, in enactivist theorizing since its origins.

As we have seen, a first consequence of the enactive conception of cognition-as-sense-making orientation is that a body is necessary for mental life—the mind is essentially *embodied*. For "body," we may in fact refer here to "two bodies," as captured by the German words *Körper* and *Leib* (see Husserl, 1950/1992; Kyselo, 2020; Moran, 2017). The first term describes the objective, physical body—a piece of the world that is measurable and that has an extension. The second term defines the body as a feeling entity—the living body that we experience in our everyday life. To say that cognition is embodied means that our mental life depends directly on these two (descriptive and felt) conceptions of the body. It is through our body-in-action (which includes the brain) that we develop viable forms of engagement with our world (perception, action, prediction); it allows us to manipulate objects and interact directly with other agents in ways that are meaningful. We do music, for example. We go to concerts, we learn how to play a new piece, we listen to our favorite track, and we move to it. We make sense of the world through our active participation in its changing demands, enacting musical conducts and nonmusical behaviors within environments, as well as inner (e.g., metabolic, emotional) configurations that are appropriate for the context and that allow for the maintenance of our organizational and structural complexity as living, social animals.

As we have also seen, the explanatory dimensions required for the study of cognition from an enactive perspective cannot be limited to the brain or the body. They must go beyond the physical boundary of the individual because active sense-making entails action, interactions, and other forms of adaptive relational behavior within the sociomaterial niche we inhabit and are thus *embedded* in. These aspects can only emerge (and be understood) in the context of the entire brain-body-world system (Beer, 2005). Likewise, participating in a musical activity involves social, cultural, and material ecological constraints that play out in various ways according to context. As such, within the brain-body-world system, cognizers encounter biological and nonbiological entities, as well as technologies and environmental features that can be helpful in achieving cognitive tasks (e.g., remembering something). Living systems establish relationships with such factors, resulting in couplings that are relevant to the organism's well-being as needs and goals are developed contextually (Malafouris, 2015).

This entails the constitution of hybrid extended cognitive systems (Menary, 2006, 2010a,b,c). For example, as one can think of how devices and objects (computers or notepads and pencils) can help us to remember an address or a password, develop ideas, or allow us to do mathematical calculations that we cannot accomplish "in the head," one can envision how musical instruments become part of the musician's cognitive domain such that they can often be experienced simultaneously as part of the performer and as part of the musical environment being enacted. Indeed, the phenomenology of engaging with tools reveals a distinction between those that merely "extend" the body and those that come to be "incorporated" by the body. The latter effectively become prostheses. This phenomenon can be observed in the mastery of a musical instrument, in which the instrument is no longer experienced simply as an object—rather, it gains a certain "transparency," as there is a sense in which the (musical) world is experienced and enacted though it.[17]

To begin to put these points in context we can think of a musician, say a drum kit player, who is situated within a certain social and material environment (figure 2.5). The ways the drummer makes sense of this environment involve engagements with all kinds of extended factors, including the machines she uses to store and create information, keep time, do calculations, and so on; music playback devices; her instrument; as well as other musicians and their instruments. More broadly, her musical activities

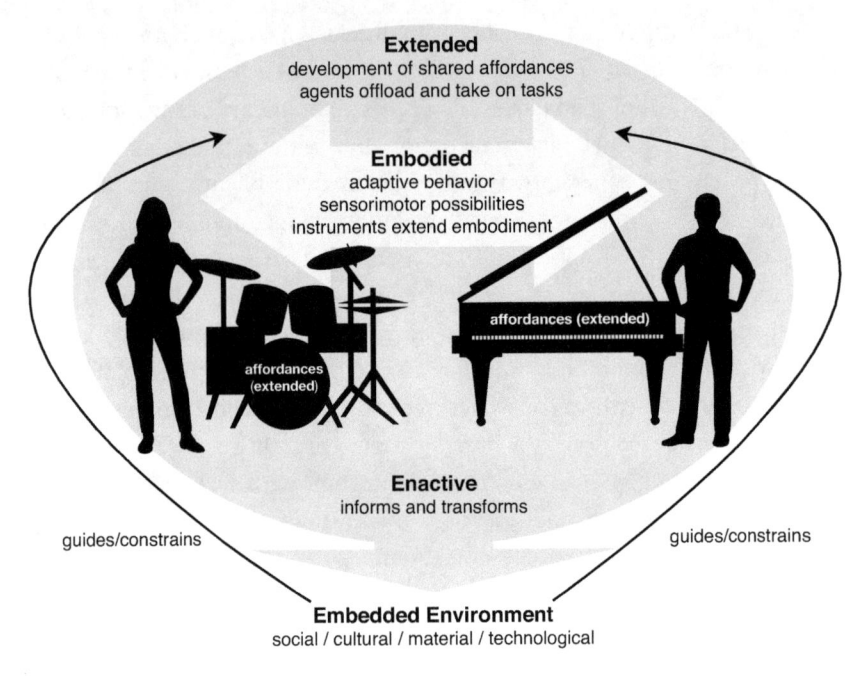

Figure 2.5
The interconnected sense-making ecology of two musical agents.

are also guided by, and affect, the social and cultural world she is embedded within. Her development within this milieu involves the ongoing enactment of meaningful patterns of action and perception in relation to environmental factors—affordances are continuously updated (sometimes very subtly) over various time scales through learning, skill development, "growth," and as she selects and adapts to new features and perturbations in the musical ecology (Reybrouck, 2001, 2005a,b; Ryan & Schiavio, 2019). These last factors can include the introduction of a novel instrument, new coperformers, a different acoustic space and/or audience, new technology, or the more immediate shifts in dynamics and timing that occur in performance.

Shifting perspective, think of a large performing ensemble, such as a symphonic orchestra. The relational dynamics of a system like this involve an array of sonic, bodily, emotional, visual, and tactile forms of communication and perception—complex sensorimotor loops (sound, vision, affect, and movement) whereby individual performers constantly take on and offload various cognitive tasks to and from the extended environment to sustain a musical world they coenact. This involves, for example, entraining with the pulse and dynamic indications provided by the visible gestures of a conductor, leading and following phrasing and timing within a section (gesture and sound), as well as the ways that musicians develop and employ the affordances of their instruments as, again, "incorporated" aspects of their embodiment. From this perspective, it is the *extended* cognitive system involving, in this case, the coupled activities of a conductor, musicians, and instruments that constitutes the explanatory field.

Here we should clarify that our understanding of this "extended" dimension of cognition is in no way "externalist." *Internalists* claim that what counts as cognition should be accounted for solely in terms of the operations that take place within the personal domain of a given agent (usually the brain). *Externalists* agree that cognition can sometimes involve processes that occur entirely in the head, but they also argue that if conditions outside the head are operationally equivalent internal processes, then those external factors should be considered as part of a cognitive process (and can even possess externally located representational content). The latter is referred to as the "parity principle" (Clark & Chalmers, 1998). As Di Paolo (2009) points out, however, the parity principle "relies both on simple prejudices about inner and outer as well as on intuitions about cognition" that are tied to the same "boundaries between inner and outer that it wishes to undermine" (p. 10). In line with such concerns, most current proponents of the extended mind thesis do not endorse the parity principle. Instead, they have adopted "integrationist" perspectives that better describe the dynamic process of structural coupling that occurs between animal and environment. Notably, this orientation connects with the so-called "second-wave" approach to the extended mind initiated by cognitive scientists John Sutton (2010, Sutton et al., 2010) and Richard Menary (2006, 2007, 2010b,c,d), which aims to address the problems associated with the parity principle by positing a more relational view, in which "extended" describes the ways organisms integrate endogenous resources with aspects of their niche. Other thinkers, such as philosopher Michael Kirchhoff (2012), have indicated the need for a "third wave" that, among other things, better accounts for the role of sociocultural practices in shaping extended multiagent cognitive systems. We will discuss these perspectives again toward the end of the book, making a few suggestions for how musical practices may inform third-wave approaches. For now, we can note that second- and third-wave perspectives are well positioned to dialogue with the co-arising view mind and world that is central to enactivism.

Since its origins, enactivism has aimed to dissolve internalist/externalist prejudices by arguing that cognition can never be reduced to inner or outer domains (cognition does not occur "inside" or "outside" of an agent). Accordingly, mind cannot be properly understood in terms of a detached external apparatus that operates *as if it were an internal process*. Rather, cognition is the "relational process of sense-making that takes place between a system and its environment" (Thompson & Stapleton, 2009, p. 4). And so, if we think about how an agent "offloads" a task to some environmental feature (e.g., using a metronome to help keep time while practicing a difficult musical passage, or using a notebook to help with memory), that feature is always a part of the sense-making activity of that agent in interaction with an environment. Likewise, when social agents take on and offload tasks to and from each other, they engage in participatory forms of sense-making that also reflect the relationally extended nature of cognition, in which they enact affordances for each other. Here again, the enactivist conception of environmental "affordances" sees them not as pregiven aspects of an external domain, but as emergent properties of the adaptive relationality between autonomous agents and the world. In brief, it is this integrative, relational approach to extended cognition that will help us to explore important aspects of musical experience, which, as we have seen, necessarily involve complex interactions between people and things.

In being embodied, embedded, and extended, cognition is necessarily *enactive*. Again, this means that living systems are not simply responders to environments; they bring forth their own domains of meaning, most fundamentally through the development of repertoires of actions that are guided by principles related to the organism's internal coherence (e.g., homeostasis, thermodynamics, regulation, nutrition, reproduction). Living systems, in other words, play an active role in shaping the extended cognitive niche in which they are embedded.

In summary, the 4E framework that guides our discussion of the musical mind is constituted by the following overlapping dimensions:

1. **Embodied:** Cognition involves the entire body of the living system. It cannot be fully described in terms of abstract mental processes (e.g., in terms of representations).

2. **Embedded:** Cognition displays layers of codetermination with physical, social, and cultural aspects of the world. It is not an isolated event separated from the agent's ecological niche. Instead, it displays layers of codetermination with physical, social, and cultural aspects of the world.

3. **Extended:** Cognition necessarily involves interactions with the features of the material and social environment. This is often examined of in terms of how agents "offload" cognitive processes to other biological beings, non-biological devices, or environmental features to serve a variety of functions that would be impossible (or too difficult) to be achieved by relying solely on the agent's own neural and/or bodily capabilities. However, the ways in which these extended processes unfold cannot be understood strictly in terms of internal and external domains. This is because these processes are always interactive and relational; they reflect the co-arising relationship between living organisms and their environments.

4. **Enactive:** Cognition is conceived of as the set of meaningful relationships determined by an adaptive two-way exchange between the biological and phenomenological complexity of living creatures and the environments they inhabit and actively shape.

We can add a further characteristic implied by this orientation. This involves the fundamentally improvisational nature of living minds (an aspect of adaptivity; Torrance & Schumann, 2019), which is echoed by Varela, Thompson, and Rosch (1991): "[T]here is always a "next step" for the system in its perceptually guided action . . . the actions of the system are always directed toward situations that have yet to become actual. Thus, cognition as embodied action both poses the problems and specifies those paths that must be tread or laid down for their solution. . . . Living cognition is like a path that is laid down in walking" (p. 205). In later chapters, we explore the possibilities this framework offers across a range of musically relevant areas, including musical emotions and empathy, the evolution of musicality in our species, musical development in infancy, music creativity, and music education.

In the next chapter, we examine the computational approach to mind in a little more detail, with a special focus on how it tackles the difficult subject of conscious experience. This will allow us to consider the advantages and shortcomings of a number of information-processing models of experience in a musical context and to explore the radical alternative offered by enactivism through the lenses of phenomenology. Since its inception, the enactivist approach has sought to integrate science, theory, and first-person experience and to develop what philosopher Maurice Merleau-Ponty refers to as an "entre-deux" between subjective and objective vantage points (see Varela et al., 1991). It therefore maintains a close affinity with the tradition of phenomenological philosophy. As we will see, investigating the phenomenology of musical experience can offer useful conceptual refinements to the 4E model, as well as a richer, embodied conception of what musical consciousness entails.

3 Music and Consciousness

How is it that a diminished seventh chord played on a piano or an open E string plucked on a guitar gives rise to the experiences it does?[1] How can we explain the feeling of closure evoked by a cadence in tonal music? How do we make sense of the sounds, rhythms, and movements produced by a Ghanaian drumming and dance ensemble, or the experience of an acousmatic[2] performance involving electronic music or *musique concrète*? These questions all evoke an aspect of our lives that is at once mundane and very mysterious: conscious experience. Indeed, while the idea that at least part of our lives should involve an awareness of our perceptions and feelings is not controversial, explaining just how this is so has proven to be remarkably difficult, so much so that the question of consciousness is often referred to as "the hard problem" (Chalmers, 1996).[3]

The enactive approach does not claim to solve the hard problem of consciousness.[4] However, it does cast the issue in a promising new light by drawing fundamental continuities between biological and mental processes. This means that consciousness can be explored not only as a product of brain function but also in terms of the interactivity of brains, bodies, and environments. Importantly, the enactive perspective decenters the traditional assumption that "linguistic access" (the ability to engage in spoken language or to access nonconscious language-like processes in brain) is a necessary component for cognition and consciousness. As we noted at the end of the previous chapter, enactive theorizing draws on insights from phenomenological philosophy, which has traditionally highlighted the role of the situated, feeling body for mental life (Thompson, 2007; Varela et al., 1991).

To better understand what an enactive-phenomenological perspective can offer for the field of music cognition, and the question of "musical consciousness" more specifically, we consider three perspectives on consciousness that are based in the computational model of mind. We argue that these models, elegant though they are, offer only limited accounts of what musical experience entails. In response to this, we then explore insights drawn from early Gestalt psychology, which inform the phenomenological and enactive orientation we develop in subsequent chapters. Before we begin, we should note that the first half of this chapter will not deal directly with musical concerns, but with outlining key aspects of the computational approach to mind and related perspectives on consciousness. We feel that this is important to add to the discussion because theorizing in music cognition traditionally has relied on information-processing assumptions, often without much precision and criticality over just what this model of mind entails.

Experience and the Computational Mind

A central concern for philosophy of mind has been to explain how mental states are physically instantiated in the brain and how these states result in intelligent behavior. In other words, one of its purposes is to describe what goes on between raw sensory input and intelligent behavioral output, or how "internal" processes are related to "external" situations and actions in the world. For some thinkers, the solution to this problem involves "symbolic computation" (see Varela et al., 1991, pp. 6–8). If symbols can be instantiated physically in the brain (through patterns of neural firing) and ascribed semantic value, they could also be subject to computational operations that function "syntactically" according to the language of the system (Fodor, 1983; see also Dreyfus, 1979; Haugeland, 1981, 1985). The representations produced by such operations at lower levels could therefore be logically manipulated (computed) to produce more complex representations and intelligent outputs that correspond with relevant aspects of the "external" environment (see Pinker, 1997). Accordingly, cognition as information processing is often understood to entail something like a "language of thought" or "mentalese" (Bechtel et al., 1998), in which the system renders "propositional" outputs that represent the world intentionally, that is, that cognize the "aboutness" of things and situations in the environment.

As mentioned in chapter 1, a central aspect of this model of mind is that the computational operations it describes must be played out at the nonconscious or "subpersonal" level. This means that not only are we not aware of such processes but that we can *never* be aware of them (Dennett,

1978; Pinker, 1997). Here, questions have been raised about how the representational outputs of information-processing mechanisms are recognized by the system beyond the (nonconscious) mechanics of system syntax (Potter, 2000; Still & Costall, 1991). This has led to philosophical concerns over where and how "experience" takes place, leading to issues with "homunculus"[5] metaphors and related problems of infinite regress (Dreyfus, 1979, 1992, 2002; Searle, 1990). Advocates of the computational mind have responded to such concerns by claiming that the use of fanciful homunculus metaphors to describe subpersonal systems is merely provisional and that these explanatory stopgaps will eventually be "discharged" by future research associated with neural networks and artificial intelligence data structures (Dennett, 1978; Pinker, 1997, p. 79).

In line with such aspirations, some computational approaches have posited models of mental content that are more closely integrated with the functionality of the system and are arguably less abstract. By this view, mental states are determined by their relations with other mental states and/or their degree of correspondence or fit with the external world (i.e., behavior and perception). A mental state is thus understood primarily in terms of its function for the cognitive economy of the system. For some, this orientation offers a way to avoid the problematic philosophical issues mentioned previously when it sees mental representations as being built into the operations of the system at a level between physical implementation and behavioral output in the information-processing hierarchy.[6] Thus, it is argued that mental states do not require a separate ontological status (à la Descartes), nor some inner observer (a homunculus). This doctrine, known as "functionalism," has been highly influential for research in cognitive science, artificial intelligence, robotics, and psychology in the last fifty years (Block, 1980; Van Gulick, 1982).

But while the functionalist approach has been the basis of many outstanding achievements in the sciences of mind, it has traditionally had little to say about the issue of consciousness. One reason for this is that consciousness and cognition are not always understood as synonymous. Generally speaking, from an information-processing perspective, cognition requires the ability to produce representations and intentional states, and conscious awareness is not necessarily a prerequisite for this to occur. Thus, accounting for phenomenal experience has not always been the prime area of investigation for many researchers in cognitive science, and notions of "access-consciousness" and "executive functions" are discussed with only rather vague suggestions of how this may correlate with consciousness as sentience (e.g., see Pinker, 1997, pp. 131–148).

The Mind-Mind Problem

Attempting to offer a purely materialist account of experience is tricky business. Thinkers who tackle this issue from an information-processing perspective often turn toward examining what is assumed to be a necessary connection between consciousness and linguistic competence (i.e., that conscious access always includes verbalizability of some kind; see Churchland, 1983). In doing so, however, "questions about the nature of consciousness itself [(i.e., the hard problem)] are left judiciously to one side" (Raffman, 1993, p. 125). Nevertheless, some supporters of the computational mind have refused to let the problem of consciousness slip away altogether. As the linguist Ray Jackendoff (1987) has argued, explaining cognition must involve more than describing the relationship between an information-processing brain and a mind that is inaccessible to consciousness (the "mind-brain" problem). One must also account for the relationship between the computational mind and the phenomenological mind—the "mind-mind problem" as he calls it (p. 20). Jackendoff attempts to deal with this issue by developing a theory of "intermediate-level representations" that are understood to support or project conscious awareness. An important implication of this approach is that it puts phenomenological constraints on computational models: "The empirical force of this hypothesis is to bring phenomenological evidence to bear on the computational theory. ... Thus, if there is a phenomenological distinction that is not yet expressed by current computational theory, the theory must be enriched or revised" (p. 25). This insight points to the need to include conscious awareness in the study of human cognition—to ensure that the structural analysis of our minds and the development of cognitive theories are continually enriched by a disciplined examination of phenomenological distinctions. As we saw, this concern is also shared by enactivism, which seeks to integrate scientific knowledge and first-person experience.

Another important insight offered by Jackendoff concerns the multiplicity of conscious experience. Our awareness of the world and ourselves is modal. We have distinct forms of consciousness that correspond to our sensory capacities: visual, auditory, tactile, and so on. His theory attempts to account for this by claiming that "each modality of awareness comes from a different level or set of levels of representation. On this view, the disunity of experience arises from the fact that each of the relevant levels involves its own special repertoire of distinctions" (1987, p. 52). What is notable here is that instead of beginning with the notion that consciousness is unified and ultimately traceable to some unique locus, Jackendoff suggests that "consciousness is fundamentally not unified and that one should seek multiple sources" (p. 52). This recognition of the fundamental multiplicity of the cognizing subject is cause for a good deal of tension as it goes against the common assumption that there should be such a thing as a stable unchanging "I" at the center of experience—an insight that also resonates with a number of enactivist and phenomenological principles (see Varela et al., 1991).

Later we consider how similar insights are developed from a phenomenological and enactive perspective and how this may shed light on the pluralistic and cross-modal nature of musical experience. For now, we note that, despite the new insights Jackendoff provides, his approach is nevertheless unable to avoid a major issue traditional computational models face when confronted with the question of consciousness. The problem, in a nutshell, is that if cognition is indeed grounded in subpersonal computations, then it seems difficult (if not impossible) to see how consciousness could have any causal influence over such processes. And indeed, Jackendoff (1987) suggests that it does not, which leads him to the rather anticlimactic conclusion that consciousness may "not be good for anything" (p. 56). In making this claim, he appears to assume that understanding the phenomenological mind should be limited to largely "mindless" forms of everyday experience (conditioned responses to environmental conditions, and so on). This is problematic because, as Varela, Thompson, and Rosch (1991) note, "[Jackendoff] considers neither the possibility that conscious awareness can be progressively developed beyond its everyday form (a strange omission given his interest in musical cognition) nor that such development can be used to provide direct insight into the structure and constitution of experience" (p. 54). Despite such challenges, philosophers of mind have continued to pursue the problem of experience, frequently invoking the notion of "qualia," which in its original usage refers to the intrinsic qualities of a subjective experience associated with a given sensory event (Jackson, 1982).

In current research, the term has come to describe subjective experience more broadly (Haugeland, 1985; Nagel, 1974), including musical varieties (see Huron, 2006; Zentner, 2012).[7]

Questioning Qualia

Competing conceptions of qualia have emerged over the past few decades.[8] However, the most prevalent approach posits that qualia should consist of a "quartet of attributes," in which qualia are understood as *ineffable, intrinsic, private,* and *directly apprehensible* in consciousness (see Dennett, 1988). Here, qualia are thought to be *phenomenal properties present in the perceiver*. The idea is that "for something to look red to someone is for it to give rise to an experience with a certain qualitative or sensational property. Its looking red consists in the fact that it gives rise to that qualitative state in a person" (Noë, 2004, p. 133). It should be noted that, from this perspective, the experience and its sensed properties are understood as two different features that may be present independently from each other. For example, although we can have a conscious experience associated with the subjective sensation of hearing a passage of music by, say, Clara Schumann, the qualities of such an experience may not be reducible to the environmental sense data associated with such an event. This is to say that the measurable physical properties of the events themselves (e.g., the changes in air pressure impacting the auditory system) are not enough to capture how the actual experience feels to the perceiver. Even the most accurate quantitative analysis of the sound waves created by a performance cannot fully capture the listener's subjective experience of it. In other words, the correlation between the physical attributes of a musical sound (e.g., its timbre, its duration, amplitude changes) and the feeling it evokes, does not appear to admit any direct causal claim: the same stimulus (or stimuli) could potentially generate different experiences in other perceivers, or even in the same perceiver over time or in a different context. Because of this, the conscious states that qualia give rise to are often understood to be "ineffable," meaning they have no direct correlation with the objective features of the stimuli and are thus not accessible to "direct" forms of description—for example, "objective" descriptions that use conventional units of measurement such as the box is 10 cm in height, 9 cm wide, 12 cm deep, and weighs 5 kg (see Roholt, 2014).[9]

A problem with this conception of qualia is that it remains rather vague about how inner mental content results in experience. It implies some internal phenomenal manifold (an "inner theater"), conjuring again the issue of who does the experiencing internally (e.g., the homunculus problem). These kinds of concerns have led some prominent thinkers to posit approaches that do away with the notion of qualia altogether.[10] Notably, the philosopher Daniel Dennett (1988, 1991) has argued that qualia-based theories maintain unresolvable (and unnecessary) issues associated with the hard problem of consciousness—which, again, revolves around explaining how qualia, whether originating in the environment, through sensory stimulation, or from neural activity, emerge into the daylight of consciousness.

A Propositional Proposal

One of the major assumptions Dennett (1978, 1979, 1988, 1991) seeks to overcome concerns the mental space, or the "inner theater," in which experience is assumed to take place. That is, again, the problem of how physical states in the brain give rise to conscious experience or how the lower-level representations are presented to consciousness. Dennett addresses this by arguing that although many of our day-to-day experiences appear to involve "momentary, wordless thinkings or convictions," it is illusory to think that this is indicative of some inner phenomenal manifold that corresponds qualitatively with the public verbal reports we issue. Dennett claims that contrary to our assumptions, there is in fact no "inner" theater at work; the raw feelings we have and the stories we tell about experience are essentially "convenient fictions."

In reality, Dennett claims, consciousness emerges from a series of propositional episodes (see Dennett, 1979) produced by nonconscious or preconscious mechanistic processes in the brain. This is an extension of the functionalist approach discussed previously: the mental processes that underpin consciousness are accomplished entirely through physical causal means via the manipulation (computation) of abstract markers (symbols) that are physically instantiated through patterns of neural firing; this allows for the computation of propositions (judgments) about a given (neural) state and how it compares to other states, or previous states of the same kind (again, representational content is built into the functionality of the system).[11] Importantly, the neural firing that results in propositional judgments about the tuning of this or that pitch, for example, do not in themselves possess the qualities of being sharp or flat. The redness of an apple does not involve some instantiation of "red" in the neural activity of the perceiver. In this way, no analogous "quality" must be presented to some internal perceiver. Put simply, Dennett argues that the inner domain involves no qualitative aspects at all. Here we find only the complex storm of firing patterns associated with the neural mechanisms that process sensory information to produce relevant judgments or "propositions" about the state of the outer world.[12]

This propositional conception of consciousness underpins Dennett's (1988) rejection of qualia as standardly conceived. His alternative is to understand conscious access to perceptual states in terms of "acts of apparent re-identification or recognition" (p. 70). He refers to the physical patterns of neural activation that result in such judgments as "phenomenal information properties"—or "pips" for short. Simply put, pips are understood as unique neural pattern-recognition devices that develop through experience. This results in "discrimination profiles" that allow properties to be identified and reidentified as causally connected to an original stimulus (or set of stimuli). Moreover, Dennett argues that because such neural activity is propositionally formatted, it should be directly translatable to other propositional formats, such as spoken and written language. Accordingly, he thinks that with enough practice we should be able to verbally describe the judgments that result from our pips and discrimination profiles *directly*—that is, without recruiting the use of metaphors, analogies, or other *indirect* forms of description. This means that Dennett sees ineffability as something that may be largely overcome with sufficient training and linguistic expertise.[13]

In all, Dennett's perspective posits a more direct relationship between brain functions and conscious experience, while still retaining an essentially computational-propositional understanding of what mental processes entail. And where Jackendoff appears to conclude that consciousness may have little relevance for survival, Dennett suggests that consciousness may have emerged in human evolution as an adaptive function associated with formation of statistical comparisons between the information stored in our pips and the sensory data received from the environment. This would entail processes of anticipation and recognition that play out at lower and higher levels of processing, in which predictions are confirmed or thwarted resulting in altered or new discrimination profiles that influence, and are influenced by, conscious experience—as Dennett (1991) writes, "all brains are, in essence, anticipation machines" (p. 177).

Whence the Nuances?

The cognitive musicologist David Huron (2006) adopts a Dennett-like perspective in his influential approach to musical experience. Here, musically generated emotional responses are understood as products of computational mechanisms associated with the satisfaction or frustration of musical expectations (essentially, statistical operations instantiated in the brain as neural firing patterns). However, others have argued that certain aspects of musical experience pose problems for Dennett's view of consciousness. Notably, the philosopher Diana Raffman (1993) claims that Dennett's approach does not permit a convincing account of *nuances*. The reason for this is that nuances are generally understood as qualities of (musical) experience that are available to conscious perception while not being explicitly conceptualizable. Indeed, experience is thought to be full of such "nonconceptual content," which is too complex, detailed, or "fine-grained" for our descriptive capacities to deal with. Raffman argues that these aspects of experience cannot be explained in terms of the propositional schema Dennett endorses.

In the visual modality, for example, we experience many more colors than we can possibly name individually (Tye, 1995, 2000, 2002)—and therefore the concept of "redness" is used to roughly describe what in fact entails a range of possible color experiences. In musical-auditory contexts, experience is often discussed in terms of nuance categories that are discrete from one another (e.g., pitch, timing, timbre, tuning, loudness). While certain aspects of such phenomena can be measured scientifically using various instruments, it is difficult to capture the experience of musical nuances through such objective forms of categorization and description. Nuances exhibit fine-grained dynamics, and the category boundaries we place them in for analysis are not explicit in most musical experience, which is more holistic in nature. Accordingly, we do not use direct forms of description to account for musical experiences in everyday speech. Instead, we employ cross-modal metaphors of space, time, texture, movement, location, narrative, and bodily-affective states: "The pitch goes *up*," "That section is *dragging*," "This chord voicing is *bright*," "The trumpet is too *soft* in that passage," "I'm *lost*," "We were really *together*."

To explain this apparent ineffability, Raffman (1993) develops a hierarchical conception of musical perception. At the "higher" schematic level (e.g., compositional structure) we may conceptualize experience and thus report on it objectively, possibly in the same way that Dennett suggests. Nuances, however, are processed before propositional schemas and therefore cannot be expressed verbally in wholly objectivist (formal, functional, or quantitative) terms because they are not directly accessible to language. As she writes, the "limits of our schemas are the limits of our language, and *qua* perceivers we are so designed that the grain of consciousness experience will inevitably be finer than that of our schemas, no matter how long or how diligently, we practice" (p. 136). Put simply, Raffman claims that although certain aspects of musical experience are schematized and thus accessible to direct linguistic description, others at the nuance level are not. Accordingly, she argues that this lack of linguistic-schematic access means that nuances cannot be remembered, reidentified, and categorized. In brief, Raffman sees the nuance level as the "shallowest" level of representation of the musical signal to which the listener has conscious access. That is, there is no prior level of representation in the information-processing chain, and thus nuance representations arrive "unheralded" to consciousness.

In computational conceptions of music cognition, like Raffman's, such nuance-level representations would have to be physically instantiated in the "hardware" (the brain in this case). This raises the issue of just how such representations could emerge unheralded directly from the shallowest level of information processing into the daylight of experience. How such nonconceptual content becomes experienced and "knowable" remains unclear. Moreover, Raffman posits a clear distinction between the structural and nonstructural features of music. This means that, although the nuances themselves are not conceptualizable, the *objectives* of nuances are. This is because they are concerned with highlighting the compositional structure of the work: for example, tuning pitches (nuance) to introduce a new key (structure) and tempo variations (nuance) to highlight cadences (structure). Beyond such structural concerns, however, nuances (on their own) appear to have no relevance: for Raffman *there are no nonstructural nuance objectives*.

The philosopher Tiger Roholt (2014) provides a critical reading of Raffman's approach, arguing that it tends to maintain Western musicology's traditional bias toward the composed "work" (see also Small, 1998). As he notes, many musics of the world do not always share this focus on structure.

For example, in Indian music, in the many varieties of African drumming, and in some forms of blues, rock, and jazz, the notion of "form" is often understood in a much more open and fluid way, where the meaning of a given performance is characterized by the dynamic moment-to-moment shifts in nuance as the performer improvises the content (see also Borgo, 2005). Moreover, these nuances can be controlled and are often repeated. In brief, these musics are characterized by *nonstructural objectives* that are enacted in performance, where the immediate qualitative aspects of musical experience are of central importance. Importantly, these musics are no less coherent than Western composed music. They are taught and discussed, and their traditions and techniques are passed on and developed with each new generation that engages with them. But if nuances are indeed nonconceptualizable and inherently ineffable—and nonschematizable and thus inaccessible to memory as Raffman suggests—how could this be possible?

Musical Consciousness as Situated Embodiment

We agree with Raffman that many key aspects of musical experience cannot be fully captured by direct, objective description and may therefore be referred to as "ineffable." However, this may *not* be the result of where nuances are processed and represented in the internal mental hierarchy, as Raffman claims. Rather, the example of musical nuances suggests that accounting for consciousness requires a perspective that extends beyond the brain to include the situated body as a primary cognitive domain. An important consequence of this perspective is that we no longer need to conceive of experience, communication, and understanding as always being dependent on some kind of preconscious quasi-linguistic access that supposedly underpins all mental activity. Effability, in other words, is no longer a necessary requirement for knowledge. Rather, it is through our direct embodied engagements with the world that we learn the fundamental repertoires of contextual movement, feeling, and expression—the "bodily knowledge"—that we develop and then communicate throughout our lives as situated social animals. This move also suggests that the forms of indirect description (e.g., metaphor, narrative) we use in everyday life to talk about musical experiences (and other complex phenomena) may in fact reflect fundamental pre- or nonlinguistic aspects of cognition and consciousness that have traditionally been downplayed or ignored by information-processing approaches.

Consider, for example, the words of jazz bassist Calvin Hill, who describes the experience of hearing a "refreshing performance" by his colleagues, the bassist Richard Davis, and the pianist Jaki Byard:

> Richard started changing things all around. At one point, everything was getting very shaky. The tempo was about to fall apart, and the drummer was trying to keep up with Richard, trying to figure out what he was going to do next, which way he was going to go. It got very chaotic for a minute as they were coming to the end of the chorus. It was like an airplane coming in for a landing that was about to crash. No one knew what was going to happen or how they were going to get out of that. At that point, Jaki was coming to the end of his solo, and he played this really strong rhythmic figure on top of what everyone else was playing, which brought all the different tempos back together and led everyone right into the "one" of the next chorus [. . .]. In that instance, Richard deliberately introduced something rhythmically into the music that made the other players feel uneasy. People will do that sometimes. They might play something that goes against the established tempo, or they might play polyrhythmic things [. . .] that make the music feel unstable. (Quoted in Berliner, 1994, p. 378)

This example illustrates well how people—including highly skilled musicians—describe musical experience through comparisons, narratives, and cross-modal metaphors that draw on (among other things) movement, space, and texture.[14] We can also consider the observations of ethnomusicologist Matthew Rahaim (2012), who examines the deep connection between hand gesture and vocal expression in Hindustani music (see also Pearson, 2013). Here, vocalists trace intricate three-dimensional shapes with their hands while improvising melody. Rahaim notes that, in learning this musical practice, students are encouraged to develop their own gestural and vocal style. Nevertheless, they inherit ways of "shaping melodic space" from their teachers: the movements of their hands and the contours of their voices are always deeply connected. He explains that this is indicative of a special kind of knowledge that is transmitted kinesthetically through generations of teachers and students. This is an excellent example of cross-modal, empathic, and corporeally based cognition unfolding over various timescales (see also Reybrouck, 2017a,b).

Likewise, everyday verbal communication also depends on the use of gesture, emotionally laden expressions, and vocal inflections in addition to the use of metaphor and narrative to connect and communicate the modalities of experience. That we do this, we suggest, is not indicative of some lack of linguistic training or awareness on the part of the experiencer-communicator (as Dennett's approach implies) but shows again that much of lived experience cannot be captured using objective forms of description. Indeed, even if one could describe the event recounted by Hill in the ways Dennett suggests, although this seems very unlikely, the amount of information that would have to be conveyed verbally would be completely impractical. Instead, as the philosopher Mark Johnson (2007) suggests, it may be that our everyday forms of communication reflect the fact that embodied sense-making in humans (and many other animals) is a fundamentally "metaphorical" phenomenon. It is important to note that Johnson's use of the term "metaphor" extends beyond its usual literary-linguistic connotations to illuminate how cognition and consciousness involve an active and ongoing synthesis of our perceptual capacities—including bodily feelings, movement and orientations, empathy, emotion, and moods—in relation with environmental factors (things and other agents).[15] We explore the implications of this metaphorical conception of mind in more detail in chapter 4. For the moment, we would like to posit that although our propositional-representational capacities do shape the ways we think, they do not "cause" consciousness—nor, as we saw, are they foundational for cognition. Our linguistic abilities *emerge* from our condition as embodied beings. Language, in other words, is another way of being embodied (Di Paolo et al., 2018).

From the enactive perspective, cognition and consciousness are instantiated by our development as multimodal perceivers, as active participatory sense-makers.[16] This reflects how the conscious mind is always situated; how it involves experiences of familiarity, transformation, and novelty; how it actively participates in constituting moment-to-moment relationships between people, things, and places; and how it is characterized by the needs, actions, feelings, and contingent desires that motivate such interactions. The suggestion here is that consciousness is not strictly localizable in the "registers" of the brain—it is, rather, an emergent property of the *relations between* embodied cognizers and the world. We argue, therefore, that the internalist mind-as-computer orientation can provide only a limited account of what musical experience entails as it is not grounded in a phenomenological reality. Indeed, conceiving of the mind as a brain-bound information-processing machine does not, and arguably cannot, address the complex ways musicians and listeners actually engage with music, how they talk about such experiences, and what these reports may reveal about the nature of musical consciousness.

In line with this, Clarke (2005) reminds us that we do not first attend to music categorically, in terms of pitch, timings, tunings, timbre, or chords. Such categories, useful though they may be in certain contexts, are products of the analyst; they are what is left over when lived experience is inhibited. In other words, the elements that characterize a given musical experience cannot be properly understood as occurrent and discrete—as objectively out there in the world—nor can they be reduced to mental processes and mechanisms in the brain. Rather, musical experiences play out in embodied-ecological contexts—they are enmeshed within the relevant interests, meanings, feelings, contingencies, and social relationships we actively live through.

From Perceptual Bundles to Behavioral Forms

In the next chapter, we further explore the embodied, multimodal, and metaphorical nature of musical experience through the lenses of phenomenology. To set this up, we conclude the current chapter by noting that there is a strong sense in which our discussion thus far reflects certain long-standing arguments that prefigure the development of the computational model of mind. Most centrally, this involves two approaches to perception that emerged in the early twentieth century, referred to as the "bundle hypothesis" and the "constancy hypothesis" (Heidbreder, 1933; Hergenhahn, 2001). The former claims that experience begins with the perception of a bundle of sense data, or "a finite number of real, separable (although not necessarily separate) elements, each element corresponding to a definite stimulus or to a special memory-residium" (Koffka, 1922, p. 533). The latter involves the related idea that each discrete or atomic sensation corresponds with some objective feature of the world. This assumes a deterministic "constancy," whereby external stimuli acting on sense organs produce the bundles of sense data that result in the experience one has (Koffka, 1922, p. 534). Consider the following passage by Dennett (1988), in which he examines the experience of plucking an open E string on a guitar and then plucking the harmonic:

> Pluck the bass or low E string open and listen carefully to the sound. Does it have describable parts or is it one and whole and ineffably guitarish? Many will opt for the latter way of talking. Now pluck the open string again and carefully bring a finger down lightly over the octave fret to create a high "harmonic." Suddenly a *new* sound is heard: "purer" somehow and of course an octave higher. Some

people insist that this is an entirely novel sound, while others will describe the experience by saying "the bottom fell out of the note"—leaving just the top. But then on a third open plucking one can hear, with surprising distinctness, the harmonic overtone that was isolated in the second plucking. The homogeneity and ineffability of the first experience is gone, replaced by a duality as "directly apprehensible" and clearly describable as that of any chord. The difference in experience is striking, but the complexity apprehended on the third plucking was *there* all along (being responded to or discriminated). After all, it was by the complex pattern of overtones that you were able to recognize the sound as that of a guitar rather than a lute or harpsichord. In other words, although the subjective experience has changed dramatically, the *pip* hasn't changed; you are still responding, as before, to a complex property so highly informative that it practically defies verbal description. There is nothing to stop further refinement of one's capacity to describe this heretofore ineffable complexity. At any time, of course, there is one's current horizon of distinguishability—and that horizon is what sets, if anything does, what we should call the primary or atomic properties of what one consciously experiences. (pp. 73–74)

The claim here is that, by plucking the harmonic, we are revealing an aspect of the previous experience of the open string that was there all along: one that was part of the bundle of stimuli that make up the experience of the open E string that we were not consciously aware of until it was pointed out to us. This implies that depending on our abilities—or, as Dennett puts it, our "horizon of distinguishability"—we could keep pointing out other bits of constituent stimuli associated with timbre, amplitude, resonance, sympathetic vibration, and so on that would eventually provide a complete objective description of all facets of the experience. Roholt (2014) notes how this approach is problematic when it assumes that listening to the open string and the harmonic separately somehow constitute the same experience, when they are in fact two different phenomena (listening to the open string after hearing the harmonic constitutes a third experience). As Roholt comments, Dennett "does not seem to realize that in the different steps of his [thought] experiment we are listening to the E-string in different ways—and when we do, a change occurs in the structure of perception" (p. 47). Indeed, it may be argued that Dennett's analysis maintains ideas associated with the bundle and constancy hypotheses mentioned earlier, according to which it is assumed that experience involves the preconscious perceptual recovery of discrete external events, which are then somehow combined into an experience and can then be individually unpacked and reassembled as constitutive of that single experience.

Here we should also point out that although bundle and constancy assumptions continue to (tacitly) influence theories and research designs, they were already being challenged early on by research in Gestalt psychology involving visual and auditory phenomena that do not maintain a direct correspondence between stimulus and experience (Koffka, 1922; Köhler, 1959; Wertheimer, 1938).[17] Such phenomena include, for example, the perception of multistable images and optical illusions (figure 3.1), the equivocal experience of pulse associated with polyrhythm, and the synthesis of the multimodal dimensions of experience (e.g., visual, auditory, bodily, spatial, emotional). Importantly, observations drawn from these studies strongly suggest that experience is not determined by some objective stimulus but requires the active participation of a situated perceiver.

The printed lines that constitute multistable images and the sonic stimuli associated with a repeating polyrhythm do not entail a constancy with the transforming experiences viewers and listeners have. Instead, these examples highlight the way a perceiver actively engages with the stimulus: the stimulus remains constant, but the experience varies (Ihde, 1977; Roholt, 2014). Likewise, it is also possible to vary the properties of the stimulus while retaining a recognizable experience: for example, when we transpose a melody, we nevertheless maintain the experience of the song (Ehrenfels, 1890/1988). In sum, this view implies a rather different conception of experience—one that involves, first and foremost, the enactment of perceptual and behavioral forms that are not simply reducible to the environmental stimuli they are

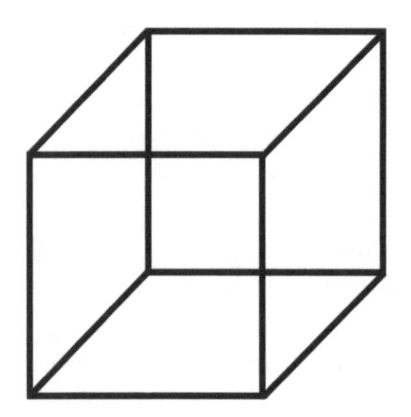

Figure 3.1
Multistable images. (Wiki Commons: Public domain)

associated with and are therefore not able to be exhaustively described in an objective way (Merleau-Ponty, 1942). Again, these observations imply a view of cognition as not always dependent on some sort of internally located symbolic or linguistic-representational access.[18] Significance, meaning, and understanding arise from the active, sense-making body in adaptive interaction with its environment, whereby recognizable and recurrent patterns of action and perception are enacted and reenacted. By this light, consciousness is not best understood as something that happens to us, nor as something we simply possess. Consciousness is something we *do in the world* (Hurley, 1999).[19]

The next chapter develops some of the examples from Gestalt psychology just mentioned (multistable images, polyrhythm, multimodal experience). In doing so, we examine how a "phenomenological attitude" is an important aspect of an enactive approach to music cognition,[20] especially as it reveals how the forms of multimodal sense-making activity (action-as-perception) enacted by the situated body constitute the very basis for what it means to be an experiencing being (Merleau-Ponty, 1942, 1945).[21]

4 Phenomenology and the Musical Body

Phenomenology is a philosophy of experience, of perception, knowledge, and being (Gallagher & Zahavi, 2008; Merleau-Ponty, 1945).[1] Most centrally, it examines the structure of consciousness and the way it is always *directed toward* things and events, including our own bodies, thoughts, and imaginations. In other words, phenomenological inquiry highlights the *intentional* nature of consciousness; consciousness involves the experience of something, and that something is always experienced in a certain way (Gallagher, 2012). This insight was developed by the Moravian mathematician and philosopher Edmund Husserl, who attempted to "suspend" or "bracket" assumptions and judgments about a given phenomenon and attend to the experience at hand in the most open and direct way possible (*epoché*). As subsequent thinkers have explored (e.g., Merleau-Ponty, 1945), this process reveals how our perceptions and understandings are often the products of habitual ways of directing our experience that have become so ingrained that they appear to take on a fixed reality of their own. This results in the development of so-called "natural attitudes" toward the things, activities, and relationships that characterize our lives, attitudes we often simply take for granted as the way things are. Phenomenology examines these natural attitudes in terms of processes of developmental (personal, cultural) *sedimentation*. This term describes how, as a river collects and deposits particles that become sedimented into formations that direct its flow, we take on ways of perceiving and acting that guide our thinking and behavior. Phenomenology can help us to better understand these processes and, in doing so, reveal new possibilities for thought and action.

The process of phenomenological inquiry introduced by Husserl begins with (1) the *what* (*noema*)[2] of experience as it appears in the nonreflective context of the natural attitude. The inquirer then attempts to identify and bracket assumptions and judgments to move from the prescriptive "literal-mindedness" of the sedimented natural attitude and better attend to the phenomena as it is given directly to experience. This leads to an examination of the *how* of experience, revealing (2) the modes of perception (*noesis*) and the way the shifting interplay of such modes may reveal new understandings and possibilities. The phenomenologist then questions back to (3) the *who* (*I-ego*) of experience, disclosing the "self" as a transforming embodied agent who plays an active role in the continuous construction of experience. This process may proceed in an ongoing way to reveal ever richer (polymorphic) ways of attending to the phenomena at hand. Figure 4.1 sketches out the basic Husserlian model: the *top arrow pointing left* describes the directed nature of experience (intentionality). The *bottom arrows* indicate the reflective processes of consciousness, which may transform the experience, resulting in new perceptual possibilities. This framework has been adapted in various ways to explore a wide range of phenomena, including musical experience (e.g., see Ihde, 1976, 1977).

Importantly, where the dominant intellectual trends associated with positivist thinking have emphasized an objectivist approach to mental life, phenomenology provides a different account. It examines experience and subjectivity not in terms of a dualist schema in which a fixed or pregiven "world out there" is represented in the mind/brain as a cause-and-response process. Rather, phenomenology explores the mind as a "circular" phenomenon, where self and world engage in a relational process of coconstitution. In brief, phenomenology highlights the exploratory, (inter)active, embodied, and bidirectional nature of perception and consciousness. It therefore plays an important role in enactivist thinking since it offers an experiential dimension to the adaptive processes associated with how living systems bring forth worlds of meaning (see Thompson, 2007, pp. 16–36).

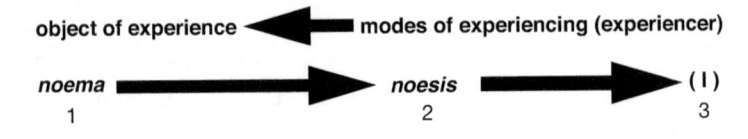

object of experience ⬅ modes of experiencing (experiencer)

noema ➡ noesis ➡ (I)
1   2   3

Figure 4.1
The recursive structure of conscious experience as revealed by phenomenological inquiry. (Adapted from Ihde, 1977)

In this chapter, we consider two phenomenological examples that advance the ideas developed in the last chapter. In the first, we examine the experience of one of the multistable images shown in the previous chapter (see figure 3.1). This will help us gain a basic understanding of how one may begin to explore phenomenological inquiry and, in doing so, develop some basic insights into the structure of experience. We then move on to discuss analogous phenomena involving a set of polyrhythms found in Ghanaian drumming. Drawing on the resulting insights, we conclude the chapter by outlining an alternative embodied and "metaphorical" approach to musical experience, one that aligns closely with the life-mind continuity that characterizes the enactive approach to cognition.

Beginning Phenomenology

One way to begin phenomenological inquiry is through the examination of the multistable images we mentioned in the last chapter. For example, figure 4.2 shows a form known as the Necker cube (Necker, 1832). This image is well known in psychological circles as an example of a bistable phenomenon; the perception of the image spontaneously reverses itself between two three-dimensional spaces (Attneave, 1971; Morris, 1971). Very often, viewers will first experience the cube as if they are oriented above it with a square on the left facing "forward" (A). This is not surprising as similar objects in our day-to-day lives (e.g., tables, bricks, chairs, boxes) are generally viewed and interacted with from such a perspective (i.e., above). However, the cube may also be perceived from "below" with a square facing toward the viewer's right (B), and when looked at for a period of time, it will often appear to shift unpredictably between the above and below perspectives. With a bit of practice, the viewer can hold onto and learn to move willfully between the above-left and below-right points of view.

The phenomenological approach allows us to develop our experiences of the image still further. By suspending (pre)judgments and explanations, and by focusing on a clear, first-person description of the phenomena at hand, the viewer may begin to develop a deeper awareness of the active nature of perception and the possibilities of experience. We have already discussed how our tendency to see the Necker cube first as a cube and not in some other way is indicative of the natural attitude we adopt in our encounters with analogous objects in our day-to-day lives (Husserl, 1960;

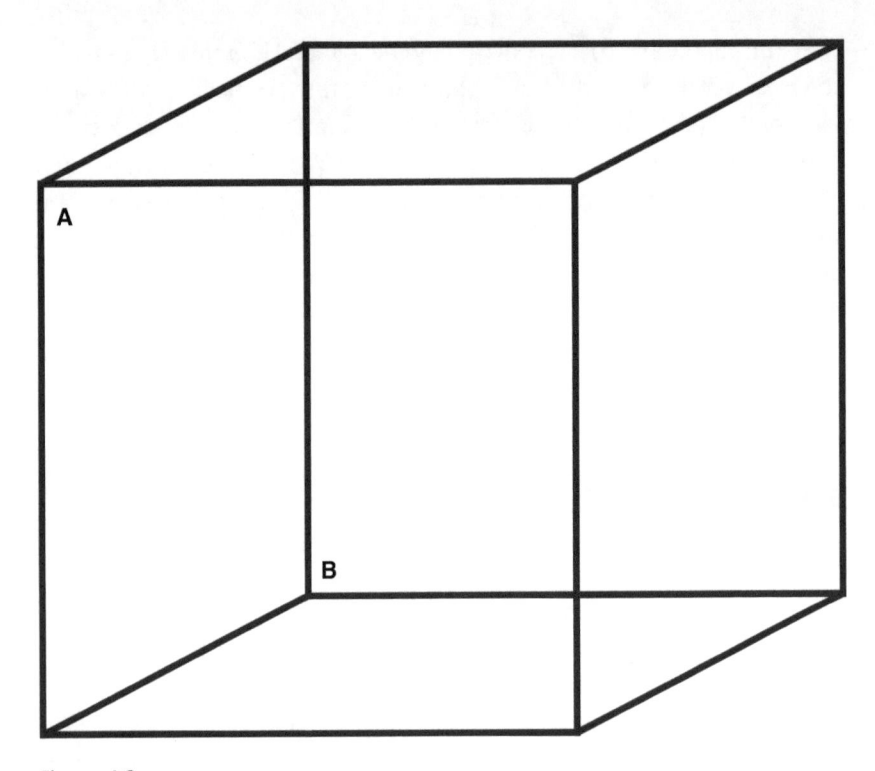

Figure 4.2
The Necker cube. (Wiki Commons: CC BY-SA 3.0)

Merleau-Ponty, 1945). However, by identifying this taken-for-granted perspective, one may attempt to suspend or "bracket" it and thus open previously unrecognized possibilities. By suspending the form's "cubeness" the viewer may explore, experiment with, and describe the object from a variety of new focal points; by moving habitual perspectives into the background, other relationships and interpretations may begin to come forward.

One possibility that emerges is that the intersections of lines may now be experienced in a two-dimensional context (Ihde, 1977). From this perspective, one may notice the vertically oriented rectangle at the center of the image, which is bounded by two right triangles, on the top left and bottom right, respectively, and trapezoidal figures above, below, left, and right. The result is a strange belt shape with a hexagonal exterior and rectangular interior; an abstract image with no deeply sedimented relationship to everyday experience. Interestingly, this "flat" perspective is "uncomfortable" and considerably harder to maintain than the cube experience. Nevertheless, the viewer can learn to make controlled movements between above-left and below-right cubes and the two-dimensional belt. One can blend the three together in in-between states or hold various features of each in place while allowing the rest of the figure and the viewer's relative position to move to other orientations. What was formerly bistable now becomes multistable and transitional states may be identified. More experimentally minded viewers may now catalogue the results of such investigations to build up a richer description of the possible experiences engagement with the Necker cube affords (see Ihde, 1977).

Although more could be said here about the Necker cube and other multistable images, what has been discussed thus far already allows us to investigate some fundamental aspects of experience that may otherwise not be considered. First, we tend toward naturalized ways of organizing the relationships that constitute experience. These are informed by our histories as embodied and ecological (i.e., located or situated) creatures—for example, the tendency to position ourselves in relation to the image (e.g., above, below, facing). Thus, a sense of movement and/or bodily orientation in space (whether actual or imagined) is a central aspect of experience (Johnson, 2007). Second, the awareness of even the simplest phenomena may be extended beyond the initial taken-for-granted perspective (the natural attitude) through a reflexive open-ended exploration of possibilities. And third, conscious experience is not simply the retrieval of a preexisting environment (the attitude of objective thought; see Merleau-Ponty, 1945)—it is an emergent, constructive, or "enactive" process where agents play an active role in constituting their perceptions of the world.

This aligns with the discussion in the last chapter when we examined the Gestalt alternative to the bundle and constancy hypotheses—where experience is not understood in first terms of the reception of packets of sense data, but more holistically as the enactment of perceptual-behavioral forms. This last point is especially relevant when we consider that from the objective perspective of direct description the Necker cube is not a cube at all but 12 lines that intersect in various ways (and that the mind projects in three dimensions). Nevertheless, from a radically empirical point of view, the *experience* of "cubeness" and variations on it we enact are real and require no further proof of validity; we can return to it repeatedly and develop it.

Intentionality and the Modes of Experience

In addition to the points outlined previously, we may now discuss a perhaps even more fundamental observation, namely that there appears to be a "directedness" to experience (Ihde, 1977, p. 43; Merleau-Ponty, 1945). Again, consciousness is always consciousness of something as something (Gallagher, 2012). This "something" could be an object in the world, something we imagine, or aspects of our own body as we interact with a musical instrument. Indeed, we began with an awareness of the "cube," which was experienced literally as such. However, an exploration of *how* we consciously attended to the cube expanded the possibilities of experience to the point where the image was no longer experienced as a cube at all. We began to move from the prescriptive "literal mindedness" of the natural attitude or taken-for-granted stance (nonreflective/constrained) toward the "polymorphic mindedness" (reflective/open to possibilities) of the phenomenological attitude.

The first thing we may note about this move is that a certain degree of self-awareness or reflexivity was needed to recognize the taken-for-granted stance and to develop alternative points of view. It brought into play *my* location in relation to the object (and/or the object's location in relation to me); it entailed *my* struggle to develop other perspectives and *my* awareness or involvement in the exploration and construction of experience. This said, it is also important to recognize that the "I" is often not explicit or "thematized" in experience; this is why it is placed in parentheses in figure 4.1 (see Ihde, 1977; Benson, 2001; Varela et al., 1991). In most mundane experience, we are often outside of ourselves in the world of "our projects" (Ihde, 1977, p. 47; Merleau-Ponty, 1945). Many of the everyday activities we engage in require the development of a repertoire of skills and ways of thinking that we come to take for granted—for example, cooking, chopping wood, playing, and listening to music. Such modes of experiencing (skillful coping) are not lacking in awareness per se but are not those in which we are explicitly self-conscious. Moreover, intense modes of experience where one is almost completely immersed in the experience itself can be pleasurable and rewarding; they may lead to positive modes of absorption that may ultimately afford an expanded and shared sense of self, as well as the cultural and therapeutic benefits that follow. By contrast, modes of experience where acute self-awareness dominates can be highly uncomfortable.

Phenomenology and the Musical Body

Such experiences may be associated with boredom, loneliness, isolation, as well as with physical and psychological pain. Extreme examples of this are found in experiences of forced confinement and torture—under such conditions one may say that the system of experience is narrowed, confined, or trapped within the self: extended consciousness is restricted; descriptive, imaginative, and narrative capacities are reduced; and a sense of dislocation or alienation from one's social, physical, or bodily milieu dominates (see Benson, 2001).

Everyday experience tends to move between the objects of experience, an awareness of the situations we find ourselves in, and a shifting sense of our own agency. We engage in alternating periods of absorbed consciousness and self-consciousness often without reflecting deeply on how such states of awareness arise or where they could lead. For example, we may consider here how a more reflexive or self-aware mode of experience may emerge out of a state of absorption because of an unforeseen occurrence. This might involve a breakdown in whatever tools one is using (e.g., a broken string on an instrument). It may also be noted how a heightened and uncomfortable level of sustained self-awareness, and even temporary feelings of alienation, and psychophysical discomfort often emerge when we are required to learn new skills or new ways of thinking and perceiving. We saw this in a very simple context when we tried to advance the possibilities of the Necker cube (more on this shortly, with a focus on a musical learning). Although such experiences may not always bring the self to mind explicitly, they nevertheless involve a shift in the mode of experience and therefore demand a further level of phenomenal reflection. Such reflection moves toward the *how* of experience and highlights the agency of the experiencer (What's going on? What am I doing here? How can I fix this? What am I doing wrong? Why is this so difficult?). In other words, phenomenological inquiry reveals consciousness as a relational process that continually moves between the poles of agent and world, where the "I" as a full-blown embodied awareness of self often comes late to experience: the situatedness of the self and its sense of being is an ongoing reflexive process; its relative significance is enacted "through its encounter with things, persons, and every type of otherness it may meet" (Ihde, 1977, p. 51). The self, therefore, cannot be pinned down as a fixed or pregiven entity but appears as an emergent phenomenon, inextricable from the embodied organism-environment interactions that give rise to it (Thompson, 2007; Varela et al., 1991).

Ghanaian Polyrhythm

In all, phenomenology provides a means by which we may begin to develop the experience of any number of phenomena through a recognition of the taken-for-granted ways we attend to them and thereby open alternative possibilities.[3] And this process may involve the integration of many more perceptual modes (e.g., auditory, tactile, bodily, emotional, social) than the almost purely visual dimension that characterized our exploration of the Necker cube.[4] Let's move on, then, to examine phenomena that engage the more explicitly embodied and intersubjective contexts associated with musical experience. Here, a relatively easy jump can be made from the simplicity of the Necker cube to the living world of musical practice, through the introduction of multistable musical phenomena such as repeating polyrhythms (figure 4.3). These are analogous to the visual example of the Necker cube in that they comprise patterns whose relations may be experienced in multiple ways. Exploring these rhythms may also require us to identify and suspend (or bracket) certain sedimented attitudes. For example, in the West we often tend toward feeling rhythmic hierarchies as fixed, as building upon a grounding pulse or figure that other pulses and rhythms work "against" or "over."[5] Ghanaian drumming does not necessarily share this conception: rhythmic hierarchies and dominant pulses shift, and their relationships are enacted and perceived in various ways. Accordingly, examining the experience of engaging with these drumming patterns can help us develop the phenomenological perspective in the more practical context of musical learning. As with our exploration of the Necker cube, readers may explore these rhythms as they follow along with our discussion.

Before we begin, we should note that notation is used here only for the convenience of conveying these drumming patterns in the format of a book. In context, these rhythms are not learned by reading them, but by doing them. That is, they are passed along socially and are developed into juxtaposable combinations via the imitation of sound and movement, and through the creative, vernacular interaction between performers over various time periods (i.e., in the moment of performance and over generational cycles of cultural evolution). It is suggested, therefore, that the best way to learn these rhythms is in collaboration with other players who have already had some experience with them. This said, our account here should at least

Phenomenology and the Musical Body 77

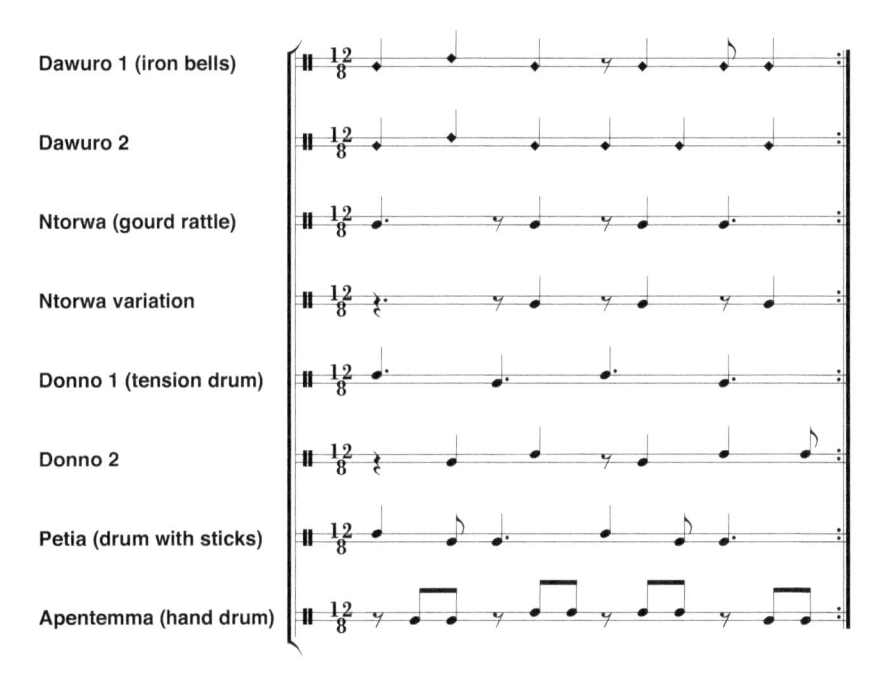

Figure 4.3
A partial representation of the Adowa rhythms played by the Ashanti people of Ghana in West Africa. (Adapted from Hartigan et al., 1995, p. 35)

give the reader a general idea of what's involved. Most readers should be able to explore the basic patterns without too much difficulty (readers unfamiliar with musical notation may ask a musician friend for help).[6]

We can start to explore these patterns by choosing pairs of them to play separately or together. One can do this by sitting on a chair and tapping one rhythm with the left hand on the left knee and the other rhythm on the right knee with the right hand. More experienced drummers can incorporate other patterns using the feet. A good combination to start with is the *dawuro 2* and the *donno 1* rhythms. Together, these create a simple 3:2 (hemiola) relationship that many will have experienced previously (the 2 pulse is a represented as dotted quarter notes in the *donno 1*, while the 3 pulse is represented as quarter notes in the *dawuro* 2). As one becomes familiar with this pattern, it is possible to shift the focus of attention between the two pulses, maintain a "neutral" position between the two, and to attend to how the experience develops. This as analogous to developing the possibilities of the Necker cube, alternating back and forth between the up and down position of the image and between the 3 and 2 perspectives of the polyrhythm and encountering both perspectives equally ("flat").

Here we may also note that each of these perspectives involves a different experiential character, both in how they feel and sound. Although the (self-produced) stimulus is consistent from an objective point of view—the timings and sound waves produced by the two hands striking the knees are essentially the same—the perceptual forms that integrate movement and sound *as an experience* are variable and depend on the agency of the perceiver. Notice too that although one could break down this experience into various objective components (e.g., timing ratios and gaps between onsets, measurements of tactile and auditory responses, neuron firing, muscular activations), the actual experience of learning and playing this pattern arguably does not involve any of this knowledge in the first instance. Rather, it reflects the fundamental organic ability to make sense of, and develop, bodily and sonic relationships in a self-organizing and goal-directed way.

The addition of a more complex rhythm, such as the *ntorwa*, takes things further. As the new rhythm emerges against the background polyrhythm, we can again actively transform the experience of it depending on which of the other pulses we choose to associate with it. For example, the longer pulse of the *donno 1* pattern (notated as dotted quarter note) with the *ntorwa* maintains the hemiola quality. However, in association with the shorter pulse of the *dawuro 2* pattern (notated as quarter notes), the relationship takes on a different duple or binary flavor. As before, we can bring the various rhythms and combinations of rhythms to the foreground (the focus) or play them on different-sounding objects.

As learners work to develop the physical coordination that will allow them to enact the various combinations, they are required to break from habitual ways of moving rhythmically. For kit drummers, this challenge involves incorporating the bodily extremities: the feet and hands must work together and independently to play separate rhythms (the voice may be included as a fifth element). Anyone who has attempted to learn a multilimb pattern on a drum set will know that this process learning is generally accompanied by a certain amount of discomfort and frustration. Interestingly, what makes this initial process of exploration so uncomfortable is precisely what makes it so informative. Here learners must shift and share the focus of attention between various bodily, auditory, and musical relationships as they play more complex three- or four-part groupings. In the process, they are confronted with a diverse range of new focal points and relationships that must be attended to and advanced reflexively without losing contact with the larger musical-polyrhythmic context from which they emerge. These are spread across the audible, visual, tactile, and proprioceptive fields of experience, and each may reveal naturalized inclinations that must be identified and developed.

The phenomenological attitude encourages the learner not to ignore or rush through this process but to attend to it carefully. Initially the experience is frustratingly disjointed, and the awareness of one's own body and its relationship to the new musical environment it is involved in creating is uncomfortable. However, a reflexive analysis of this state may help learners become more aware of their bodies' proclivities—its *sedimented* ways of moving—and thus develop more nuanced and flexible ways of experiencing their embodied musicality. In the process, one begins to see that what was once taken for granted as "naturalized" may be better understood as a historical process of embodied and conceptual sedimentation and that new possibilities (bodily-instrumental-sonic affordances) can emerge with sustained work. Such forms of embodied learning result in the enactment of new self-made musical-rhythmic worlds in which one may come to feel increasingly comfortable or "at home."

Things become even more interesting when one begins practice these patterns in social groups. In line with our observations in the previous chapter, this context requires the musician to engage in an adaptive and relational way with the sounds and actions (the behavioral forms) produced by other agents. Importantly, in such situations the musical ecology is enacted and maintained by the self-organizing activity of the ensemble, where autonomous individuals engage in shared acts of participatory sense-making (De Jaegher & Di Paolo, 2007) that allow for new musical affordances to develop between musicians and between musicians and their instruments.

Revisiting Representation

The example of learning Ghanaian drumming patterns shows how phenomenological and enactive perspectives can offer insights into the corporeal, situated, and intersubjective nature of musical cognition that are arguably not available from the traditional computational perspective. This said, the experience of learning these rhythmic patterns can involve types of representation. After all, we used notation to represent the rhythms we explored. And in the absence of a teacher, one could figure out, for example, the 3:2 relationship between the *dawuro 2* and the *donno 1* patterns by representing it in a mathematical way. However, we suggest that conflating the experience of listening to, learning, and performing these patterns with such representations may be misleading. The reason for this, and in line with our discussion in previous chapters, is that the kinds of representations we have just discussed are in no way primary or subpersonal. Rather, they are consciously created as learning tools.

A critic may respond to this by arguing that the fact that we can make representations of these rhythms is indicative that there should be some corresponding algorithmic computational process occurring in the brain that allows us to parse the relationships in this way, and that such internal mental content is what permits us to understand, perform, and represent these patterns, or to experience them at all. Additionally, the critic may ask about the possibility of finding neural activity that correlates to the stimuli of each rhythm as perceived and/or performed, or even patterns of firing that align with other underlying ratios of these rhythms. This, they may argue, would imply that the experience of such stimuli is reducible to mental content—perhaps a form of qualia, or (à la Dennett) the output of some propositional schema produced by information-processing in the brain.

We would respond to this by noting that even a brief description of learning these patterns shows that they cannot be reduced to an internal computational process; this would turn a highly visceral experience to an abstraction. Recalling our discussion in chapters 2 and 3, corresponding neural firings may be alternatively understood not as the cause of bodily outputs and perceptions, but rather as one aspect of an integrated dynamical system involving body, brain, and environment. Likewise, and as we mentioned previously, in most contexts learning such patterns does not involve the use of notation. Instead, the drumming movements and sounds are encountered directly and are transferred through mimetic cross-modal processes whereby learners strive to bring their bodily sound-making activities into coherence with those of the musicians they learn from. Note, too, that such processes do not require theoretical descriptions, nor do they appear to reflect or be dependent on high-level linguistic competence. Rather, learning occurs by watching, listening, feeling, and moving in adaptive and goal-directed ways through interactions with the other agents and sound-making objects (e.g., drums,

Phenomenology and the Musical Body

bells, rattles) in the environment. We suggest that, in the first instance, these cognitive processes depend on an awareness of the shifting states of the lived body in relation to its environment—homeostasis, the balance of stability and instability, and the corporeal-affective dimensions that guide (participatory) sense-making on the most basic levels.[7]

To be clear, we are not saying that there is no relationship between linguistic and rhythmic capacities.[8] We simply wish to note that acquiring knowledge of these rhythms does not first involve detached, analytical, fact-based, or conceptual forms of learning, nor some form of depersonalized representational retrieval of a pregiven set of stimuli. Rather, musical learning begins in a direct corporeal engagement with the sociomaterial environment—musical development, understanding, communication, expression, and meaning making are manifest first in the self-organizing forms of action-as-perception enacted by environmentally situated musical bodies. Further, in this context notational representation can be seen for what it is—a heuristic for learners who are unable to have this kind of direct social experience (who do not come from the sociocultural environment in which this manifestation of musical sense-making emerges in the day-to-day environment). A moment's reflection on the experience of learning these patterns "from the page" will confirm that, here too, a complex range of corporeal-sonic relationships must be enacted in which the body places itself in new behavioral configurations. From a phenomenological perspective, it is these bodily and affective dimensions that are most present—the various feelings that arise as one struggles and finally succeeds at learning the rhythms, and as one gains increasing fluidity in performing, perceiving, and improvising with the sounding forms of bodily action these patterns involve.

Although a traditional computational perspective may be able to account for certain aspects of such experiences, it would arguably become overtaxed when confronted with the full range of corporeal-environmental factors involved. Think, for example, of the overwhelming amount of information that would have to be described if every phenomenon involved—every bodily movement and feeling, every relational transformation of sound and action, every experience of agency—were formatted propositionally, as Dennett's approach implies. Alternatively, the intercranial dynamics involved in the experience of musical learning may be better accounted for by recent nonreductivist trends in neuroscience. Here, brain activity that correlates with bodily and environmental events is not necessarily assumed to represent those events. Rather, behavioral coherences between body, brain, and world are discussed in terms of the oscillatory dynamics between the nonlinearly coupled components of an integrated system. This perspective—sometimes referred to as "dynamical attending"—explores brain, body, and environment as inseparable aspects of an evolving dynamical system, where the cognitive load is distributed across the corporeal and sociomaterial ecology (Jones, 2009; Large & Jones, 1999; Large et al., 2015; McGrath & Kelly, 1986).

We will have more to say about this approach. For the moment, we can note that the phenomenological account of learning Ghanaian polyrhythm aligns closely with the fundamental self-organizing and relational characteristics of (participatory) sense-making we touched on previously. Likewise, these experiences also bring the 4E framework to mind: where the processes involved in musical learning can be explored in terms of the *embodied* dynamics of musical development (the enactment of new muscular linkages and coordinated patterns of action); the *embedded* social, material, and cultural environments in which such actions play out; the *extended* dimension of instruments and other agents that allow for various cognitive processes to be offloaded and taken on (e.g., where another drummer plays a pulse so that one can learn to play another rhythm with or against it); and the ways such processes involve the active shaping or *enactment* of a meaningful musical environment. This orientation also recalls the discussion in the last chapter over the cross-modal nature of experience and the "metaphorical" nature of the embodied human mind. We now turn to consider these ideas in more detail, exploring what they reveal about the nature and origins of musical experience.

Embodied Mind as a "Metaphorical" Phenomenon

Several phenomenologists (Gallagher, 2005; Johnson, 2007; Polyani, 1969) have noted that the general tendency to ignore the corporeal dimensions of mind is exacerbated by the fact that in nonreflective day-to-day life, the body tends to "hide out." That is, our "biological self" tends to retreat to the background as our intentionality is directed "out into the world," while nevertheless tacitly providing the very means and context by which all our perceptions and engagements take place—on a primordial level, all experience is "passively motivated" by the sensing body.[9] It is important to understand that, in this context, the term "passive" is not synonymous with "inactive." Rather, it refers to the fundamental "openness" to the world that characterizes all life, to how the lived body, even in preconscious or minimally conscious states, actively constitutes its structurally coupled relationship with the environment through affectively motivated (valenced) activity, resulting in patterns of behavior, dispositions and moods, recurrent affective episodes (feelings, moods, emotions), memories, habits, and so on (Colombetti, 2014). This recalls the relationship between autonomy, relationality, and sense-making discussed in chapter 2, where the biocognitive nature of living organisms is at once *operationally closed* and *thermodynamically open*, such that the organism maintains itself (its differentiation and "identity") through its interactivity with the environment that it inhabits and shapes. Phenomenologists sometimes use the term "passive synthesis" to describe this multisensory process, which can help us understand how the object-directed or intentional structure of experience discussed earlier (*noema-noesis-I*) emerges from more primordial modes of sense-making that do not always entail an explicit subject-object structure.

Figure 4.4 outlines the continuity between conscious experience and the more fundamental forms of biocognitive organism-environment relationality (structural coupling) associated with "passive synthesis" (see Thompson, 2007, pp. 28–31). As we noted in chapter 2, although affordances are shown here to be associated with the niche side of the organism-environment system, they are in fact inextricably linked with, and are shaped by, the activity of organism with its sensorimotor coupling (abilities) and the cross-modal forms of action-as-perception this enables (see also Gallagher, 2012; Merleau-Ponty, 1945; Thompson, 2007, pp. 28–31).

Importantly, these insights highlight the essentially relational and cross-modal nature of embodied sense-making, which involves an ongoing synthesis of sight, sound, touch, smell, movement, feeling and emotion, as well as interoceptive and proprioceptive capacities (Johnson, 2007; Sheets-Johnstone, 1999). Support for this view comes from the work of neuroscientists Antonio Damasio (1994, 2003) and Joseph LeDoux (2002), who have demonstrated that all cognitive processes depend on the basic bodily systems that allow us to maintain a state of well-being: metabolism, basic reflexes, the immune system, pain and pleasure responses, basic drives, emotions, and feelings (Damasio, 2003). This research highlights the apparent autonomy of our sensory and metabolic systems while embracing

Socio-material environment

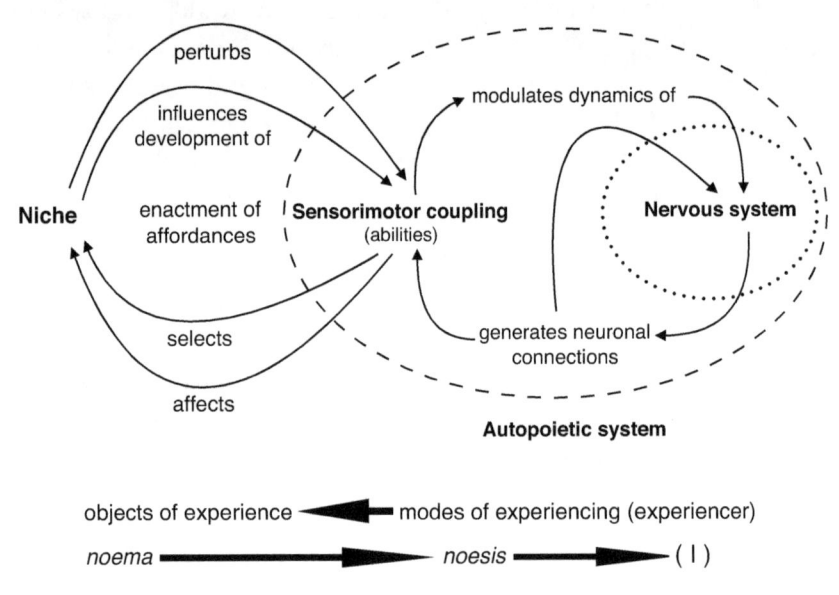

Figure 4.4
The "circular" structure of experience revealed by phenomenology aligns with the bidirectional structure of sense-making associated with the enactment of affordances and niche construction in biological systems.

how they develop codependently, how meaningful bodily configurations and motivations for action and perception for action arise relationally.

Further support comes from empirical research that strongly suggests the development of shared cognitive resources for action and perception. For example, developmental studies show that infants have a remarkable proclivity to imitate actions and sounds they have never experienced before (Meltzoff & Moore, 1977; Meltzoff & Prinz, 2002). This proclivity, it is argued, permits the prelinguistic bodily origins of communication (Bateson, 1975; Trevarthen, 1998, 1999). Such processes have recently been investigated in light of the discovery of mirror neurons, which, as we saw, activate both in the brains of those performing actions and in those of perceivers, and appear to function intermodally (Gallese & Goldman, 1998; Rizzolatti et al., 2002). In connection with this, neuroscientist V. S. Ramachandran's (2011) research in mirror neurons and synesthesia has led him to posit that the brain areas that develop in association with specific cognitive functions engage in "cross-activation." He argues that cross-activation does not involve the neat sharing of resources via the inputs and outputs of discrete cognitive modules and mechanisms, but a more plastic process whereby neural "webs" develop in complex overlapping ways that allow for deep and sometimes highly idiosyncratic relationships to form between different modalities of experience—for example, between color and number, or movement, sound, shape, and feeling.

Cross-activation in sense-making is suggested in a cross-cultural study done by Ramachandran and Hubbard (2001) that replicated previous work by the Gestalt psychologist Wolfgang Köhler (1929). Here the researchers showed the image in figure 4.5 to participants who spoke different languages and asked them, "Which of these is shapes is Bouba, and which is Kiki?" Almost unanimously, participants assigned the name "Bouba" to the shape on the right, and "Kiki" to the shape on the left. This result suggests that the human mind ascribes meanings to shapes and sounds in a nonarbitrary way. Not only do the shapes appear to coincide with the "soft" and "hard" sounds of long vowels and short consonants, but also with the physiology of producing the names associated with them. Pronouncing "Bouba" requires a "rounded" mouth shape. "Kiki" requires more angular and tighter shapes, as well the "percussive" action of the tongue to produce the "sharper" dorsal consonants. In brief, this simple experiment strongly implies that meaning making involves the cross-activation

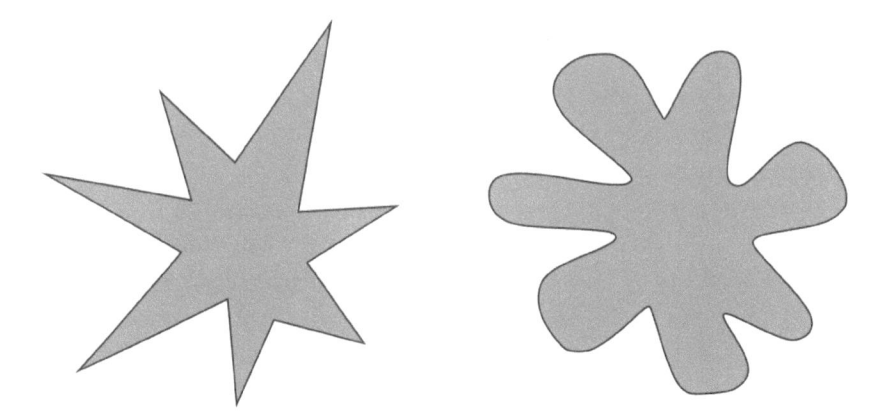

Figure 4.5
The Bouba/Kiki effect. (Wiki Commons: CC BY-SA 3.0)

of spatial, visual, textural, sonic, and physiological dimensions (see Ramachandran, 2011).

It is also suggested that neural cross-activations evolve as a result of experience and environmental alterations, highlighting again the inseparability of brain, body, and environment. Ramachandran claims that these aspects are crucial for understanding the complex diversity of human cognition and experience and how they unfold through learning and development. For example, he posits that the cross-activation of neural areas associated with sight, sound, and movement may allow for the development of speech. He also suggests that this cross-modal perspective might reveal useful insights into the evolution of language in our species, which could have a biocultural origin in the ritualization of bodily gestures involved with the practical aspects of our ancestor's daily lives. Ramachandran's work also gives biological grounding to the "metaphorical mind" notion introduced earlier. As we discussed, this idea goes deeper than the common linguistic-conceptual usage of the term "metaphor" to describe the prereflective and nonlinguistic processes that allow us to enact meaningful experiences through the development of cross-modal relations, beginning with the most basic (and sometimes minimally conscious) forms of human sense-making (passive synthesis).

This metaphorical approach to human sense-making is explored by philosopher Mark Johnson (2007), who draws on the pragmatic psychological-philosophy of William James and John Dewey, research in cognitive linguistics, developmental psychology, neuroscience, and the phenomenological tradition associated with Merleau-Ponty (e.g., Sheets-Johnstone, 1999). Following Dewey's "principle of continuity," Johnson examines cognition beginning with basic bodily processes, movement, and organism-environment couplings—where "rational operations grow out of organic activities, without being identical with that from which they emerge" (Dewey, 1991, p. 26). In connection with this, he also discusses the notion of "vitality affect contours" introduced by the psychiatrist and psychoanalyst Daniel Stern (1985). This concept describes how, as infants, we strive to create a secure, coherent, and meaningful world through non- or prelinguistic embodied means. This process, Johnson argues, is based on the developmental coupling of the organism and the environment through action; it allows us to recognize and create "metaphorical" relationships between cross-modal perceptions, and between emotional-empathic engagements and feelings as we develop through time and space. Indeed, Stern (1985) describes these fundamental relationships (vitality affect contours) in terms that relate equally well to visual, auditory, tactile, emotive, or kinetic modes of experience—for example, *surging, fleeting, fading away,* and *drawn out.*

Developing these ideas, Johnson argues that human sense-making is rooted in a basic proclivity to enact embodied and cross-modal relationships within the environments we inhabit. It is this primordial capacity that grounds the origin of meaning and mind. Likewise, Johnson claims that these corporeally based forms of meaning making continue to shape the contours of our experience and guide how we meaningfully orient ourselves in the world even as we grow up and engage in more abstract or propositional ways of thinking. Importantly, this recognition of the metaphorical nature of experience strongly suggests the possibility of a "corporeal intentionality" (Leman, 2007), in which knowledge of the world (including other agents) begins in movement, corporeal articulations, embodied states of being, and the enactment of organism environment relationships through multimodal forms of action-as-perception (Reybrouck, 2017b). Indeed, while cognition may be based first in "manifest" (visible, physically sensed) experience, our metaphorical capacities may also drive sense-making in "covered" ways "that relate physical corporeal activity with imagination and emotion" (Cano, 2006, p. 6). As Johnson (2007) puts it, "[O]ur experience of meaning is based, first, on sensorimotor experience, our feelings, and our visceral connections to the world; and, second, on various imaginative capacities for using sensorimotor processes to understand abstract concepts" (p. 12). The core idea, then, is that even in our most physically inactive and reflective states, our embodied existence appears to guide our ability to attribute valences, goals, and intentions within the environments we inhabit.[10] And this allows us to engage in the ongoing process of enacting the meaningful relationships with the people, activities, ideas, and things that constitute the complex fabric of our lives as sociocultural animals.

A metaphorical approach to human cognition can also help us think about musical sense-making, which involves cross-modal, situated-embodied experiences that are both explicit (dancing, performing) and covert (e.g., listening and the experience of movement, space, texture, contour).[11] By this light, the environmentally situated, feeling body is the primary cognitive domain for musical experiences—which are not, most fundamentally at least, representational, propositional, or conceptual phenomena. Accordingly, Johnson (2007) argues that the assumptions behind music-as-language analogies can lead to distorted and reduced understandings of both linguistic and musical communication. He therefore suggests an approach that examines music first in terms of the actual experiences it affords. Such experiences, he claims, are grounded in the basic logics of space, time, and movement that, via the cross-modal, embodied, and affective nature of human cognition, give rise to the ways we get involved with music within the extended physical and sociocultural ecology. Johnson argues that we are drawn to music because "it appeals to our felt sense of life" (Johnson, 2007, p. 236).[12] Thus, as he suggests, we would do better to describe musical experience first in terms such as "moving music," "moving times," "musical landscape," and "music as moving force"; or image schemas that describe paths of motion such as "source-path-goal" (Cox, 2016; Johnson, 1987, 2007; see also Lakoff & Johnson, 1980, 1999).[13]

Some Implications for Future Research and Theory

This shift away from internalizing and rationalizing perspectives on human meaning making is not just a sort of theoretical exercise. It may help researchers implement new models and strategies that will shed light upon the embodied roots of music cognition. In qualitative research, for example, phenomenologically guided frameworks could aid scholars in identifying particular metaphorical passages in interviews, highlighting the relational and evolving nature of musical experience nonreductionist terms (see Høffding, 2018; Ravn & Christensen, 2014; Ravn & Hansen, 2013). At the same time, it could help the interviewer develop richer questions that point more directly to the core of the problems of musical experience, without assuming an initial dichotomy between inner subjectivity and an objective world. Asking performers and listeners to discuss their personal engagements with music (historically and culturally) in conjunction with their moment-to-moment descriptions of musical experience—involving cross-modal correspondences, embodied metaphors, and so on—could lead to interesting new insights. In quantitative research, a shift away from information-processing accounts could also inspire the development of new experimental tools and designs to address the question of musical experience in new ways.

With regard to this last point, we should make it clear that, although the enactive approach highlights the centrality of the environmentally situated body for cognition, it is not dismissive of the use of functional Magnetic Resonance Imaging (fMRI), Transcranial Magnetic Stimulation (TMS), Magnetoencephalography (MEG), or other techniques that enable the examination of the brain. Although we maintain that cognition is not simply "in the head" and should be understood in an embodied and ecologically situated way, brain research is, of course, an incredibly valuable resource if we want to understand how experience works. Claiming that cognitive processes go beyond the boundaries of the skull does *not* mean that the brain does not participate in it. In this sense, then, our approach may be situated within the recent nonreductionist trends in critical neuroscience mentioned previously, which emphasize the role of bodily and extraneural factors in driving cognitive processes (see Colombetti, 2014; Fuchs, 2011; Slaby, 2014; Slaby & Gallagher, 2015; Thompson, 2007).[14]

In sum, the phenomenological attitude reveals that musical experience unfolds cross-modally through the integration of patterns of feeling, emotion, movement, and sound. These patterns and their meanings are enacted and reenacted in manifold ways—both shared and personal—over various time scales and in different contexts by the people, communities, and cultures involved. There are, of course, many examples that can help demonstrate how this is so. For the sake of brevity, we ask the reader to consider only one of them—namely, guitarist Jimi Hendrix's groundbreaking performance of the "Star Spangled Banner" at Woodstock in 1969 (an example also considered by Clarke, 2005). This performance radically reframes a cultural icon (the Anthem) by placing it in the context of a sonic enactment that expresses the brutality of the Vietnam war and the hypocrisy of the hegemonic American foreign policy that supported it. Here, Hendrix creates a violent (and ironic) soundscape where the symbol of patriotism incarnate in the Anthem is deconstructed through its interpenetration with chaotic noise—out of which sonic episodes arise that evoke gunfire, falling bombs, and screaming. What is so remarkable about this performance is not only how Hendrix enacts new understandings of what musical performance and guitar playing entail but also how his return to the primal cross-modal ("metaphorical") origins of music in sound, affectivity, movement, space, and empathy allows the evocative "musical" episodes to be experienced as more than simply symbolic. Rather, they may be witnessed as transformative phenomena that are lived through—visceral evocations for critical, emotional, and compassionate imagining that emerge from, and simultaneously inform and transform, the interpenetrative system of performer, listener, and the (dissenting) sociocultural ecology being enacted. This example goes well beyond an understanding of music as reproduction and representation, revealing the deeply communal nature of musical activity.

5 Music and Emotion

In this chapter, we develop the ideas introduced previously by exploring the relationship between emotion and musicality through the lenses of phenomenology and the 4Es. We first examine some prominent perspectives on the subject, arguing that although many important insights have emerged in recent decades, much current research and theorizing tends to maintain the problematic internalist bias, in which emotional engagements with music are explained in terms of appraisals, affect programs, component processing, and forms of (nonconscious) statistical analysis that play out within the brain. In response to this, we draw on a range of recent interdisciplinary research to explore musical emotions as dynamical and emergent phenomena that span bodies, brains, and the environments they are embedded in.[1]

Searching for Musical Emotions

Although recent decades have seen a rapid growth of interest in the topic of musical emotions from musicologists, philosophers, computer scientists, psychologists, neuroscientists, and music therapists, among others (Juslin, 2019; Juslin & Sloboda, 2010), a direct connection between music and emotion has not always been universally accepted. As we mentioned in chapter 1, some nineteenth century musicologists argued that musical expressivity should be approached as a purely intellectual phenomenon: musical beauty involves the perception and understanding of the formal characteristics encoded into the musical work by the composer. As such, any emotional response is merely a by-product of subjective experience and has little to do with the music itself (Hanslick, 1891). This view, referred to as "formalism," understands music as a profound and decorative art but denies the relevance of the "extra-musical world of concepts, actions, emotional states, and character" (Meyer, 1956, p. 1). Other scholars, by contrast, have claimed that musical works do in fact possess emotional content as a central aspect of their communicative function. This so-called "expressionist" position argues that the structural relationships inherent to a given piece of music are intended to produce a predictable and universal set of feelings and emotions in listeners. For instance, the musicologist Deryck Cooke (1960) claimed that the minor second expresses "spiritless anguish," while the major second equates to "pleasurable longing" (p. 90).

Debates over various formalist and expressionist orientations constituted a major part of philosophical musicology in the first half of the twentieth century. Despite their differences, however, both perspectives coincide in a common assumption—namely, that whatever music expresses should be found *outside* of the listener. This was driven by traditional beliefs central to Western musicology, where until relatively recently the putatively autonomous (and superior) status of the composer and the score tended to be taken for granted (see Bohlman, 1999; Cook, 2001; Goehr, 1992). Because of this, the answer to the question of where (or from whom) musical emotions originate was deemed self-evident: musically expressed emotions belong to composers, who encode their feelings into scores so that trained performers can reproduce them and educated listeners can decipher them (Martin, 1995).[2]

This view was questioned by philosophical musicologists of the late twentieth century (Budd, 1989; Davies, 1994). It was recognized, for example, that it is not necessary for composers and performers to be in a specific corresponding emotional state to produce emotionally expressive music. Likewise, the suggestion that the locus of musical emotions should reside in the "music itself" (the score and/or the performance thereof) is made problematic by the observation that musical works are not sentient beings, therefore making it difficult to see how they may feel and produce the emotions they are thought to express. Additionally, while the expressionist position rests on the idea that specific musical relationships (like a minor third) produce specific responses (like sadness), the Western canon is nevertheless full of instances in which compositions in minor keys produce music that cannot be properly described as inherently "sad": for example, Bach's Concerto for two violins in D Minor, BWV 1043. These kinds of concerns prompted the development of new theories aimed at addressing this "where/whose" problem.

Some writers have posited idea that musical experience may involve the listener's projection of a "virtual persona" to whom the emotions belong (Cochrane, 2010; Cone, 1974; Levinson, 1996). Here emotions are thought to be perceived as "possessed by" the persona in music, but not necessarily as "owned" by the composer. Others have argued that emotions ascribed to the music are in fact aroused in perceivers and that experiencing emotions as "in the music" is therefore a form of misattribution: emotions should therefore be attributed to listeners (Matravers, 1998; Nussbaum, 2007). Still others (Davies, 1994, 1997; Kivy, 1980) have claimed that emotional experiences associated with music do not require positing a subject who possesses them. The thinking here is that because musical sounds often resemble emotionally expressive human behaviors (e.g., vocal utterances, bodily movements, gestures), these similarities should be sufficient to initiate emotional engagements with music. Note that these philosophical approaches often tend to imply a close connection between the emotional and empathic dimensions of musical experience, a theme we explore in more detail later.

Routes and Mechanisms

Where philosophical approaches have been mostly concerned with accounting for the "whose" or "where" of musical emotions, psychological theories have been more focused on how emotions are aroused—or "induced"—in listeners in response to musical sounds. This has often been framed in terms of two "processing routes" associated with the psychology of emotions more generally (e.g., Chaiken & Trope, 1999). The first route posits that the elicitation of an emotion involves an "appraisal" by the perceiver (Lazarus, 1982; Scherer, 2005; Solomon, 1976). Such appraisals are thought to depend on information-processing mechanisms selected by evolution: for example, because they may have permitted features of the environment to be evaluated quickly (and preconsciously) in terms of their relevance for the goals and continued survival of the animal (i.e., to what degree a stimulus is good/bad, safe/threatening, expected/unexpected, attractive/repellent).

Now, it has often been assumed that music has no immediate biological connection to human well-being, in both day-to-day and evolutionary contexts (Juslin et al., 2010; Pinker, 1997; Scherer & Coutinho, 2013). It is argued, however, that the perception of music nevertheless employs appraisal mechanisms that evolved to process other forms of survival-relevant stimuli and that this allows us to form mental representations of the music we encounter. According to David Huron (2006), such processing involves statistical forms of learning that, over time, permit the listener to make predictions about what is likely to come next in a piece of music that follows the formal rules he or she has become accustomed to hearing (e.g., enculturated harmonic progressions, rhythmic structures).[3] The appraisals initiated by the contrast between statistically based expectations and the way the music actually unfolds generate different emotional states like anticipation, tension, surprise, relief, and disappointment. Others (e.g., Panksepp & Bernatzky, 2002) have added that the emotional experience of music also involves more "primitive" mechanisms that entail the triggering of innate, reflex-like sensorimotor responses associated with subcortical brain areas that produce appraisals of danger or urgency. This implies that (musical) appraisal processes may occur over various timescales and involve different "levels" of cognitive processing, where even a single stimulus input may produce "values for one or more appraisal variables as its output" (Moors, 2013, p. 133). Thus, more complex approaches see appraisals as performed both by basic and quick mechanisms (e.g., the "novelty check" that is produced in less than 500 ms by primitive brain areas, including the amygdala) and by slow mechanisms (like those associated with the aesthetic evaluation of a piece of music), which are thought to depend on complex, propositional forms of processing in the cortex (see Clore & Ortony, 2000; Sander et al., 2005).

The second route explores musical emotions, not in terms of appraisals but as products of mechanisms involved with "associative processing." Here, emotional responses are induced in a listener because the music he or she perceives has previously been associated with an emotional state. This route can involve forms of preconscious or minimally conscious conditioning in which emotional responses are generated because, in the past, the music was experienced as coinciding with events that were positively or negatively valenced. But listeners may also be completely aware of the associations that drive emotional experiences with music—for example, when a piece of music evokes episodic memories of specific emotionally laden events in one's life (Juslin & Västfjäll, 2008).

Other approaches do not fall neatly into appraisal nor associative routes. One of these involves the idea of "activation spreading." Here the hypothesis is that emotions are organized as networks of nodes in the brain, which are connected to each other by pathways that become strengthened over time. When one of these nodes is activated, responses are triggered across the network (Innes-Ker & Niedenthal, 2002). This is thought to underlie the mechanisms of "rhythmic entrainment" and "emotional contagion" associated with musical experience. The first of these describes how a listener's movements and physiological rhythms synchronize (or "entrain") with the periodicity of the music (the pulse or the beat); this, in turn, increases arousal and engagement with the music and may induce mood shifts and feelings of pleasure (Labbé & Grandjean, 2014). The second term, "emotional contagion," refers to the ways listeners unconsciously mirror the emotional expression of the music, and how this mimicry leads to the contagious induction of the emotion in the listener (Scherer & Zentner, 2001). It should be noted that some approaches to musical emotions incorporate both routes, as well as many of the other mechanisms considered previously. We later consider two prominent examples of these. First, however, it will be useful to introduce another important appraisal-based perspective that is currently gaining ground in music psychology.

Are Musical Emotions Basic?

The idea of "basic emotions" coincides with a long-standing assumption in folk psychology—namely, that emotions come in a group of predefined types, such as happy, sad, and angry. Basic emotion theory (BET) formalizes this view by positing the existence of a small group of basic emotions that evolved in response to the challenges our ancestors faced in their everyday life (Ekman, 1980). It is argued that each basic emotion is defined by a dedicated, innate, and distinct neural network that controls the affect programs that give rise to them (Tomkins, 1962, 1963). The term "affect programs" refers to the evolved brain mechanisms that produce "the patterns for these complex organized responses, and which when set off directs their occurrence" (Ekman, 1980 p. 82). Such programs, and the responses they trigger (i.e., basic emotions), are thought to be culturally universal in humans and are thus understood to function as auto appraisals selected by evolution (Ekman, 1992, 2003; Ekman & Friesen, 1971, 1986). It has also been suggested that because basic emotions are biologically fixed, they form the building blocks from which more complex emotions emerge (Prinz, 2004a,b).

Although BET proper has only recently begun to make an appearance in music psychology (see Juslin, 2013a,b), research in this area has traditionally been framed by the idea that musical emotions come in discrete categories associated with the ways specific neural mechanisms respond to musical stimuli. These categories tend to vary from study to study. Some have focused on more "basic" emotions (e.g., joy and sadness), while others have introduced categories that might be better described as moods (e.g., tenderness or nostalgia). Recent approaches have developed more nuanced models that introduce complex "aesthetic" categories such as wonder and awe (see Eerola & Vuoskoski, 2013; Juslin & Laukka, 2004; Juslin & Västfjäll, 2008; Trost et al., 2013). In all, studies of musical emotions guided by BET have produced some interesting insights into the relationship between perceived emotions and physiological changes (Lundqvist et al., 2008), as well as how emotional categories may be recognized consistently across listeners (Fritz et al., 2009). Because of this, some researchers argue that BET should be adopted more widely in musical emotion studies[4]—this includes the music psychologist Patrik Juslin (2013b), who advises that we should build our theories of complex emotions around the "core layer" of "iconically-coded basic emotions."

It should be noted, however, that BET is currently a topic of debate in the domains of affective science and philosophy of emotion. Some critics have raised the question of how, precisely, alleged basic emotions relate to nonbasic ones (Ortony & Turner, 1990). There is also an ongoing disagreement between those who argue that basic and nonbasic emotions are continuous (Clark, 2010; Plutchik, 2001; Prinz, 2004a,b), and those who insist that the two involve distinct processes and should thus be studied separately (Damasio,1994; LeDoux, 1996). Other authors (Barrett, 2006; Russell, 2003) question BET's lack of context sensitivity: emotions can be expressed and experienced in various ways, and their meanings shift depending on the situation. Attention has also been drawn to the BET methodology, which often employs a forced choice approach (generally between images facial expressions and a given scenario). Research has shown that quite different responses may be given when participants are not constrained by emotional labeling (Russell, 1994). It has also been pointed out that the close correlations that are expected to occur between specific brain areas, autonomic nervous system activity, and the putative affect programs associated with basic emotions are often less consistent than hypothesized (Barrett, 2006).

While advocates of BET have responded to such criticisms in various ways (see Ekman & Cordaro, 2011), another more general problem remains concerning the rather arbitrary way basic emotion categories were originally introduced. This concerns the work of Silvan Tomkins (1962, 1963), who posited the existence of nine primary affect categories (1962, pp. 111–112). As the philosopher Giovanna Colombetti (2014, pp. 36–40) notes, while no sustained justification was given by Tomkins for why these categories were chosen, various subsets of them have nevertheless been assumed to compose the basic responses that guide our emotional engagements with the world.[5]

BET is also problematized in musical contexts by studies that show that the physiological changes associated with the induction of musical emotions do not always align clearly with those exhibited in association with basic emotion categories (Krumhansl, 1997, p. 351; see also Scherer & Zentner, 2001). Additionally, cross-cultural research associated with the perception of musical emotion suggests that while there do appear to be certain aspects that are shared between cultures—especially with regard to tempo and pitch, where faster tempo and higher pitch correspond to happiness and slower tempo and lower pitch to sadness (Eibl-Eibesfeldt, 1989; Juslin & Laukka, 2003)—there are also many differences. For instance, when listening to Javanese music that is meant to express joy, Westerners often interpret the (close to) minor third interval in the *pélog* scale as being indicative of sorrow (Gregory & Varney, 1996). And even where there does appear to be intercultural agreement, precisely *how* the emotion is meaningfully interpreted may differ from culture to culture, over time, and from one individual to the next (Becker, 2010; Finnegan, 2012).[6]

These issues have led some to posit that musical emotions may be somehow different from—or perhaps "impoverished" versions of—"real" (i.e., basic) emotions. Music psychologist John Sloboda (2000) describes the situation in the following way: "Very often we feel that there is an emotion present . . . but we cannot quite tie it down. In such a state of ambiguity . . . we may well expect the profound and semi-mystical experiences that music seems to engender. Our own subconscious desires, memories, and preoccupations rise to the flesh of the emotional contours that the music suggests. The so-called 'power' of music may very well be in its emotional cue-impoverishment. It is a kind of emotional Rorschach blot" (p. 226).

In response to these kinds of concerns, advocates of BET have developed complex theories that integrate aspects of the routes and mechanisms discussed previously to explain how basic emotional responses ground complex musical experiences. Notably, Juslin's (2013a; Juslin & Västfjäll,

2008) approach integrates a range of factors including *B*rainstem reflexes, *R*hythmic entrainment, *E*valuative conditioning, emotional *C*ontagion, *V*isual imagery, *E*pisodic memory, *M*usical expectancy, and *A*esthetic judgement (*BRECVEMA* for short). Basic emotional responses are influenced and focused by these mechanisms, allowing for the experience of so-called "aesthetic emotions," such as wonder, transcendence, nostalgia, tension, or awe (Zentner et al., 2008).[7]

The Component Process Model

Other researchers have posited alternatives to BET-based approaches. Notably, the psychologist Klaus Scherer has offered an intriguing perspective based on his own component process model (CPM), which was first developed to explain emotions more generally (2001). Briefly, the CPM offers a way of accounting for emotional responses through a set of criteria, including physiological symptoms, motor expression, action tendencies, and cognitive appraisal. These components are thought to be integrated multimodally and are continuously updated largely through nonconscious processing. This results in representations, "parts of which may then become conscious and subject to assignment to fuzzy emotion categories which may then lead to labeling with emotion words, expressions, or metaphors" (Scherer & Coutinho, 2013, p. 123). The CPM thus allows for the differentiation of affective phenomena into several classes that are not reducible to the discrete categories associated with BET, where "the meaning of emotion words for the respective classes may be difficult to define and are multiply interrelated" (p. 128). Among these classes, three are understood to be properly emotional. The "utilitarian" describes appraisals that are relevant to our goals and well-being. The "aesthetic" refers to responses associated with experiences of beauty, wonder, and awe. And the "epistemic" concerns the role of affective responses and appraisals in the construction of knowledge.

To explain musical emotions, Scherer and Coutinho (2013) draw out a "multifactorial" approach based on the CPM. In doing so, they consider a wide range of interacting elements, including the structural elements of the music, performance dynamics, listener experience, and contextual features. This leads them to posit five possible mechanisms or routes that afford "the production of emotions in listeners": appraisal, memory, entrainment, emotional contagion, and empathy. These processing components are understood to interact with each other in complex ways that span preconscious and conscious levels.

Scherer and Coutinho acknowledge that musical experiences can involve many different types of affective states. However, their analysis is restricted to what they refer to as "properly emotional reactions to music" (p. 139). These reactions are categorized in terms of the utilitarian, aesthetic, and epistemic classes mentioned previously. According to Scherer and Coutinho, utilitarian emotions should have little to do with musical experience. This is because they see this type of emotion as driven by factors that have immediate personal relevance for the listener—musical experience, it is assumed, has none. They write,

> A major difference between utilitarian emotions on the one hand and aesthetic and epistemic emotions on the other is the fact that appraisals concerning goal relevance and coping potential involve different criteria (such as different goals and coping mechanisms) in aesthetic or epistemic emotions as compared to utilitarian ones. In other words, an aesthetic or epistemic emotional experience is not triggered by concerns with the immediate relevance of an event for one's survival or well-being, nor with how well one can cope with the situation. Rather, the appreciation of the intrinsic qualities of a piece of visual art or a piece of music or the degree of discovery or insight one achieves through novel and complex stimulation in different modalities is of paramount importance. This corresponds in many ways to Kant's well-known definition of aesthetic experience as "interesseloses Wohlgefallen" (disinterested pleasure; Kant 1790/2001), a definition which insists on the need for a complete absence of utilitarian considerations. (2013, p. 125)

Thus, CPM places special emphasis on so-called aesthetic and epistemic emotions for explaining musical experience because they correspond to, again, the supposed *lack* of personal relevance music has for human goal making and well-being. This, Scherer and Coutinho argue, explains the ambiguity found in traditional attempts to align emotional responses to music with nonmusical basic or everyday emotions (which are relevant to our goals). In all, then, Scherer and Coutinho (2013) posit an essentially Kantian conception of musical emotion that rests upon detached perceptual processes. Despite references to context and culture, the overriding assumption is that musical emotions are caused by external structural antecedents intrinsic to the music itself, acting on specific internal processing mechanisms—whereby *music creates emotions in listeners*.

The Emotional Meaning of Musical Experience

Like Juslin's BRECVEMA model, CPM offers a rich, multifaceted approach to musical emotions. It may even provide more nuance, seeing emotions emerging through the dynamical interaction of neural components as opposed to being built on discrete emotion categories. Nevertheless, as we saw, CPM does posit its own classes of response, only two of which (aesthetic and epistemic) are applicable to musical emotions. Additionally, while aesthetic or epistemic emotions are not seen as completely disembodied, the corporeal dimensions of musical experience are understood to be limited to "diffuse responses" (goose pimples, shivers, tinkling on the spine, or moist eyes) that "do not serve any obvious adaptive purposes" (Scherer & Coutinho, 2013, p. 125).

It should be noted that the assertion that music has no utility in everyday life has been seriously questioned in recent years. For example, the neuroscientist Stefan Koelsch (2013) argues that current work in musical emotions has been guided by the following group of problematic assumptions:

- Music does not involve goal-oriented responses.
- Music is not directly relevant to human well-being.
- Music does not elicit real emotions.
- Musical emotions have little material (i.e., biological) effect on listeners.

He counters this by considering the deep social significance of musical activity, outlining interrelated functions that serve empathy, communication, group cohesion, and social cognition. In doing so, Koelsch posits that music *is* goal-directed as it helps fulfill social needs, noting that "music-evoked emotions are related to survival functions and to functions that are of vital importance to the individual" (p. 232). He reinforces these claims from a neuroscientific perspective by tracing the neurophysiological correlates of music-evoked positive emotions, showing how these align with survival relevant interactions in the social environment.

The shift in perspective offered by Koelsch connects with a more relational and developmental orientation toward the musical mind. A range of research and theory (e.g., Cross, 1999, 2001; Krueger, 2013; Trevarthen, 2002) has revealed the importance of musicality for developing embodied, prelinguistic, and emotional-empathic forms of understanding, communication, and social cognition, beginning with the first interactions between infants and primary caregivers (and perhaps before birth; see Parncutt, 2008).

Musicality is also understood to play a major role in the processes of joint action and participatory sense-making in developmental and social contexts (Loaiza, 2016; Schiavio & De Jaegher, 2017), which are associated with the coconstruction of meaningful social and cultural worlds.

This social and developmental perspective paints a rather different picture of what music means for the human animal. Indeed, by this light, one of the main issues that drives appraisal-based theories—the assumption that music is not relevant for human goals and well-being—loses its significance. In line with this, it may be argued that the CPM model downplays or ignores important dimensions of musical experience when it restricts musical emotions to aesthetic (in the Kantian sense) and epistemic types. Additionally, while CPM presents itself as a dynamical[8] and "emergent" model of emotion, the interaction of the components involved are "driven by the appraisal of the eliciting event" (Scherer, 2007, p. 115). In Colombetti's (2014) critique of CPM, she notes that this aspect moves the model away from the kind of recursive dynamics associated with self-organizing systems—it implies, in fact, a more linear causal schema that is driven by the appraisal component, which assumes executive properties (see also Camras & Witherington, 2005, p. 336).

Similarly, many of the other theories discussed previously tend to examine musical experience in a cause-and-response context, reflecting the tacit acceptance that the emotional experience of music should begin with the internal appraisal of external events. BET approaches reinforce this in-the-skull orientation when they ground the emotional experience of music in the reactions produced by evolutionarily determined affect programs. Likewise, prediction-based models, such as Huron's (2006), conceive of emotional responses to music in terms of mechanisms that produce statistical inductions of environmental regularities through algorithmic processing, in which listeners' experiences of musical anticipation, satisfaction, surprise, and so on are caused by "weighted sums" drawn from many representations.[9]

We should note here that BRECVEMA, CPM, and Huron's model do not deny that the body plays an important role in the emotional experience of music. BRECVEMA and CPM, especially, highlight the importance of entrainment for musical experience: our ability to become synchronized with the rhythm of the music plays an important part in how we start responding to it emotionally. But here again, these accounts tend to place the body in a kind of mediating, responsive role between inner and outer realities, where cognition as such is restricted to the brain. Additionally, we argue that although the approaches discussed previously capture important aspects of what musical emotions involve, they also miss crucial dimensions of what musical experience entails when listeners are essentially taken to be *responders* to musical stimuli in the environment and where, accordingly, musical perception is examined mostly in terms of how individuals react to specific (and often isolated) musical events.[10]

As considered in previous chapters, it has been shown that even in contexts that appear to be passive (e.g., listening in a concert hall), people actually play active roles in shaping their engagements with musical environments (Clarke, 2005; Krueger, 2014). And such active processes of musical meaning making draw on those primordial corporeal sense-making capacities that allow us to construct metaphorical, cross-modal, and narrative relationships between various temporal, spatial, textural, bodily, social, ecological, sonic, and affective dimensions of experience (Johnson, 2007). These types of embodied-emotional sense-making are, again, based first in adaptive corporeal experience within a contingent sociomaterial milieu—they depend on various levels of interactivity in which patterns of behavior and perception self-organize and transform as a history of structural coupling between experiencer and the environment.

Now, all of this does not negate the observation that emotional engagements with music can involve recurring psychophysical patterns of behavior that can sometimes bear striking similarities across individuals and groups and that may indeed be nameable. It does indicate, however, that we need to develop new approaches and tools for studying such phenomena. Accordingly, we suggest that a more dynamical and enactive perspective on musical emotions is needed—one that can better address the social, developmental, embodied, ecological, and active-creative dimensions of musical experience.

Musical Emotions as Dynamical and Emergent

Research in affective science based on dynamical systems theory (DST) holds that emotional episodes may not be reducible to pregiven response mechanisms in the brain (and corresponding behavioral categories). Rather, emotions are also guided by developmental processes that span the integrated body-brain-environment network (Freeman, 2000; Lewis, 2000, 2005).

Recalling our introductory remarks on DST in chapter 2, this approach examines emotions in terms of the agent's history of structural coupling with a given environment, where a range of factors combine over different time scales to produce patterns of activity and meaningful organism environment relationships. These relationships can become stable over time ("basins of attraction" in DST speak) but can also be disrupted to form new patterns and relationships—for example, through growth, learning, social interactivity, and the mutual influence of other exogenous and endogenous factors. As we considered, such phenomena can be described in terms of the enactment of areas of attraction and repulsion, of stability and entropy, which unfold across the body-brain-world network and constitute the developing phase portrait of the organism-environment system.

Colombetti (2014, pp. 57–82) suggests three mutually informing levels of inquiry that can help to capture how such dynamical processes play out across body, brain, and environment. The first involves the development of the muscular linkages and coordinative structures associated with emotional expression (e.g., facial, vocal, limb, and hand movement). This aligns with the developmental research mentioned previously, which shows that the appearance of recurrent and meaningful patterns of expressive behavior in infants is not best understood wholly in terms of predetermined genetic programs but as emergent properties of the interaction of a range of environmental and bodily factors as infants actively reach out to and make sense of their sociomaterial worlds (Camras & Witherington, 2005; Thelen & Smith, 1994). The second level draws on existing work in affective neuroscience that highlights the plasticity of neural structures (see Freeman, 1999, 2000). Here, emotional episodes are explored as patterns of convergence in the neural trajectories of an agent that may both stabilize and transform as a result of contingent shifts in the global constraints of the brain-body-world system. The third level focuses on environmental concerns, exploring how agents enact meaningful emotional engagements (i.e., recurrent patterns of relational behavior) with the things and people they interact with and the situations they live through (Hsu & Fogel, 2003; Laible & Thompson, 2000).

Notably, from this perspective, emotional episodes are not limited to the skull nor the skin of the organism. There is a strong sense in which they "extend" into environments where agents coenact meaningful worlds through forms of participatory sense-making (Krueger, 2014; Krueger &

Szanto, 2016; Slaby, 2014). DST considers these kinds of processes as recursive, or "circular," phenomena in which all parts of the extended system—body, brain, and environment—influence each other in a nonlinear way. Indeed, at each level described earlier, adaptations are constantly needed for a living agent to maintain coherent relationships with the world and to develop new ones as needed. For example, growth means that a new set of bodily and environmental affordances becomes available to an animal, which must then be developed in conjunction with new skills that enhance its sense-making capacities. This entails the development of new neural trajectories, as well as the enactment of novel relationships with things and other people, which, in turn, involve the development of new sets of meaningful movements and gestures by which an embodied agent interacts with, makes sense of, and communicates with its world. This, again, influences possible neural states of the agent, and so on.

Put in simple terms, the enactive-DST perspective explores emotions as emergent properties of living systems—as "patterns" of adaptive sense-making behavior that may recur in conditions with similar sets of constraints but that may also evolve in various ways because of shifts in bodily and environmental conditions. These insights align well with the fundamental enactive principles of autonomy and autopoiesis. As we discussed, the world of salience that is brought forth (enacted) through the ongoing interactivity between a self-organizing living system and its environment necessarily involves a primary conative-affective drive associated with survival and well-being (Reybrouck, 2001). The sense-making activities that shape and are shaped by such processes are always relevant to the life-world of the organism and are thus emotionally motivated. From the "primordial affectivity" of simple organisms to the more complex individual and sociocultural self-organization of humans, emotionality lies at the core of mental life and therefore grounds cognitive processes (Colombetti, 2014; Damasio, 1994).

In all, this orientation offers a way to address the *active* and *self-organizing* nature of emotional experiences—how they develop, stabilize, and transform through the interaction of bodily, neural, and ecological dimensions. An enactive-DST approach can therefore help us explore emotions as simultaneously plastic, patterned, and recurrent, without having to first posit the existence of pregiven emotion categories and the fixed neural mechanisms that give rise to them. It also suggests that while emotional episodes can be unique to an individual organism, they may also emerge and develop ways that are shared between agents with similar biosocial needs and histories of interaction with the environment.

Importantly, this emergent conception of emotion casts the insights offered by the previously discussed theories of musical emotions in a new light. For one thing, it suggests that while a musical experience may be described as "sad" or "angry," for example, that sadness or anger cannot be divorced from the developmental trajectory and current situation of the musical agent within a particular sociomaterial environment. At the same time, it also embraces the fact that emotional experiences are often enacted that are not easily labeled with words but that are nevertheless recognized, reenacted, and developed by individuals and musical communities, implying that the states of being we refer to with specific emotional signifiers may be far more complex, contextual, and idiosyncratic than is suggested by the language we use. For example, what we categorize as "fear" in a given instance may in fact involve a complex range of relational entailments that make this or that fear unique to its context and the person experiencing it.[11] Additionally, this perspective shows that the "predictive" or "anticipatory" aspects of musical experience need not be understood solely in terms of complex internal computations (e.g., statistical processing). Instead, they can be traced to the dynamical patterns of action and perception associated with the idea of (participatory) sense-making. This suggests that how agents engage in musical appraisals may be rooted first in the repertoire of affectively driven body-environment relationships (structural coupling) they have enacted throughout their lives and how musical experiences coincide with, push against, or otherwise engage such body-based proclivities for action and feeling (Reybrouck & Eerola, 2017). In contrast with Dennett's notion of pips (see previous discussion in chapter 3), this perspective sees acts of prediction and recognition not as limited to (preconscious) neural mechanisms, but rather in terms of the enactment and re-enactment of corporeal-environmental configurations that reflect the adaptivity of the organism, basins of attraction across neural, bodily, and environmental trajectories, whose fulfillment may be achieved or thwarted.

Music, Emotion, and Social Life

In subsequent chapters, we consider how such self-organizing dynamics play out in musical developmental contexts, what they can tell us about the nature of musical creativity, and what impact they may have for our understanding of musical learning and performance. We can summarize by noting that, from an enactive-DST perspective, emotional engagements with music are intimately linked with our sensorimotor faculties and our histories as self-making, social, cultural, and environmentally situated beings. This means that the experiences of resonating with musical environments, preparing for musical action, and enacting personal and social musical worlds cannot be reduced to in-the-skull information-processing mechanisms.

For improvisers, composers, listeners, and interacting performers, musical experience emerges through affective-motivational processes, which play out in unique ways depending on how musical environments interact with the developmental histories of the participants involved. Modes of engaging with the same piece of music can differ among individuals, and listeners can display diverse emotional experiences, despite having similar cultural backgrounds and similar levels of musical expertise. Likewise, differences also emerge with the specific sensorimotor interactions adopted to engage with the affordances and demands of changing musical environments. Musicians explore and play with the processes of sense-making in a wide range of ways, sometimes adjusting their performance and expressions to produce consensus between performers or shared embodied states between interacting listeners (e.g., dancers). At other times, they initiate radical shifts that demand new emotional-bodily-cognitive relationships and a heightened adaptability to the sonic environment (e.g., free improvisation). And while the measurable physiological effects of the emotions involved in such diverse settings may cover a relatively limited range of parameters, the actual experience of such events may take on a wide range of characteristics and meanings given the situatedness of the musical agent. That is, while musical emotional episodes may bear striking physiological similarities to one another, they can also involve important phenomenological differences that reflect the contingencies of existence and adaptation.

Accordingly, we believe that music is not something we simply react to. And, likewise, emotions are not just responses, or things that we (or musical works) possess. Rather, emotionality and musicality are intimately related aspects of human life that are central to how we actively constitute and make sense of the world. In line with our exploration of music and consciousness in previous chapters, musical emotions involve relational forms of action and perception that play out in a variety of ways over different timescales and across a range of contexts. Throughout our discussion we have also been emphasizing the deep connection between emotional experiences with music and our nature as social beings. As we began to consider, musicality plays an important role in the development of lasting emotional bonds and preverbal or nonverbal (i.e., embodied-affective) modes of communication between infants and primary caregivers (Johnson, 2007; Krueger, 2013; Trevarthen, 2002). Likewise, research shows that children who engage in coordinated musical activities display heightened levels of cooperation and empathic understanding with their peers in other contexts (play and group learning) (Kirschner & Tomasello, 2010; Rabinowitch et al., 2012, 2015). Musical factors such as interpersonal entrainment and the enactment of mimetic forms of sonic and corporally based communication are also thought to account for the power of music in cultural contexts that involve more complex social coordination—work, ceremony and ritual, play, athletics, and more (see Clayton et al., 2012; Hove & Risen, 2009; Small, 1998; Stokes, 2010; Stuart, 2012; Tomlinson, 2015).

Moreover, while the relationship between music, emotion, and sociability is perhaps most obvious in the real-time interactions between performing musicians and in situations where a listening audience is present,[12] an empathic dimension also seems to guide the experience of music in more solitary contexts. This is implied by some of the philosophical perspectives we considered previously—for example, those that suggest the idea of a "musical persona" or that musical sounds reflect emotionally expressive movements and utterances that a listener may empathize with. Similar phenomena have been documented in recent studies that show that people who are experiencing sadness sometimes listen to sad music because it gives them the sense that someone is empathizing with their feelings (Eerola et al., 2016; Lee, Andrade, & Palmer, 2013; Van den Tol & Edwards, 2013).

Because musical expressions possess increased complexity, temporal range, subtlety, and force when compared with everyday gestures and facial expressions (Cochrane, 2010; Roberts, 2015), musical environments can offer affordances for feeling, thought, and action that are not always available from other environmental resources. Music can therefore help people to explore their feelings within a field of possibilities that goes beyond normal modes of expression—to work through powerful emotions and social experiences, such as grief or anger. These uses of music are reported by the participants of research done by the musicologist Marie Strand Skånland[13] (2013). Below are examples from two participants in her study:

> Sometimes I can use [music] to investigate [my mood] a little. . . . In a way, it can help me to find out what mood I'm in, and to feel it. . . . And then maybe manage to understand why, and to do something about it.

> I actually use music to amplify the mood I'm in already. . . . Instead of just being angry I'm able to distance myself from the feeling and monitor the feeling, in a way . . . examine the feeling, touch it and study it a little. (pp. 7–8)

These comments are indicative of how people use extended musical environments to develop richer understandings of their emotional and social lives (Clarke, 2019; Peters, 2015). Music, in other words, can afford increased *epistemic access* to our affective states (see also Reybrouck, 2017a,b). This can happen in interpersonal situations and in solitary contexts where, again, people sometimes experience music in terms of a virtual persona who empathizes with their feelings. Let's now turn to explore some possibilities for understanding the social dimensions of musical experience though the lenses of embodied music cognition and the 4E framework.

6 The Empathic Connection

As with emotion, conceptions of empathy have been influenced by long-standing assumptions in folk psychology.[1] Most centrally, this has involved the idea that empathy entails a kind of mind reading whereby we "enter into" the thoughts and feelings of another person, or otherwise mentally place ourselves in their position (Laurence, 2007; Lipps, 1907; Smith, 1759). To explain how this is so, scholars have developed an influential hypothesis referred to as theory-theory, which, as the name suggests, posits that empathy entails the construction of theories about the minds of others (Baron-Cohen, 2011; Marraffa, 2011). Because such theory making would require knowledge of the other person's intentions and beliefs, as well as other contextual factors, empathy is often thought to rely on complex cognitive processes involving deduction and imagination.[2] Taking this further, some researchers have suggested that we need to define what really counts as empathy to distinguish it from other aspects of human thought and feeling. Notably, the psychologist Amy Coplan (2011) has argued that a clear distinction should be made between the lower-level perceptions, feelings, and bodily states associated with emotional contagion and imitation and the higher-level processes involved in constructing representations of the experiences and intentions of other people. The term "empathy," she argues, should refer to the more complex cognitive operations involved in the latter.[3] Additionally, she posits that empathy is not best understood in terms of the construction of theories as such, but as a simulation process whereby the mental states of others are replicated in the mind of the empathizer, while a clear differentiation between oneself and the other person is conserved (2011, pp. 5–6).

Although these perspectives do capture aspects of what empathy entails in certain circumstances, we suggest that they may be too restrictive in musical contexts. Consider, for example, the communicative musical (or music-like) interactions that occur between infants and primary caregivers. In these situations, interpersonal understanding does not appear to be driven by the kinds of higher-level processes involved with theory-theory or Coplan's notion of empathy-as-simulation. Here, infants arguably would *not* be engaging in the construction of complex mental simulations, theory making, and clear self-other distinctions. It also seems difficult to separate the emotional-corporeal experiences the caregiver has in interaction with the infant from the processes involved in understanding the infant's feelings and desires. Should it follow, then, that these kinds of experiences do not involve empathy? Perhaps it is only a specific aspect of the caregiver's higher-level cognizing that is properly empathic. Another example is a situation in which one "loses" oneself in the music as a performer or listener, or in which a powerful musical event appears to break down distinctions between self and other within a group (e.g., between performers, audience members, and performer and audience). Are these so-called "immersive"[4] musical experiences also not empathy involving, even if they evoke powerful feelings of belonging, cohesion, and shared conceptions of identity (Vuoskoski et al., 2017)?

In brief, we suggest that both theory-theory and Coplan's approach may place too much focus on complex internally based mental processes and that they therefore do not fully capture the corporeal and affective-emotional aspects of social cognition—which, as we have seen, appear to be central to musical experience. In line with this, we argue that it may be problematic to posit strict distinctions between more "basic" forms of social empathy driven by emotional and bodily engagements and more "complex" or "higher" varieties associated with imagination, deduction, and other more abstract forms of thought (although such higher-lower distinctions could be useful as heuristics).[5] As we began to consider in chapter 4, the latter aspects of human sense-making are deeply continuous with the former, and musical experience appears to engage both simultaneously and to varying degrees. With this in mind, let's now consider two alternative approaches that may offer a way forward for thinking about musical empathy in a manner that highlights the continuities between perception in action, feeling, and social understanding.

Embodied Simulation Theory

Like Coplan, the perspective referred to as "simulation theory" in cognitive neuroscience bases social cognition in the capacity to form internal simulations (Gallese & Goldman, 1998; Goldman, 2006). Unlike Coplan, however, the simulations posited by recent accounts of this theory do not rely first on higher-level cognitive processers but on the automatic construction of preconscious representations of the bodily/emotional states of the person one is empathizing with (Gallese, 2005; Gallese & Sinigaglia, 2011). In a nutshell, the idea is that we ascribe meaning in our social encounters by recruiting our own bodily states and by functionally attributing them to others (Gallese, 2014, 2017). Accordingly, this approach is sometimes specified as "embodied simulation theory."

The supporting evidence for this view comes from recent research in neuroscience associated with the functional properties of the mirror neuron mechanism. As we began to consider in chapter 2, mirror neurons become active both when observing a goal-directed action performed by another agent (e.g., picking up a drumstick) and when performing a similar action. Importantly, such neural activation reflects the degree of motor expertise of the perceiver and is modulated by the goal of the action witnessed rather than only by perception of bodily movement. These points are shown in an elegant (and oft-cited) functional magnetic resonance imaging (fMRI) experiment carried out by Buccino and coworkers (2004). Here, participants were asked to watch silent videos in which a man, a monkey, and a dog perform actions associated with eating and communication, respectively. The goal-directed action associated with eating (biting food) is shared by all three animals. Communication, however, entails different forms of action for each: talking for humans, lip smacking for monkeys, and barking for dogs. When participants watched the actions associated with eating, fMRI results showed a clear activation of areas associated with the mirror neuron mechanism—"two sites (a rostral and a caudal) in the inferior parietal lobule as well as the posterior part of the inferior frontal gyrus and the adjacent precentral gyrus" (Rizzolatti & Sinigaglia, 2008, p. 132). However, the video with communicative actions resulted in very different outcomes: activity in the neural areas associated with mirroring was weaker when watching the monkey lip smacking and stopped completely when participants were shown the dog barking. Indeed, this kind of action, barking, is not present in the human motor repertoire. Although we can understand what a dog is doing while barking, this kind of understanding is not primarily defined in motor terms. This action is not understood "from within" because it cannot be properly simulated. Just like watching a dog wagging its tail, we simply cannot reuse our bodily states to simulate this action because it does not resonate with any motor program we use in our everyday life.

The importance of goal specificity for social understanding is summarized in the following passage:

> By mapping actions in terms of their motor goals and intentions, the cortical motor system is able to represent them as such without needing to specify all the kinematic parameters. Interestingly, these goal-related motor neurons are somatotopically organized so that when they activate, they instantiate a motor representation of their corresponding bodily part (e.g., hand, mouth, etc.) as accomplishing a given motor goal (e.g., grasping, biting, etc.). In other words, our body is mapped within the cortical motor system as a manifold of possibilities for action. (Gallese & Sinigaglia, 2011, p. 129)

The process of simulating and understanding an observed action, therefore, does not simply rely on the perceived kinematics—it depends on the degree to which the observing agent is able to perform the action in question and perceive its contextual purpose (its goal-directedness) (Cappuccio, 2009, p. 62). Consider, for example, how an expert drummer may simulate the witnessed movements of a fellow drummer: the perceiving drummer is sensitive to different subtle nuances observed in the performance and can, for example, predict and understand in a more accurate way than a nonexpert how a change in the hand grip of the drumstick will lead to different musical outcomes.[6]

Mirroring mechanisms are also thought to be involved in how we share emotions with others (Gallese, 2003a,b, 2009; Wicker et al., 2003). It has been shown, for example, that when we observe other individuals expressing disgust, or pain, the same brain areas become active when we feel disgust or pain ourselves (Hutchison et al., 1999; Jackson et al., 2005; Likowski et al., 2012; Pfeifer et al., 2008; Singer et al., 2004). This research has led a number of theorists to conclude that the brain mechanisms associated with social understanding are grounded in the automatic activity of sensorimotor neurons and that we empathize with others by activating "the same neural circuits underpinning our own emotional and sensory experiences" (Gallese, 2009, p. 523). Such insights may help to explain how many of our social engagements involve body-based forms of understanding that appear to arise spontaneously in the moment—that is, too quickly for complex forms of mental deliberation and theory making. Neuroscientist Vittorio Gallese (2003a) explains it in the following way, "Whenever we are exposed to behaviors of others requiring our response, be it reactive or simply attentive, we seldom engage in explicit and deliberate interpretative acts. The majority of the time our understanding of the situation is immediate, automatic, and almost reflex-like" (p. 520).

Additionally, while theory-theory remains influential in developmental psychology,[7] scholars in neuroscience and philosophy of mind have noted that the "theories" of other minds it posits cannot be directly observed, which draws its scientific status into question. Now, all of this does not negate the role that more complex mental operations may play in certain forms of empathizing. However, these insights strongly suggest that theory making (and the like) are better understood as secondary aspects of social cognition, which is guided by the more fundamental capacity to integrate the sensorimotor (and emotionally expressive) acts of others with those we enact ourselves.

Put simply, embodied simulation theory posits that social cognition is based primarily on the processing of information associated with the perception of bodily action—which, as Gallese (2001) notes, "considerably deflate[s] the role played by abstract theorizing when ascribing mental states (at least some mental states) to others" (p. 42). For a growing number of researchers, then, embodied simulation theory offers a more parsimonious and empirically grounded account of the foundation of social cognition in humans and many other animals. These advantages have been recognized in recent research associated with embodied music cognition (Leman & Maes, 2014; Leman et al., 2018; Maes et al., 2014; Novembre et al., 2012, 2013). Supporting research (Kohler et al., 2002) indicates that the perception of sounds associated with goal-directed action may be enough to trigger a subset of mirror neurons, which are referred to as "trimodal," meaning they are sensitive to auditory, visual, and motor contingencies.[8] Accordingly, a number of studies have shown that musical experience appears to activate the motor cortex even in situations in which listeners are not observing or engaging in bodily movement (see Molnar-Szakacs & Overy, 2006; Overy & Molnar-Szakacs, 2009). This suggests that simulation-like mechanisms may also (partially) account for empathic, emotional, and "metaphorical" experiences in listening contexts in which no explicitly correlated visual information is present.

Additionally, recent music research suggests that empathic processing may involve interacting routes that result in two main forms of musical empathy (Fan et al., 2011; Gazzola et al., 2006; Wallmark et al., 2018). The first is referred to as "emotional empathy," which describes the preconscious ability to share emotional states. The second, "cognitive empathy," describes the ability to consciously deduce and understand the internal states of others.[9] These forms interact with each other through the integration of bottom-up (emotional contagion and the preconscious simulation of intentional action and affective gestures in others) and top-down processing (complex deductive processes). Both cognitive and emotional empathy depend, to varying degrees, on the mirror mechanism and the ability to create the internal neural simulations that are thought to be fundamental to social cognition. Here it is also interesting to note studies that suggest that the influence of top-down processing can be minimal, allowing listeners to integrate the more imaginative and reflective aspects of musical experience with the (preconscious) bottom-up, corporally based simulations associated with neural mirroring (see also Schubert, 2017).

Among other things, an embodied simulation theory approach to musical empathy provides for a more integrated view that includes primary corporeal-affective forms of interaction as a central aspect of social cognition. It could also offer insights into how listeners attribute emotional and corporeal states to performers and composers—from the mirroring of basic bodily affective states to more complex imaginative representations of their lives and intentions (Cox, 2016). Likewise, embodied simulation theory may also help to explain how the emotional experiences we have with music in more solitary listening contexts (where no explicit person-to-person interactivity is present) are nevertheless guided by empathically driven motor simulations (imagining-feeling musical movement, space, narrative, or, again, the presence of a virtual persona).

Interaction Theory

Embodied simulation theory appears well poised to offer important insights into the social nature of music cognition. However, it has been argued that this approach may offer an account that remains too individualistic when it sees empathy in terms of internal responses that occur within the psychophysical domain of the perceiver. Accordingly, some thinkers (De Jaegher et al., 2010; Gallagher, 2008a,b, 2011, 2020; Gallagher & Varga, 2013) are exploring an alternative approach that adopts a more dynamically relational perspective, in which the neuronal activity described earlier is seen as one component of a larger cognitive system that spans interacting bodies (including brains) and other developmental and environmental factors.

This orientation—sometimes referred to as "interaction theory"—has antecedents in the work of a number of phenomenological philosophers. Most notably, Edith Stein (1989), Edmund Husserl (2006), Maurice Merleau-Ponty (1945), and Max Scheler (1954) challenged the assumption that the mental states of others are not directly observable and that we must therefore always employ indirect means—such as inference, theory making, or simulation—to access them. These thinkers argued that the meaningful experience of most social interactions is dependent on bodily movements and facial gestures, various emotionally laden and goal-directed conducts (forms of behavior) that are perceived directly and that therefore form the front line for social understanding. The main aspect that differentiates these thinkers from the embodied simulation theory approach just discussed is that they see no need to first posit the existence of internal simulations to explain such experiences: our emotions, intentions, and desires are not hidden inside us but are manifest in our actions and interactions (Gallagher, 2001, 2020). Social cognition is, most fundamentally, a direct process of embodied interactivity within a shared world. Philosophers Shaun Gallagher and Daniel Hutto (2008) write, "In most intersubjective situations, that is, in situations of social interaction, we have a direct perceptual understanding of another person's intentions because their intentions are explicitly expressed in their embodied actions and their expressive behaviors. This understanding does not require us to postulate or infer a belief or a desire hidden away in the other person's mind. What we might reflectively or abstractly call their belief or desire is expressed directly in their actions and behaviors" (pp. 20–21).

The phenomenology of encountering a face of another person who is scowling or smiling, for example, does not involve an inference that the internal state of that person is angry or happy. Rather, as the philosopher Dan Zahavi (2011, 2014) notes, the emotion is directly perceived as an aspect of the world embodied on the face of the person one is engaging with.[10] This philosophical orientation is now referred to as "direct social perception." In line with the previous discussion of the extended, relational conception of autonomy associated with the enactive approach, direct social perception highlights the continuity between mind, body, and the social environment, arguing that we experience ourselves and others, first and foremost, as a dynamical psychophysical unity that evolves over time (Krueger, 2018a). Now, we do sometimes hide our emotions or desires behind behavior: for example, by suppressing actions and expressions related to sadness, joy, or excitement or that point toward intentions and goals. Understanding what someone may be experiencing in these situations could require processes of inference, deduction, theory making, and simulation. However, these contexts are arguably not indicative of more primary social interactions as they involve interventions (from both parties) that may be better understood as secondary rather than foundational for social cognition.

Psychological Precursors to Interaction Theory

A main psychological precursor to interaction theory can be found in the concept of "primary intersubjectivity" developed by psychologist Colwyn Trevarthen (1979, 1999, 2002). This notion describes the early development of meaningful sensorimotor interactions between infants and primary caregivers, including (proto) musical varieties (Malloch & Trevarthen, 2009).[11] In line with this, Vasu Reddy and colleagues (2013) have shown that infants play active roles in social engagements with caregivers and other people they encounter, enacting specific bodily movements and vocal expressions that facilitate forms of contextually relevant actions by the other agent (figure 6.1). Interestingly, it appears that, in these situations, the needs, desires, intentions, and goals of both infant and caregiver are not best understood as hidden behind such communicative activity—that is, simply as causes and responses confined to the body and brain of each agent. There is instead a strong sense in which communication and understanding are enacted in the patterns behavior that arise communally *between* both agents, which are "shaped and adjusted as the interaction unfolds" (Fantasia et al., 2014, p. 6).

The development of these interactive capacities is thought to lead to more complex forms of cooperative behavior referred to as "secondary intersubjectivity" (Hubley & Trevarthen, 1979; Trevarthen & Hubley, 1978). Here, infants begin to observe the actions of others in different contexts,

The Empathic Connection

The emotional-empathic frontier between child and caregiver

interacting neural trajectories

interacting neural trajectories

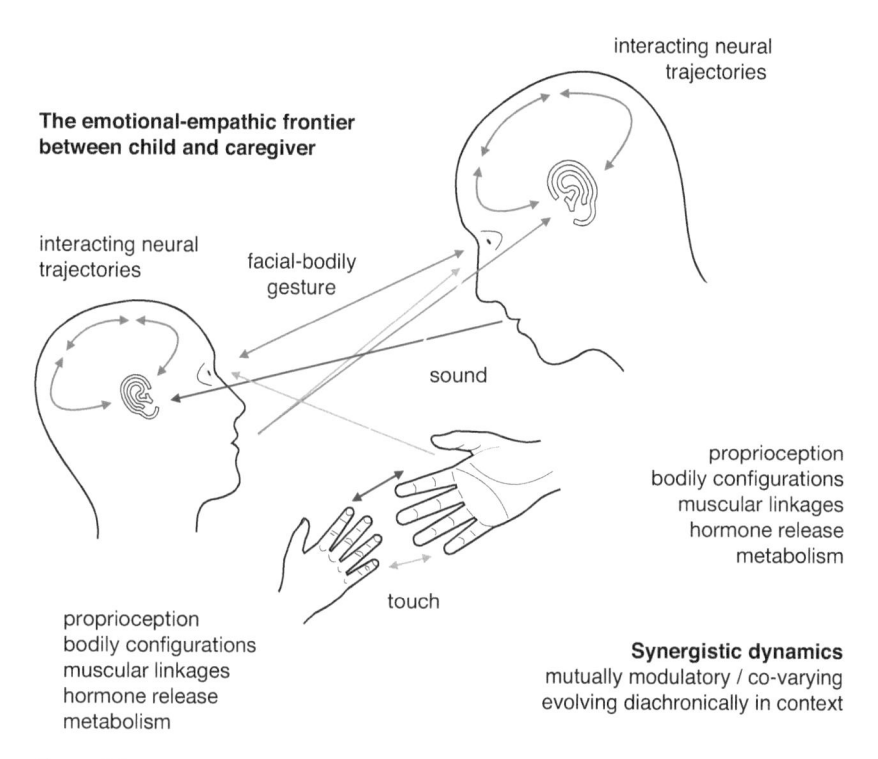

facial-bodily gesture

sound

touch

proprioception
bodily configurations
muscular linkages
hormone release
metabolism

proprioception
bodily configurations
muscular linkages
hormone release
metabolism

Synergistic dynamics
mutually modulatory / co-varying
evolving diachronically in context

Figure 6.1
Primary intersubjectivity: social cognition emerges at the shared frontier of perceptual relationality that develops between embodied agents as they interact over various timescales. (Adapted from Trevarthen [2017])

developing a pragmatic understanding of the meaning of a given action in terms of what the other agent wants or needs within a specific situation. This allows for more complex forms of joint attention and joint action to emerge. Importantly, such forms of social understanding appear to be best described in terms of direct perceptions of a shared world and the coordinated forms of action associated with the realization of common goals.

This work resonates with the direct social perception orientation just discussed, as it shows how our earliest forms of social understanding are based in the development of shared repertoires of movement, touch, and facial and vocal expression rather than in abstract processes of simulation associated with inferring the otherwise hidden mental states of other agents. It also aligns with the work in emotion research discussed in the previous chapter, which suggests that emotions are not only foundational for cognitive processes but that they are also highly dependent on developmental processes involved with embodied social interaction (see Colombetti, 2014). As we saw, these dynamical processes entail the emergence and interaction of the muscular linkages and coordinative movement associated with emotional expression (Camras & Witherington, 2005; Fogel & Thelen, 1987); the self-organization of neural structures (see Freeman, 1999, 2000); and the way in which these aspects develop, stabilize, and transform over various timescales as the agent engages with the social environment (Hsu & Fogel, 2003; Laible & Thompson, 2000).

The philosopher Matthew Ratcliffe (2017) develops similar insights in more complex social situations associated with clinical psychology, in which he examines how communication and understanding develop between patients and therapists. In these contexts, empathy entails socially interactive processes that unfold over time: forms of communication, understanding, and trust emerge that are special to each relationship. These are manifest first in the coenactment of patterns of behavior that are experienced directly as shared environmental affordances. Here it should be noted that deduction, inference, and simulation may have an important role to play at certain points in the development of these relationships: for example, when a therapist develops a provisional theory about why a client may be repressing certain experiences. However, the more fundamental dynamics of these relationships are not based in such detached processes. Social understanding develops through the direct interactivity of the two parties, involving the negotiation of bodily, emotional, and narrative factors toward a unique domain of shared understanding. Accordingly, Ratcliffe argues that while simulation may play a role in certain forms of empathy, it is *neither necessary nor sufficient* for empathic relationships to occur. What is required is that agents are able to interact directly and thereby coenact meaningful repertoires of gesture, emotional expression, and speech that develop diachronically.[12] This decenters the traditional focus on the rationalizing internal domain of an individual empathizing agent. Such processes are arguably better described by the shared, "we" perspective offered by interaction theory.

In line with the direct social perception orientation, the interaction theory approach sees empathy as primarily determined by the processes of reciprocal interaction that occur between embodied agents in specific environmental contexts (De Jaegher et al., 2017; Gallagher, 2001, 2008a,b; Krueger, 2018a; Zahavi & Michael, 2018). Because of this, an interaction theory view of musical empathy aligns closely with the enactive orientation we presented in the opening chapters and offers important new possibilities for thinking about social cognition. In particular, we suggest that this orientation may provide an extension to the insights associated with embodied simulation theory—especially as it explores how social cognition (including musical varieties) can occur directly. Let us now develop interaction theory and direct social perception within a musical context by considering notions of "musical scaffolding" and "empathic space."

Musical Scaffolding

As we have seen, musical environments afford a range of "extended" cognitive processes. And we use this musical capacity to "scaffold" the emotional-social environments we inhabit, thereby shaping experience and guiding behavior in various ways (DeNora, 2000). Again, we can trace this phenomenon to the earliest stages of human development. Humans are born with limited capacities for self-regulating attention and emotion and are therefore highly vulnerable to environmental perturbations (Posner & Rothbart, 1998). The musical scaffolding facilitated by caregivers is therefore an important developmental aspect of their environments. This is shown in research that demonstrates how infants' engagement with music can modulate and stabilize an "array of physiological states and microbehaviors associated with instability into an array associated with stability—stable heart rate, blood pressure, color, feeding, changes in posture, muscle tone, less frantic movements, rhythmic crying, cessation of grimacing, and an ability to sleep or become animated and intent" (DeNora, 2000, p. 81). In this way, entraining and resonating with extended musical environments can help infants develop endogenous resources for psychophysical regulation and begin to enact the social-empathic configurations that are conducive to their well-being to create new stable patters of interactivity between neural, corporeal, and environmental dimensions.

These primary forms of (proto)musical sense-making lead to a basic "knowledge" of our corporeal and emotional selves and of the social and cultural environments we are embedded in. Such knowledge does not simply involve the acquisition of pregiven information and facts; it develops from our actions and interactions in the world. Biologically speaking, this is reflected by a more plastic and relational conception of the endogenous (corporeal-neural) systems involved, such as the mirror mechanism or the proposed components of emotion processing. As we considered in the previous chapter, it may be reductive to think of these "components" as the loci or cause of empathic and emotional (musical) experiences. Rather, they stand as parts of an evolving dynamical system that includes body, brain, and world (which, of course, includes other embodied minds).

Here, we should be careful not to lose sense of the enactive conception of autonomy. As we saw in chapter 2, this implies a kind of "interactional asymmetry" in the organism-environment system, whereby the living sense-maker brings forth an adaptive "point of view" but can only do so through its relational history of coupling with the environment it inhabits. This is reflected in the following quote by Evan Thompson (2007) regarding the nature of information and the formation of meaning:

> [I]nformation, dynamically conceived, is the making of a difference that makes a difference for some-*body* somewhere. Information here is understood in the sense of *informare*, to form within. An autonomous system becomes informed by virtue of the meaning formation in which it participates, and this meaning formation depends on the way its endogenous dynamics specifies things that make a difference to it. (p. 5)

To understand this in a musical context, we can think of an interacting ensemble, such as a jazz trio, a string quartet, or a group of musicians performing the Ghanaian polyrhythms discussed in chapter 4.[13] In each case, the musicians involved must develop the endogenous resources (neural and muscular linkages, emotional and motivational resources, cross-modal perceptual relationships, and conceptual capacities) required to participate in, and meaningfully transform, their musical environment. However, these factors emerge and are given meaning within the context of extended material (instruments, acoustic spaces) and social world in which the musician participates. As Eric Clarke (2019) writes, "Empathy . . . is not only the basis of intersubjective feeling and understanding, but is also the foundation for our grasp of our own subjectivity and consciousness."

Participatory musical sense-making engages synergistic processes of mutual specification that involve the enactment of tightly coupled, self-organizing feedback loops. These loops play out at neural, muscular, behavioral, body-instrument levels and engage self-regulative processes across personal and interpersonal domains (figure 6.2). Interestingly, this evokes again the seeming paradox between the idea of autonomy and the

The Empathic Connection 121

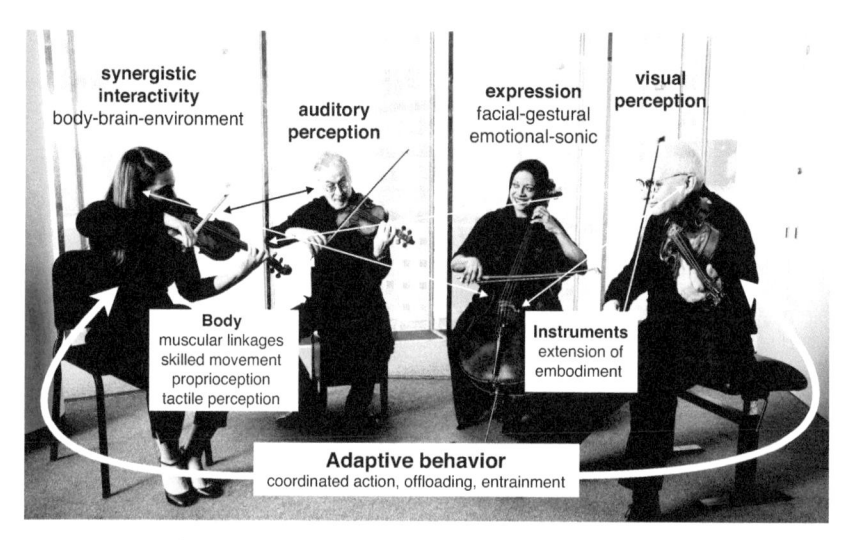

Figure 6.2
Musical empathy in performance. (Photo of the Julliard String Quartet by Claudio Papapietro; used with permission of the Julliard String Quartet)

interactive or "co-arising" relationship between organism and environment we discussed in chapter 2. And once more, this can be overcome through a revised understanding of what "autonomy" entails in human social contexts. The received view handed down from Enlightenment thinking posits a conception of autonomy that involves "primordially lone individuals extending their cognitive reach" (Urban, 2014, p. 4). By contrast, the enactive approach offers a relational perspective that highlights the origins and potential fluidity of self-hood as it arises within the embodied, contextual, and adaptive-cooperative processes associated with participatory sense-making (De Jaegher & Di Paolo, 2007). Again, this "dialectical" conception of autonomy entails the kind of relational thinking associated with understanding the complex, recursive dynamics of self-organizing living systems.

This relational perspective on autonomy and selfhood can help us recognize how the musically scaffolded environments enacted in performance involve experiences that are at once highly personal and collectively shared. Musical agents rely on each other (on each other's endogenous resources) to adaptively maintain the extended musical environments they coenact by "taking on" and "offloading" various tasks (entraining with a beat provided by a drummer; leading phrasing, intonation, or dynamics). This engages synergistic processes that entail the enactment of tightly coupled, self-organizing feedback loops that play out at neural, muscular, behavioral, body-instrument levels and that engage self-regulative processes across personal and interpersonal domains (Krueger, 2014).

Participatory musical sense-making, then, involves the various ways we directly and collaboratively engage with rich, cross-modal networks of bodily, emotional, sonic, social, cultural, technological, and material scaffolding that support and constrain intelligent (i.e., creative, adaptive, skillful, goal-directed) behavior. In line with our discussion of interaction theory and direct social perception, the ways this plays out cannot be properly explained only in terms of internal simulations. It is more fully described, we argue, in terms of an unfolding process in which agents come to know and understand each other directly through histories of direct embodied interaction that occur within the shared (and extended) sociomaterial milieu they are embedded in.

We should note too that the ability to scaffold extended musical environments also drives musical experience in more solitary listening contexts: for example, when we use music to regulate our moods, to relax, or to stimulate ourselves (Skånland, 2013). We also engage in musical scaffolding when we use music to enhance the everyday worlds we live through—for instance, when we use a personal listening device to create a soundtrack for the daily journey to work or school, or while wandering through a station while waiting for a train (see, e.g., Bull, 2000, 2008). These situations also involve extended interactions with the material environment (technologies and spaces) as the experiencer offloads various self-regulative and relational dimensions to the sonic world they enact. Here too, then, musical experience is not something that can be limited to the personal domain of listeners and performers—it is an environmentally extended phenomenon.

As Clarke (2005, 2013) reminds us, human musicality is seamlessly materialized across a wide range of technologies and artifacts, as well as various social, cultural, and acoustic environments that afford manifold forms of musical practice and engagement.[14] Music, like emotion and empathy, is something we experience *in the world*, which necessarily includes bodies, other people, cultures, and technologies. As such, the social and emotional aspects of the musical mind cannot be abstracted away from the embodied, embedded, extended, and enactive dimensions that give rise to it. Moreover, it appears that musical experiences always entail—in different ways and to varying degrees—an integration of social-empathic and emotional-affective dimensions. And even in those instances in which the empathic connection may not be explicit, our ability to engage with and make sense of musical environments nevertheless appears to be rooted in our early development as situated, social beings who structure sonic environments as a means of personal and interpersonal regulation, communication, and meaning making. As the as the philosopher Joel Krueger (2011a, 2015) points out, music is something we do things with.

Empathic Space

Let us now consider one of the main things we do with music as participatory sense-makers: the enactment of what Krueger (2011c) refers to as "empathic spaces." If the idea of musical scaffolding highlights the embodied and extended aspects of human sense-making with music, then empathic space draws our attention more toward the embedded dimension. That is, it examines the kinds of social environments afforded by musical scaffolding and how we use these environments to connect, create, and share meaningful experiences with others. Indeed, music plays a central role in how we enact and manipulate the dynamics of the manifold social environments that give our lives meaning. It guides meaning making and behavior in specific contexts, and this plays out in different ways across human cultures. Consider, for example, the ways music is used in religious ceremonies, where various empathic spaces are evoked throughout an event to guide the various phases of the ritual (Clarke et al., 2010, pp. 1–8; Wynn, 2004). We also use music to create empathic spaces that afford feelings of cooperation and affiliation and to evoke a shared sense identity or patriotic feelings.[15] To take a common everyday example, we can think of how we may use music at a dinner party to help establish the right kind of shared environment for intimate and friendly conversation or to change the mood toward more energetic postmeal interactions (e.g., dancing, singing, game playing).

In brief, musically scaffolded empathic spaces engage those shared corporeal, emotional, and cross-modal (or "metaphorical") cognitive capacities (mimicry, behavioral synchronization, affectively motivated movements, and so on) that emerge early in life and that guide our development as embodied participatory sense-makers.[16] These spaces afford the alignment of feelings, cross-modal perceptions, and behaviors between multiple agents, thereby enhancing attentional and motoric coordination and strengthening cohesion and affiliation. This also connects with recent research, which suggests that when musical agents synchronize to the same beat or melody, they also synchronize respiration, heart rate, and brain wave activity (see Cross, 2007).[17] Importantly, musically scaffolded empathic spaces unfold literally *in between* individuals as they jointly negotiate (or enact) musical environments and subject positions.[18] As Krueger (2011c) discusses, the "mutually-modulatory" expressions afforded by musically scaffolded environments highlight the dynamics of the second person "we-space"—that is, the shared embodied space of *direct engagement* associated with direct social perception and interaction theory.

Toward an Enactive View of Musical Empathy

As we have suggested throughout the previous chapters, musical experience involves a complex set of factors associated with how people enact meaningful relationships within the contingent social, cultural, and material environments they inhabit and actively shape. Music spans and integrates developmental, corporeal, emotional-affective, and more abstract imaginative processes in various ways. Accordingly, restricting musical empathy to representations or simulations produced at "higher" levels of mental activity may offer accounts that are too dependent on linear and hierarchical models of cognition that downplay the deeply interactive, embodied, and emotional-affective aspects that characterize musical experience. The corporeally driven conception of simulation increasingly associated with embodied music cognition offers one way to examine musical empathy from a more embodied perspective. By considering both lower and higher forms of processing as relevant to empathic experience, embodied music cognition offers a coherent alternative to perspectives that attempt to define empathy more narrowly.

The interaction theory orientation, for its part, takes this further by extending empathy out of the brain and body of the individual into the world. It suggests that while simulations (and other representations or theorizing) may play a role in some forms of empathic experience, they are neither required nor sufficient for empathic relationships to occur. What *is* required, first and foremost, is that embodied agents can engage directly with each other in specific sociomaterial contexts. It therefore aims to explore how situated (ecologically embedded and socially extended) interactivity gives rise to forms of empathy that are unique to context and that unfold diachronically. It follows, then, that while "simulation" is an important factor to consider, it may be only one aspect of a more relational and ecological process that plays out differently in various contexts.[19]

This aligns with our discussion in the previous chapter, in which we began to consider how musical emotions are not best understood as responses dependent on pregiven neural mechanisms but rather as dynamical and emergent phenomena that arise, stabilize, and transform through the history interactivity (structural coupling) between embodied agents and the extended environments they are embedded in. For humans and other social animals, such states emerge in infancy and develop through histories of (positively and negatively) valenced experiences—resulting in "basins of attraction" that are shared with, and influenced by, the activity of all those involved in the social environment (Sheets-Johnstone, 2010, 2012). In this way, emotional-empathic experiences may be considered as both plastic and patterned-recurrent—that is, as dynamically emergent phenomena that may bear likeness to previous states of being and to episodes experienced by others who share similar metabolic needs, physiologies, and interactions (Colombetti, 2014). This is all to say that, like music, emotional and empathic experiences do not inhere in the individual but develop relationally (Fogel et al., 1992; Laible & Thompson, 2000); this often involves the coenactment of behavioral patterns that afford the ability to recognize shared states of being. Therefore, musical emotion and empathy also must be studied as "extended" phenomena (see Krueger, 2011c, 2014, 2018b). In line with this, we suggest that the set of overlapping dimensions associated with the enactive approach to mind can help us better understand the emotional and empathic experience of music within a more pluralistic and interactive framework.

As we have seen, the concepts of musical scaffolding and empathic space begin with the idea that musical cognition is deeply rooted in our capacities as embodied, emotional, and social sense-makers. As such, the experience of music fundamentally involves living bodies that are embedded within specific sociomaterial environments and that use the social and material affordances of these environments to (cooperatively) extend their cognitive domains. In doing so, they enact forms of meaningful behavior and shared worlds of meaning (i.e., culture). These aspects, in turn, may feed back into the extended cognitive system, producing new constraints and possibilities to which agents must adapt, both individually and collectively. This perspective can help us think about the deep connections between the empathic and emotional aspects of musical sense-making, showing how they reflect the fundamentally self-organizing or autopoietic nature of living autonomous cognitive systems as they continually strive to enact and maintain relationships with the environment within ranges that are conducive to survival and well-being. For example, although collective music making does not usually involve situations that are life threatening in the literal sense, this activity nevertheless demonstrates a continuity with the adaptive dynamics associated with sustaining more basic organism-environment relationships. As we touched on in chapter 4, the phenomenology of participatory musical sense-making involves the adaptive maintenance of (musically) relevant behavioral linkages between body and world—the ongoing coenactment of viable balances between the corporeal, instrumental, emotional, social, and sonic factors required to keep the musical event "alive" and "flourishing."

Therapeutic Interventions

The insights and ideas discussed previously can also help us to think about the therapeutic nature of musical experience in everyday and clinical contexts. The therapeutic effects of music have been recognized for thousands of years. However, music therapy as a contemporary healthcare profession began after the Second World War (see Bunt, 1994). It has since developed a wide array of approaches to use music in the treatment of, among other conditions, emotional-behavioral or mood disorders (Layman et al., 2002; Magee & Davidson, 2002), brain damage (e.g., stroke recovery and Alzheimer's disease; Nayak et al., 2000; Sacks, 2007), physical and cognitive disabilities, as well as in developing self-esteem and sociability (Henderson, 1983). Neuroscientist Aniruddh Patel (2010) discusses how music therapy treatment can result in positive and long-lasting changes to brain structures and functions for stroke patients and in the recovery of verbal fluency in aphasia (largely, he suggests, through neuroendocrine effects and mechanisms of brain plasticity). Music therapist Concetta Tomaino (2009) writes, "Singing may serve as a priming element for speech . . . stimulating either peripheral language areas or compensatory areas in the right temporal lobe" (p. 216). And indeed, we can note the remarkable effects of clinical music therapy in the case of American congresswoman Gabby Giffords as she struggled to regain the faculties of speech after being shot in the head (Michaels, 2012).

By fostering musically scaffolded environments that afford the development of new muscular, neural, and environmental linkages, music therapists provide effective treatment for people suffering from acquired and degenerative speech and motor disorders (e.g., Parkinson's disease). Through singing and musically entrained rhythmic movement and speech, patients can develop better regulation of bodily movements—gait, breathing, oral articulation and intonation for speech, and so on (Schiavio & Altenmüller, 2015; Tamplin & Baker, 2017). These kinds of treatments capitalize on the various ways musical activity can help patients form new corporeal-neural connections through carefully guided engagements with the extended musical environment, which includes the unique history of emotional-empathic interaction between patient and therapist.

Extended musically scaffolded environments can also be discerned in clinical contexts focused on social and emotional regulation. Consider, for example, the case of Gary, a young man unable to see or speak but who, through music therapy, is able to engage in "co-ordinated activity with another person . . . develop his sense of self, his presence to self and other(s)" and enjoy "interaction capable of producing pleasure [and] security" (DeNora, 2000, pp. 14–16). Music therapy has afforded Gary the tools for "stabilizing" his environment and himself and for developing the capacity to enact an empathic space with his caregivers, whereas before he could only resort to shrieking, biting, and scratching to express distress, activities that would only further alienate him from his own social and physical existence. Other researchers have considered the positive effects of group music therapy for individuals who suffer from shared or similar traumatic experiences. For example, in contexts involving grief, bereavement, and other forms of trauma among teenagers, various forms of musicking (e.g., improvising, songwriting, drumming) can provide shared empathic spaces where complex emotional dynamics can be "let out" and explored in various ways (McFerran, 2019; McFerran et al., 2010).

As mentioned previously, research has also revealed the beneficial effects of music on neonates who are in a profoundly disorganized state of being (Anderson & Patel, 2018; Haslbeck, 2014; Haslbeck & Bassler, 2018;

Kaminski & Hall, 1996). The introduction of music into the neonate's world masks other potentially stressful noises of the hospital environment and aids in creating a regulated calming environment with which the infant may become entrained, thereby regaining some of the auditory stability reminiscent of the intrauterine environment (e.g., maternal movements, breathing, heartbeat) (Collins & Kuck, 1990; DeNora, 2000; Leonard, 1992; Parncutt, 2006, 2008). Additionally, music therapy interventions involving parents of neonates have resulted increased levels of emotional bonding in situations in which other forms of connection are not possible because of the medical condition of the baby: "parents' perceptions of their baby's responsive behavior during musical moments scaffolded their own ability to understand and construct their new role as a parent to their baby" (McLean et al., 2019, p. 4). The therapeutic power of music is also employed outside of clinical contexts. For example, DeNora (2000) considers the case of Lucy, who in the course of everyday life "self-administers" music (in the form of Schubert's Impromptus) to move from a state of stress to one of calm. Similarly, Standley (1995) has shown how music may be successfully introduced into medical and dental environments to reduce anxiety and pain in patients (see also Bunt, 1997). There are, of course, many more instances of how we use music to scaffold environments that promote healing and well-being. Nevertheless, the brief review provided here begins to demonstrate how music's therapeutic powers, in both clinical and "everyday" contexts, depend on the comprehensive way musical activity engages the embodied, embedded, and extended dimensions of human life and how it enables us to enact environments that can both strengthen existing relationships and foster new ones across corporeal, neural, and social domains. We suggest, therefore, that fields of music therapy and enactive cognitive science have a great deal to offer one another (see Maiese, 2020) and look forward to future work inspired by 4E and dynamical frameworks.

A central aspect of this book involves exploring the ways we use music to organize (and reorganize) the dynamical relationships that imbue our lives with meaning. In the current chapter we have highlighted the deeply social nature of musical experience—how we use music to enact shared meanings and scaffold social spaces for pleasure, ritual, physical-emotional regulation, healing, and more. In later chapters, we extend our discussion further into the areas of infant musical development, musical creativity, and music education. Before we do that, however, we would like to explore what an enactive orientation may reveal about another contentious issue in interdisciplinary musicology: the question of why and how we evolved into musical animals in the first place. As we will see, the ideas and arguments we have considered thus far will offer interesting counterpoints and extensions to many current perspectives in the field referred to as "evolutionary musicology." Additionally, we will introduce new concepts drawn from archeology, theoretical biology, and evolutionary theory that will inform our discussions in later chapters. So, let us now expand our view of human musicality into the ancient past by considering how protomusical behavior emerged and developed through the interactivity of embodied minds and material resources in the social spaces enacted by our prehuman ancestors.

7 The Evolution of the Musical Mind

The inquiry into the origins of musicality in our species draws on a fascinating array of knowledge from across the sciences and humanities.[1] But despite the diversity of perspectives on offer, evolutionary musicology has often tended to adhere to a traditional "adaptationist" view of biological evolution, a perspective that has come under increasing scrutiny in recent decades (Tomlinson, 2015). The central problem is illustrated by Varela, Thompson, and Rosch (1991):

> The dominant orthodoxy in evolutionary thinking over the last few decades saw evolution as a "field of forces." Selective pressures ... act on the genetic variety of a population producing changes over time according to an optimization of the fitness potential. The adaptationist or neo-Darwinian stance comes from taking this process of natural selection as the main factor in organic evolution. In other words, orthodox evolutionary theory does not deny that there are a number of other factors operating in evolution; it simply downplays their importance and seeks to account for observed phenomena mainly on the basis of optimizing fitness. (p. 18)

Importantly, this orientation implies a rather strict separation between the products of natural selection (i.e., adaptations) and those of culture.[2] As a result, evolutionary musicologists have been faced with something of a dichotomy: music tends to be seen *either* as a naturally selected adaptation that has contributed directly to our survival as a species (i.e., to fitness potential) *or* as a product of culture with little or no direct connection to our biological heritage.[3]

The computational model of mind has also exerted considerable influence here. As we have seen, this approach has focused research and theory in music cognition toward a complex information-processing hierarchy implemented by the brain (Deutsch, 1999; Huron, 2006; Levitin, 2006; Sloboda,

1985). Likewise, in evolutionary contexts, the mind is often understood in terms of innate information-processing modules that have been naturally selected to perform specific tasks related to the survival of the species (Barrett & Kurzban, 2006; Coltheart, 1999; Fodor, 1983, 2001; Pinker, 1997; Tooby & Cosmides 1989, 1992).[4] Among other things, this approach has led some scholars to suggest 1:1 mappings between brain regions and musical functions (Peretz & Coltheart, 2003). An ongoing debate exists over whether these brain areas evolved to support musical functions, or if they were selected to perform other roles and were later co-opted for music perception.

Recent research has tended to weaken the modular hypothesis by emphasizing the plastic and self-organizing properties of the (musical) brain (Altenmüller, 2001; Jäncke et al., 2001; Lappe et al., 2008; Large et al., 2016; Münte et al., 2002; Pantev et al., 2001). Additionally, the past two decades have also seen the development of a "biocultural" approach to the origins and nature of the musical mind that looks beyond the traditional nature-culture dichotomy implied by a strict adaptationist framework (see Cross, 1999, 2003; Killin, 2013, 2016a, 2016b, 2017; Tomlinson, 2015). This perspective draws on a range of research in theoretical biology, neuroscience, archeology, embodied and ecological cognition, and dynamical systems theory, positing a more integrated model that sees biological and cultural dimensions of the human phenotype as aspects of the same evolving system. Put simply, the biocultural orientation explores the origins of music in terms of the cycles of interactive behavior that arose within the social and material environments of our prehuman ancestors.

In this chapter, we contribute to the discussion over the biocultural hypothesis by exploring it through the lenses of the enactive approach to cognition. We begin by providing a brief overview of some key positions in the field of evolutionary musicology, examining how many tend to adhere to the "nature-or-culture" dichotomy mentioned previously. We then outline the biocultural hypothesis, placing a special focus on the approach developed by the musicologist Gary Tomlinson (2015) as, for us, it represents the current state of the art in the field. We then attempt to show how an enactive perspective could be used to support and refine Tomlinson's claim that the origins of the musical mind should be sought in the embodied dynamics of coordinated action that occurred within the developing sociomaterial environments of our ancestors and not first in terms of cognitive processes involving (quasi-linguistic) representational mental content.

To conclude, we consider some tentative possibilities for how a 4E framework may help guide future research and theory.

Music and the Dichotomy of Adaptation

An important point of discussion in evolutionary musicology concerns whether musicality can be considered as a bona fide adaptation, or if it is better understood as a product of culture (Davies, 2012; Honing et al., 2015; Huron, 2001; Killin, 2016a, 2017; Lawson, 2014). Some researchers (including Darwin, 1871) have drawn on comparisons with music-like behavior in other animals, suggesting an adaptive function for music in mate selection and territorial display in our prehistoric ancestors (see Miller, 2000). It has been argued, however, that although music-like behavior in nonhuman animals (e.g., bird song) may well be a product of natural selection, these traits are not homologous with human music making but are merely analogous (Hauser & McDermott, 2003; Pinker, 1997). Accordingly, it is claimed that comparative studies involving more phylogenetically distant species may not have great relevance for understanding the biological origins of human musicality (McDermott & Hauser, 2005; but see Fitch, 2006). Additionally, evidence of musical behaviors in our closest primate relatives is often understood to be sparse. For some scholars, this suggests there was no properly musical phenotype before modern humans in the hominin line (Huron, 2001; Justus & Hutsler, 2005; Patel, 2008).

Such arguments have been traditionally used to support claims that music should not be conceived of as an adaptation, but rather as a product of culture (e.g., Pinker, 1997; Sperber, 1996). Here it is posited that music is largely dependent on cognitive structures (instantiated by brain modules) that evolved to support properly adaptive functions in our ancestors—language, auditory scene analysis, habitat selection, emotion, and motor control (see Trainor, 2015). The strongest version of this approach is offered by Steven Pinker (1997), who argues that music is an "invention" designed to "tickle" these naturally selected aspects of our cognitive and biological nature. Music itself, however, has no adaptive meaning: from an evolutionary point of view, it is the auditory equivalent of a "cheesecake"—a cultural invention that is pleasurable, but biologically useless. In line with this, it is suggested that music may be a kind of "exaptation," in which the original (i.e., adapted) function of a trait becomes co-opted to serve other purposes

(Davies, 2012).[5] Thus, as cognitive scientist Dan Sperber (1996) posits, music may be understood as "parasitic on a cognitive module the proper domain of which pre-existed music and had nothing to do with it" (p. 142).

By contrast, other researchers have suggested the existence of cognitive modules that appear to be specialized for musical functions. For example, music psychologist and neuroscientist Isabelle Peretz's (1993, 2006, 2012) research in acquired *amusia* has led her to (cautiously) posit an innate music-specific module for pitch processing, suggesting that music may be as "natural" as language (Peretz, 2006). Such claims have been countered by cognitive neuroscientist Aniruddh Patel (2008), who has argued that evidence indicating the existence of adapted music specific modules may in fact be explained by (ontogenetic) developmental processes, whereby cortical areas become specialized for certain functions through experience (e.g., via processes of "progressive modularization"; see Karmiloff-Smith, 1992). However, while Patel (2008, 2010) maintains that musicality in humans is not a "direct target" of natural selection, he also acknowledges the profound biological and social benefits associated with musical activity, claiming that music is a powerful "transformative technology of the mind" (Patel, 2008, pp. 400–401). Here, Patel discusses how musical experience may lead to long-lasting changes in brain structure and processing (e.g., though neuroendocrine effects). He also notes a number of behavioral and biological factors that may be indicative of adaptations that supported the emergence of both language and vocal music. These include the phenomenon of infant babbling, the anatomy of the human vocal tract, and the fixation of the *FOXP2* gene, which is involved in building circuits that sequence motor behavior (and cognition) in speech and musical rhythm (Patel, 2008, pp. 371–372). However, he suggests that because language appears to emerge more quickly and uniformly in humans, and because the lack of musical ability does not appear to entail significant biological costs, these factors are better understood to support the adaptive status of language. In brief, Patel posits here that musical processing is a "by-product" of cognitive mechanisms selected for language and other forms of complex vocal learning (see also Patel, 2006, 2010, 2012).[6]

These last claims are questioned by those who argue that they may reflect a rather narrow perspective on what musicality entails—for example, the assumption that musicality necessarily requires special forms of training, or that music is a pleasure product to be consumed at concerts or through recordings (for discussions see Cross, 2003, 2010; Honing et al., 2015; Small, 1998). Ethnomusicological and sociological research has revealed musical activity around the world to be central for human well-being; it is inextricable from work, play, social life, religion, ritual, politics, healing, and more (Blacking, 1976, 1995; DeNora, 2000; Nettl, 1983, 2000). Moreover, in many cultural environments, music is highly improvisational in character: the acquisition of musical skills begins in infancy and develops rapidly, often without the need for formal instruction (Blacking, 1976; Cross, 2003; Solis & Nettl, 2009). It has also been suggested that because certain physical and cognitive deficits need not hinder survival and well-being in modern Western society, certain "musical" impairments may go almost completely unnoticed. Likewise, music's relevance for human survival across evolutionary time has been considered in terms of its importance for bonding between infants and primary caregivers and between members of social groups (Benzon, 2001; Dissanayake, 2010; Dunbar, 2012; Tolbert, 2001). Moreover, musical developmental processes appear to begin very early on in life (Parncutt, 2006), and researchers have demonstrated the universal and seemingly intuitive way caregivers create musical (or music-like) environments for infants through prosodic speech and lullabies (Dissanayake, 2000; Falk, 2004; Trehub, 2003b,d,e). And as we began to consider in the last chapter, Trevarthen (2002) has proposed that humans possess an innate "communicative musicality" that serves the necessity for embodied intersubjectivity in highly social beings such as ourselves (Malloch & Trevarthen, 2009).

In all, it is argued that the wide range of activities associated with the word "music" have immediate and far-reaching implications for survival and socialization for many peoples of the world, as they may have had for our prehistoric ancestors (see Blacking, 1976; Mithen, 2005). And indeed, the archeological record shows evidence of musical activity (i.e., bone flutes we mentioned in chapter 1) dating back at least 40,000 years (Higham et al., 2012; Morley, 2013). Such concerns drive the "musilanguage" theory put forward by archeologist Steven Mithen (2005) and others (e.g., Brown, 2000; Lawson, 2014), in which both music and language are understood to have developed from a "protomusical ancestor" that evolved because of selective pressures favoring more complex forms of social behavior: for example, enhanced types of communication associated with foraging and hunting, mate competition, increased periods of child rearing (soothing infants at a distance), and more complex forms of coordinated group activity (Balter, 2004; Bannan, 2016; Cross, 1999, 2003; Dunbar, 1996, 2003, 2012; Falk, 2000, 2004). Here it is also suggested that musical behavior may have contributed to the development of shared intentionality in modern humans, which in turn permitted the rapid development of cultural evolution and the emergence of modern human cognition (Tomasello, 1999; Tomasello et al., 2005).

While many fascinating accounts have emerged on both sides of this debate, we suggest that the dichotomous nature-or-culture perspective that frames this discussion renders both sides somewhat problematic. On one hand, arguing that music is primarily a product of culture may tend to downplay its deep significance for human well-being, as well as the rapid and intuitive ways it develops in many cultural contexts. As we have just considered, these manifold developmental and social factors are taken to be indicative of the biological relevance of music for the human animal. On the other hand, arguments for music as an adaptation often tend to posit a singular adaptive status for what is in fact a complex phenomenon that spans a wide range of biological, social, and cultural dimensions.

A Biocultural Perspective

In connection with the concerns just outlined, a number of biologists and musicologists have reasserted the importance of comparative perspectives that examine music-related (or music-like) behaviors in cross-species studies (Fitch, 2006; Ravignani et al., 2016b; Wallin et al., 2000). This research involves drawing analogies with human musicking (e.g., with bird and whale "song"), or suggesting potential homologies (e.g., the bimanual drumming behaviors of great apes). Among other things, comparative approaches have revealed distinctions between learned (music, language, and some forms animal song) and unlearned (laughter and various animal cries) vocalizations and auditory signals. By exploring the convergent evolutionary development of these traits across different species, comparative research offers useful insights into the "constraints influencing the evolution of complex signaling systems (including both song and speech)" and it suggests that "ape drumming presents a fascinating potential homology with human instrumental music" (Fitch, 2006, p. 1). Importantly, the comparative orientation recognizes that no one selective force is sufficient to account for the complexity of human musicality and thus decenters the traditional focus on explaining the origins of music in terms of specific sets of adaptations.

Likewise, other scholars (Cross, 1999, 2001, 2003; Currie & Killin, 2016; Killin, 2013, 2016a) have offered alternative "biocultural" approaches to the nature and origins of human musicality—where the question of whether *either* biology *or* culture should account for deeply social and universal human activities that require complex cognitive functions (e.g., music) is replaced by a perspective that integrates the two. For example, musicologist Ian Cross (1999) suggests that musicality is an emergent activity—or "cognitive capacity"—that arises from a more fundamental human proclivity to search for relevance and meaning in our interactions with the world. It is claimed that because of its "multiple potential meanings" and "floating intentionality" music provides a means by which social activity may be explored in a "risk free" environment, affording the development of competencies between different domains of embodied experience and the (co)creation of meaning and culture (Cross, 1999, 2003). Tomlinson (2015) develops similar insights, arguing that what we now refer to as "language" and "music" began with more basic forms of coordinated sociocultural activity that incrementally developed into more sophisticated patterns of thought, activity, and communication (see also Morley, 2013). Moreover, such activities are understood to have transformed environmental niches over time (Killin, 2016a, 2017; Sterelny, 2014) and with them the behavioral possibilities (affordances) of the hominines who inhabited them through recursive cycles of feedback and feedforward effects.

Over the past few years, the biocultural perspective has gained recognition from a wider range of researchers, including Patel (2018; see also Podlipniak, 2017; and Savage et al., 2020). One of the reasons for this growing interest is that the biocultural orientation suggests a way through the problematic nature-or-culture dichotomy discussed previously. In doing so, however, it necessarily draws on models of evolution and cognition that differ from those that have traditionally guided evolutionary musicology. Indeed, Tomlinson's (2015) approach develops Neo-Peircean perspectives in semiotics (e.g., Deacon, 1997, 2010, 2012), exploring how embodied and indexical forms of communication may in fact underpin our linguistic and musical abilities both in evolutionary and ontogenetic terms. As we discuss next, this is further supported by work in theoretical biology associated with developmental systems theory, studies of musical and social entrainment (rhythm and mimesis), and insights from ecological psychology and enactive cognitive science.

Looking beyond Adaptationism

Tomlinson (2015) argues that although music-as-adaptation perspectives reveal important aspects of why music is meaningful for the human animal, they are also problematic when they tend to assume a "unilateral explanation for a manifold phenomenon" (p. 33; see also Killin, 2016a). That is, because music takes on so many forms, involves such a wide range of behavior, and serves multiple functions, it seems difficult to specify a single selective environment for it. And thus, these traits sit "uneasily side by side, their interrelation left unspecified" (Killin, 2016a). To be clear, this does not in any way negate the claims regarding the social and developmental meanings of music. These biologically relevant traits do exist, but they are just too numerous and complex to be properly described in terms of an adaptation (at least not in the orthodox sense of the term). Because of this, Tomlinson (2015) claims that we must be careful about how we frame evolutionary questions—and especially those regarding complex behaviors such as music and language—lest we fall into the reductive theorizing associated with "adaptationist fundamentalism." In brief, he argues that dwelling on the question of the adaptive status of music has had the effect of "focusing our sights too narrowly on the question of natural selection alone—and usually a threadbare theorizing of it, at that" (p. 34).

We should make it clear that the idea of a genealogy of species is not in question. Rather, it is the mechanism by which this process occurs that is contentious. Darwin did not believe that adaptation through natural selection should be the sole force driving evolution. And indeed, it has been argued that natural selection (i.e., the constraints of reproduction and survival) may not be sufficient to shape genomes and organisms toward optimal fitness and that "survival of the fittest" may not be the "goal" of evolutionary processes after all (Fodor & Piattelli-Palmarini, 2010; Gould, 2002; Gould & Lewontin, 1979; Ho & Saunders, 1984; Lewontin, 1983; Sober, 1984, 1993; Varela et al., 1991).

With this in mind, the "developmental systems" approach to biological evolution posits a compelling alternative that aligns closely with enactivist principles (see Oyama, 1985; Oyama et al., 2001). In contrast to the one-directional schema that characterizes more traditional frameworks (in which evolution is understood to involve *adaptations to* a given environment), developmental systems theory presents a more relational view, in which organism and environment are understood as mutually influencing aspects of the same integrated system. Here evolutionary processes do not only entail the adaptation of a species' phenotype to a fixed terrain but also involve "a dynamic interaction where other species and the non-living environment take part" (Tomlinson, 2015, p. 35). In other words, this approach explores the complex ways genes, organisms, and environmental factors—including behavior and (sociocultural) experience—interact[7] with each other in a recursive way. Such interactions are understood to lead the formation of phenotypes and the construction of environmental niches, which influence the further development of the organism in an ongoing way (genome ⇔ cells ⇔ environment ⇔ phenotype) (Jablonka & Lamb, 2005; Laland et al., 2010; Moore, 2003; Richerson & Boyd, 2005; Sterelny, 2014). This view therefore eschews the classic nature-nurture dichotomy, preferring instead to examine the interaction between organism and environment as a recursive or "dialectical" phenomenon (Lewontin et al., 1984; Pigliucci, 2001), in which no single unit or mechanism is sufficient to explain all processes involved.[8] "DNA short sequences, genes, whole gene families, the cell itself, the species genome, the individual, 'inclusive' groups of genes carried by different individuals, the social group, the actually interbreeding population, the entire species ... the ecosystem of actually interacting species, and the global biosphere" (Varela et al., 1991, p. 192).

This perspective also moves away from the logic of fitness optimization that drives the orthodox adaptationist position and toward a recognition *viability* as an enacted process. Optimality implies that the biocognitive functions and environmental interactions of a given organism should be highly prescribed: what is not permitted by the constraints of optimization is forbidden. The logic of viability, however, opens a more flexible perspective in which what is not impossible may occur. By this light, natural selection can be seen to operate but in a "modified sense: selection discards what is not compatible with survival and reproduction" (Varela et al., 1991, p. 195). That is, evolution is no longer understood as a prescriptive drive toward optimal fitness, but rather in terms of more general ecological constraints that are associated with maintaining the life-world of the organism and its lineage, thus highlighting "the richness of self-organizing capacities in biological networks" (Varela et al., p. 197). This perspective is reflected in the words of the biologist Richard Lewontin (1983): "Our central nervous systems are not fitted to some absolute laws of nature, but to laws of nature operating within a framework created by our own sensuous activity. [. . .] We do not further our understanding of evolution by general appeal to 'laws of nature' to which all life must bend. Rather we must ask how, within the general constraints of the laws of nature, organisms have constructed environments that are the conditions for their further evolution and reconstruction of nature into new environments" (p. 163).

Importantly, this view aligns with many principles at the core of enactivism. Most centrally, it highlights the insight that the organism plays an active role in shaping the environment it *coevolves with*: the activities of the organism alter the selective pressures of the environmental niche. This, in turn, affects the development of the organism, resulting in a dynamical coevolutionary feedback cycle. As illustrated in figure 7.1, sociocultural developments add additional epicycles involving patterns of behavior that can sometimes hold stable over long periods of time.[9] These are passed on

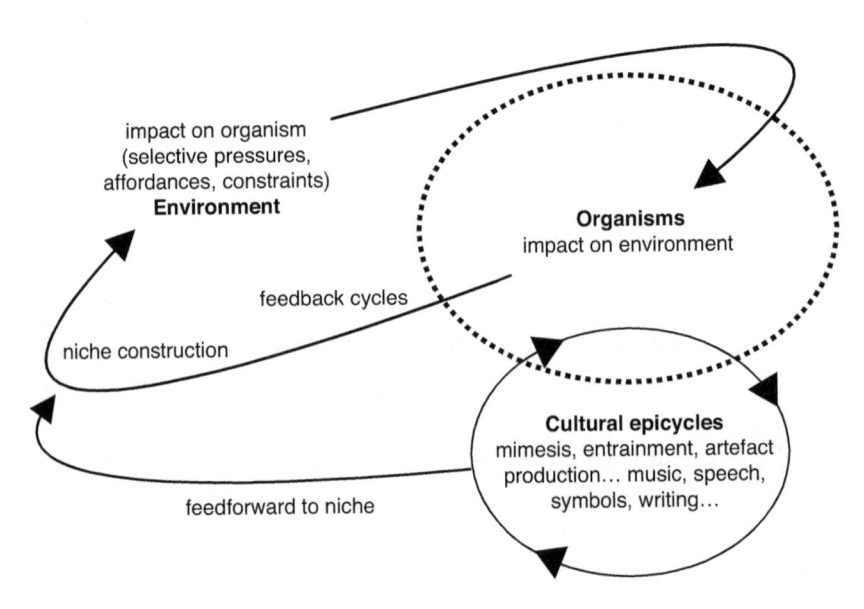

Figure 7.1
The cyclical process of biocultural coevolution. (Adapted from Tomlinson, 2015, pp. 46–47)

inter- and intragenerationally through embodied mimetic processes (more on this later; see also Sterelny, 2012). While such epicycles necessarily emerge from the coevolution cycle, they may, once established, develop into self-sustaining patterns of behavior that develop relatively independently. However, the effects of these cultural epicycles inevitably feed forward into the broader coevolutionary system, resulting in additional alterations to environmental conditions and shifts in biological configurations (e.g., gene expression and morphological change; see Killin, 2016a; Laland et al., 2010; Skinner et al., 2015; Wrangham, 2009).

The making and use of tools is offered as a primary example of what such cultural epicycles may entail. The archeological record contains many examples of biface stone hand axes that were made by our Paleolithic ancestors. These tools are remarkably consistent in their functional and aesthetic qualities, implying method and planning in their manufacture (Wynn, 1996, 2002). However, it is now thought that the production of these axes entailed a "bottom-up" process based on the morphology and motor possibilities of the body, unplanned emotional-mimetic social interaction, and the affordances of the environment (Davidson, 2002; Gamble, 1999). In other words, it is argued that the emergence of Paleolithic technologies did not involve abstract or representational forms of thought (e.g., a mental template, or "top-down" thinking)—a capacity these early toolmakers likely did not possess (but see Killin, 2016b, 2017). Nor were they the result of genetically determined developmental programs. Rather, tool making behavior is thought to have originated, developed, and stabilized primarily through the dynamical interaction between our pre-human ancestors and the material environments they inhabited and shaped (Ingold, 1999). It is suggested that such self-organizing forms of (mimetic, rhythmic, coordinated) social-technological behavior provided the grounding from which more complex cultural activities, like music and language, emerged later (Tomlinson, 2015).

The Rhythmic Origins of the Musical Mind

One way to think about how these developments could have occurred is to consider the mimetic-empathic dimensions of these prehuman social environments and how they may offer insights into the origins of musicality in coordinated rhythmic behavior. In social animals, attention tends to be turned toward the world and the activities of others (McGrath & Kelly, 1986). This entails the capacity to observe, understand, and emulate the actions and feelings of conspecifics and thus enact the empathic spaces we considered in the previous chapter. In line with this, it is suggested that simple mimetic processes allowed our Paleolithic ancestors to engage in increasingly complex chains of actions that were passed on from one individual or generation to the next (Gamble, 1999; Ingold, 1999; Leroi-Gourhan, 1964/1993). Notably, these early forms of social scaffolding involved the enactment of culturally embedded "action loops" (see Donald, 2001; Tomlinson, 2015) that depended on a basic proclivity for forms of social *entrainment*.

The phenomenon of entrainment may be observed in many ways and over various timescales in both biological and nonbiological contexts (Becker, 2011; Clayton et al., 2005; de Landa, 1992; Knight et al., 2017). Most fundamentally, it is understood in terms of the tendency for oscillating systems to synchronize with each other. As we considered in chapter 2, a simple example of this is found in how wall-mounted pendulums mutually constrain one another, resulting in synchronization or, indeed, "entrainment" over time (see Clark, 2001). However, biological and social systems can also be conceived of as dynamically interconnected structures formed by oscillating components—from metabolic cycles to life cycles, from single neuron firing to regional patterns of activity in the brain, from individual organisms to social groups and the broader biological and cognitive ecology (McGrath & Kelly, 1986; Oyama et al., 2001; Varela et al., 2001; Ward, 2003). The components of such systems influence each other in a covarying or synergistic way (Chemero, 2009). Moreover, the development of coupled systems is guided by local and global constraints that allow the system to maintain stability, to be resistant to perturbations or to regain stability once a perturbation has occurred. The ability to adaptively engage with such factors is crucial for living systems, which, as we saw, must maintain metabolic functioning within certain parameters if they are to survive. Such self-organizing processes result in "emergent properties"—relationships, structures, and patterns of behavior that may remain consistent over long periods or that may be subject to transformation because of shifts in local and global constraints of the organism-environment system.

The mathematical techniques associated with dynamical systems theory (DST) have aided researchers in modeling such phenomena. As we began to explore in chapter 2, patterns of convergence (stability) in the state of the system are contrasted with areas exhibiting entropy (i.e., gradual decline into instability; see de Landa, 1992). Again, this is often represented as a topographic "phase-space" that describes the possible states of a given system over time—periods of convergence in the trajectories of the system are represented as "basins of attraction" (Abraham & Shaw, 1985; Chemero, 2009). A "phase transition" occurs when new patterns of convergence arise (i.e., new attractor layouts). And indeed, researchers associated with developmental systems theory (discussed earlier) use DST methods to model the evolutionary trajectories of coupled organism-environment systems, mapping dynamical patterns of stability and change as functions of constraint parameters (see Oyama et al., 2001). As we also considered, DST is used to examine how social animals bring their actions in line with those of other agents—and with other exogenous factors—by "dynamically attending" to the environment through sight, sound, movement, and touch (Large & Jones, 1999; McGrath & Kelly, 1986). This results in the enactment of coordinated forms of behavior that can occur both voluntarily and involuntarily. Emotional-empathic aspects may also come into play here. For example, when a stable pattern is disrupted, entropy emerges in the system and a negative affect may result. The (living) system then self-organizes toward regaining stability, resulting in a positive affect (feeling, emotion). It is suggested that the action loops associated with Paleolithic toolmaking emerged from these forms of social entrainment, in which dynamical couplings between various trajectories in the social environment led to increasingly stable patterns of behavior (basins of attraction) in the cultural epicycle. This permitted the mimetic transmission of cultural knowledge without the need for symbols, referentiality, or representation (see Tomlinson, 2015, p. 75).

Interestingly, the idea of dynamical attending mentioned previously has been explored empirically in the context of musical (i.e., metrical, rhythmic) entrainment (Jones, 2009; Large & Jones, 1999; Large et al., 2015). In line with this, Tomlinson (2015) suggests that such dynamical models may help to reveal the distant origins of musical rhythm in the mimetic, emotional, and sonic-social environments jointly enacted by the coordinated (entrained) motor patterns of early toolmakers. These possibilities are supported by a range of recent comparative research into the evolution of rhythmic behavior (Fitch, 2012; Merchant & Bartolo, 2017; Ravignani et al., 2016 a,b; 2017). Indeed, evolutionary musicology has often tended to explore the origins of music in terms of its vocal dimensions (i.e., music as pitch/song production and its relationship to spoken language) and has thus had to wrestle with the issues associated with complex vocal learning and its apparent absence in other primates. The focus on rhythm, however, has shown similarities between nonhuman and human behavior (Bannan, 2016; Fitch, 2010; Iversen, 2016; Merchant et al., 2015; Patel & Iversen, 2014; Wilson & Cook, 2016).

Several recent contributions have also explored the deep relationship between rhythmic behavior and social cohesion in both human and nonhuman subjects (e.g., Knight et al., 2017; Large & Gray, 2015; Tunçgenç & Cohen, 2016; Yu & Tomonaga, 2015). Additionally, recent research led by the cognitive scientist and biomusicologist Andrea Ravignani and colleagues (2016a) has attempted to empirically model the cultural evolution of rhythm. As the study shows, participants presented with random percussive sounds tend to develop structured and recurrent rhythms from such information; these patterns continue to develop through subsequent generations of participants who are asked to imitate the rhythms of previous generations.[10] These observations align with the conception of cultural transmission based on mimesis and entrainment just discussed. They also imply that the enactment of musical (or music-like) behavior may not be traceable solely to the genome but arises as a result of a more general propensity to structure acoustical experience in certain ways (see also Fitch, 2017).

Here it should be noted that the biocultural approach also develops a theory about the origins of vocal musicality, albeit one that is deeply connected to the rhythmic factors just described. This entails the development of a repertoire of "gesture-calls" like those found in modern primates and many mammalian species (grunts, pants-hoots, growls, howls, barking, and so on; see Tomlinson, 2015, pp. 89–123). These calls do not involve the abstract, symbolic-representational, and combinatorial properties employed by modern languages. Rather, they are tightly coupled with the same mimetic, emotional, and embodied forms of communication that likely characterized prehuman toolmaking. It is suggested, then, that the vocal expressions associated with these gesture-calls reflected the sonic aspects (rhythmic and timbral) of prehuman environments, the motor patterns of production, as well as the gestural and social rhythms (e.g., turn taking, social entrainment) that developed within the cultural ecology. In line with this, studies show connections between rhythmic capacities and the development of vocal forms of communication, including language (Bekius et al., 2016; Cummins, 2015; Cummins & Port, 1996; Ravignani et al., 2016b). As an aside, it is also posited that the process of knapping may have resulted in specific forms of listening (Morley, 2013, p. 120) and that the resonant and sometimes tonal qualities of stones and flakes may have afforded music-like play with sound (Killin, 2016 a,b; Zubrow et al., 2001).[11]

In brief, these rhythmic forms of behavior may have led to protomusical and protolinguistic forms of communication that arose simultaneously. However, as Tomlinson (2015) notes, "half a million years ago there was no language or musicking" (p. 127). While many music-relevant anatomical features were in place by this period, there is no evidence that these hominins possessed the more complex forms of combinatorial thinking required for the hierarchical structuring of rhythm, timbre, and pitch associated with musical activity—the kind of thinking that is also needed to build tools that employ multiple components in their planning and manufacture (e.g., those specifically intended for musical use, such as bone flutes). Rather, it is posited that protomusical and protolinguistic communications were initially limited to deictic copresent interactions (in-the-moment face-to-face encounters that integrated gesture and a limited number of vocal utterances) that incrementally developed into more complex sequences of communicative behavior. Over time, this led to the enactment of increasingly sophisticated forms of joint action and social understanding (Dunbar, 1996, 2003; Knoblich & Sebanz, 2008; Sterelny, 2012).

Such developments in the cultural loop fed forward into the coevolutionary cycle, allowing the environmental niche to be explored in new ways, affording previously unrecognized modes of engagement with it. This, in turn, altered selective pressures, leading to incremental phase transitions in the dynamics of the system, in which previous constraints were weakened, and new behavioral-cognitive phenotypes became possible. By the Upper Paleolithic period, the growing influence of the cultural epicycle favored an enhanced capacity to understand the actions and intentions of others and the related capacity to think "offline," "top down," or "at a distance" from immediate events (Bickerton, 1990, 2002; Carruthers & Smith, 1996; Tomasello, 1999). These developments allowed for the marshaling of material and social resources in new ways, leading to the creation of more complex artifacts (e.g., musical instruments), as well as more sophisticated types of cultural activity (e.g., ritual) and communication, including the hierarchical and combinatorial forms required for language and music as we know them today.[12]

Plastic Brains

Because biocultural approaches see (musical) cognition as an emergent property of situated embodied activity within a developing sociomaterial environment, it requires a different view of mind than the information-processing model associated with an adapted (modular) brain. If evolutionary processes do not involve adaption to a pregiven environment but instead require the active participation of organisms in shaping the environments they *coevolve with*, then "selection" and "adaptation" can be understood in a contingent and dynamically cyclical context. And this suggests, more generally, that cognitive processes may not depend wholly on genetically programmed responses or be reducible to a collection of fixed information-processing mechanisms in the brain. Rather, they may entail more plastic and perhaps nonrepresentational characteristics that reflect the cyclical integration of brains, bodies, objects, and sociocultural environments.[13]

In line with such concerns, scholars are questioning whether the notion of modularity continues to have much relevance for understanding the complexities of the human brain and cognition (e.g., Anderson, 2014; Doidge, 2007; Uttal, 2001).[14] For example, it is suggested that brain regions that appear to consistently correlate with specific processes, such as Broca's area and syntax, represent vast areas of the cortex that may in fact develop multiple overlapping or interlacing networks, the manifold functions of which may appear evermore fine grained and plastic as neural imaging technology becomes more refined (Grahn, 2012; Hagoort, 2005; Poldrack, 2006; Tettamanti & Weniger, 2006). In relation to this, recent research suggests the existence of "global systems" that function in a flexible and context-dependent manner (see Besson & Schön, 2012, pp. 289–290). These do not work independently of any other information available to the brain and are thus nonmodular (i.e., they are not discrete). Additionally, research into various levels of biological organization is showing that biological and cognitive processes develop in interaction with the environment: for example, that epigenetic factors play a central role in the expression of genes and that the formation of neural connections unfolds as a function of context (Lickliter & Honeycutt, 2003; Panksepp, 2009; Sur & Leamey, 2001; Van

Orden et al., 2001).[15] In short, the idea that brain and behavior are best understood as linear systems decomposable into discrete modules and corresponding functions is being replaced by more plastic and dynamically interactive perspectives.[16]

Such insights have contributed to the growing view that music cognition is the result of *nonmodular* cognitive developmental processes that are driven by a more general attraction to coordinated forms of social behavior (Trehub, 2000; Trehub & Hannon, 2006; Trehub & Nakata, 2001–2002; see also Drake et al., 2000). Accordingly, recent decades have seen researchers turn to "connectionist" models to account for essential cognitive functions such as (musical) perception and learning (see Clarke, 2005; Desain & Honing, 1991, 2003; Griffith & Todd, 1999). Likewise, Tomlinson (2015) discusses the connectionist approach as a way of understanding how the embodied-ecological processes of mimesis and social entrainment contributed to the development of music and language. As we considered earlier, the connectionist strategy does not rely on the idea of fixed modules but on the fact that when simple devices (such as individual neurons) are massively interconnected in a distributed way, such connections may change and grow through experience—when neurons tend to become active together, their connections are reinforced and vice versa (Hebb, 1949). Such connectivity is thought to result in the emergence of complex subsystems of activity as well as global convergences that produce system-wide properties. This is often modeled using DST and can also be understood in terms of the oscillatory dynamics mentioned previously (see Chemero, 2009).

Evolving Embodied Minds

While the connectionist approach was initially seen as an alternative to the computational orientation, more recent modeling has revealed the ability of complex connectionist networks to simulate syntactic, representational-symbolic, and combinatorial cognitive processes (see Bechtel, 2008; Smolensky, 1990), that is, those required by the "adapted brain" hypothesis. Such developments are attractive for some researchers because they allow for the assumed computational-representational nature of cognition to remain while accommodating the growing evidence around brain plasticity and dynamism (Chalmers, 1990; Clark, 1997; Dennett, 1991; Smolensky, 1990; see also van Gelder, 1990). As we have seen, however, others maintain that because the brain's connectivity cannot be separated from its dynamical history of coupling with the body and the environment, living cognition is not best understood as strictly limited to in-the-brain computations and representational content, even in a revised connectionist sense (Chemero, 2009; Hutto & Myin, 2012; Thompson, 2007).

To better understand what this means for the biocultural approach to music's origins, it may be useful to consider Tomlinson's (2015, pp. 129–139) reading of Cheney and Seyfarth's (2008) research into the social lives of baboons. As Tomlinson notes, observations of baboon vocal and gestural interactions lead Cheney and Seyfarth to suggest that the social behavior of these animals is indicative of an underlying hierarchical and syntactic-representational cognitive structure, one that is continuous with the Fodorian notion of "the language of thought" or "mentalese" (again, a process of non- or preconscious symbolic manipulation in the brain according to syntactic rules). This, they suggest, may reveal a deep evolutionary connection between linguistic processing and social intelligence, in which linguistic-computational processes are thought to underpin social cognition even if no spoken or symbolic language is present (as with baboons and our pre-human ancestors; cf. Barrett, 2018). However, Cheney and Seyfarth also hint at an alternative possibility, in which a more plastic and dynamical connectionist framework comes into play. The idea here is that once a system learns to organize itself in various ways, the patterns it develops can be recognized by the system in association with various things and relationships and thus may be said to "represent" them.[17] For this reason, connectionist processes are sometimes thought to be "subsymbolic" in that they provide a link between biological processes at lower levels and representational processes at higher ones (Smolensky, 1988; Varela et al., 1991, p. 100). In line with this, Cheney and Seyfarth (2008) suggest that, as animals engage with their environments, neural networks could be reinforced, leading to multimodal forms of "distributed neural representation" (p. 241; see also Barsalou, 2005). As Tomlinson (2015) points out, this implies something more concretely embodied and ecological:

> [A] *quite literal re-representing, a solidifying, affirming, salience-forming set of neural tautologies.* There is no reliance on abstracted social identities such as those humans conceive, on a mysterious language of mind that does the representing, or on baboon comprehension of causality, proposition, and predication. In their place are the accretion of intrabrain and interbrain networks and the responses

they enable in face of situations that are both familiar and less so. Networks are, within sheer biological constraints, products of environmental affordances, forged through the repeated patterns of an organism's interaction with the socio material surroundings. . . . All the intricacy Cheney and Seyfarth find in baboon sociality may well be explained . . . without recourse to anything like mentalese. (pp. 135–136; italics original)

Similarly, when Tomlinson refers to the mimetic nature of the developing protomusical environments, he clarifies that the action loops associated with this may be understood as "representational," but not in the sense of mental templates or propositions. Following the neuroanthropologist and cognitive neuroscientist Merlin Donald (2001), Tomlinson comments that the notion of "representation" employed here may entail little more "than the rise to salience of an aspect of a hominin's environment—in this case an enacted sequence of physical gestures imprinting itself in neural networks that fire again when repeated. Or . . . a set of interconnected neural oscillations" (pp. 73–74). It is suggested that this revised conception of representation may be more conducive to understanding cognition across a wider range of developmental and phylogenetic contexts. Indeed, the problem with applying the more traditional approach associated with computational psychology is that it tends to encourage a kind of "reverse engineering, retrospectively projecting human capacities onto earlier hominins or onto nonhuman species understood as proxies for our ancestors" (Tomlinson, 2015, p. 138). Likewise, the psychologist Louise Barrett (2011) discusses how our tendency to construct highly anthropomorphic views of other life forms can lead to false understandings, not only of their cognitive capacities but also of the nature and origins of human minds. To counter this, Barrett draws on enactivist theory, 4E cognition, and DST to describe the various ways insects and animals (including humans) use the morphologies of their bodies to enact the environmental affordances that allow them to realize various cognitive tasks. Importantly, Barrett's critique aligns with our previous discussions, in which we explored how the traditional assumption that cognition necessarily involves some form of linguistic competence (syntax, propositional thought, symbolic representation, and other forms of abstract "mental gymnastics") tends to overshadow the more primary aspects of human sense-making (Johnson, 2007). This, as we saw, extends to music, which over the past three decades has been examined with a special emphasis on its relationship to linguistic capacities in cognitive and evolutionary contexts (Patel, 2008; Rebuschat et al., 2012).

Now, all of this is not meant to imply that research into the cognitive and evolutionary relationship between music and language should be abandoned. This is an important area of inquiry and should continue to be investigated. However, other developmental and sociocultural factors are receiving growing attention from researchers. This includes accounts that explore the dynamical, ecological, and embodied nature of musical experience (e.g., Godøy et al., 2016; Krueger, 2013; Large & Jones, 1999; Reybrouck, 2005a,b). Indeed, while music and language both involve hierarchical and combinatorial forms of thought, it may be that both emerge from more domain-general capacities and proclivities related to agent-environment relationships. For some scholars, this implies that the symbolic-representational and propositional forms of cognition associated with language may be derivative rather than primary (see Hutto & Myin, 2012, 2017). As such, the origins of cognition may be found in the self-organizing dynamics associated with biological development itself—in the cycles of action and perception that are directly linked to an organism's ongoing history of embodied engagement with its environment. This recalls the coevolution cycle discussed previously, but it may also be considered in the context of ontogenesis—again, in how infants enact meaningful realities through embodied and affective interactivity with their sociomaterial niche (Bateson, 1975; Dissanayake, 2000; Reddy et al., 2013; Service, 1984).

Such insights are not lost on Tomlinson (2015), who highlights the continuity between the embodied activities of Paleolithic tool makers and cognition as such—where, he argues, cognition may in fact be rooted in interactions with the environment that over time result in increasingly complex extensions of individual embodied minds into the broader cognitive ecology (e.g., via mimesis and social "rhythmic" entrainment). Tomlinson also entertains the possibility that the self-organizing (or "self-initiating" as he sometimes refers to it) nature of the activities discussed earlier may not have to be understood in representational terms at all. However, he does not go much further than this general suggestion. This is perhaps somewhat surprising because he does, here and there, draw on the notion of "affordances" and the field of ecological psychology it is associated with—an explicitly nonrepresentational approach to cognition in its original version (Gibson, 1966, 1979).

Once Tomlinson outlines the deeply embodied, ecological, and socially interactive precursors of musical behavior, he then turns to explain music cognition using generative (e.g., Lerdahl & Jackendoff, 1983) and prediction- or anticipation-based models (e.g., Huron, 2006) that focus on the (statistical-grammatical) processing of musical stimuli and the behavioral responses they lead to. These perspectives are relevant as they focus on the more abstract and combinatorial ways the modern human mind may process musical events. We would like to suggest, however, that further insights into the origins of musicality could be gleaned by aligning the biocultural approach with enactive cognitive science.

Enactivism Meets the Biocultural Perspective

Enactivism and the biocultural perspective both draw on a plastic and self-organizing conception of the genome/phenotype relationship. This highlights the circular and coemergent view of organism and environment central to the enactive approach, but now over a deep evolutionary timescale. And because the enactive approach traces cognition to the fundamental biological concerns shared by all forms of life, it may help avoid some of the anthropomorphizing tendencies noted previously (e.g., imposing language-like capacities on non- or prehuman animals). Here, an interesting perspective is offered by recent theory associated with "radical enactivism" (Hutto & Myin, 2012). While this approach argues that so-called "basic minds" do not deal with any form of representational content, it also suggests that culture and language impose certain constraints that result in cognitive activities that may be understood as content bearing (Hutto & Myin, 2017). The explanatory advantages of this approach are currently a subject of debate. Nevertheless, the insights that arise from this discussion may shed new light on the cultural epicycles discussed previously. As Tomlinson (2015) points out, although musical activity is not fundamentally symbolic or representational, it necessarily occurs and develops within cultural worlds of symbols and language. Put simply, the debate surrounding radical enactivism could offer new perspectives on how, over various developmental periods, cultural life may simultaneously constrain, and be driven by, the nonsymbolic, social-affective, and embodied forms of cognition that characterize musical activity.

In previous chapters, we also discussed how researchers are using DST models to examine biocognitive processes in terms of the nonlinear couplings that occur between:

- The body: the development of muscular linkages and repertoires of corporeal articulation
- The brain: the emergence of patterned or recurrent (i.e., convergent) trajectories in neural activity
- The environment: the enactment of stable relationships and coordinated behavior within the sociomaterial ecology

We suggest that this perspective may be employed in conjunction with existing knowledge of early hominin anatomical and social structure, evidence from the archeological record, as well as comparative studies with other species and existing musical activities. This could include studies of how musical environments and behavior affect the expression of genes and gene groups and how this may recursively influence behavioral and ecological factors (see Bittman et al., 2005, 2013; Kanduri et al., 2015; Laland et al., 2010; Schneck & Berger, 2006; Skinner et al., 2015).

A 4E framework may also contribute to the biocultural approach in terms of describing the interacting dimensions of the emerging musical mind. Here, the *embodied* dimension explores the central role the body plays in driving cognitive processes. This is captured, for example, in the description of the early Paleolithic tool-making societies, where the reciprocal influences of sight, sound, and coordinated movement led to the production of artifacts with specific characteristics. Such forms of embodied activity also formed the basis from which more complex forms of thought and communication emerged. As we also considered, the biocultural model explores how such embodied factors arise in specific environments, leading to stable and recurrent patterns of activity where bodily, neural, and ecological trajectories converge. This highlights the *embedded* dimension, which concerns the ecological and sociocultural factors that coconstitute situated cognitive activity. The biocultural model explores this in terms of the sonic, visual, tactile, and emotional-mimetic nature of the niches enacted by our early ancestors, as well as the growing influence of the cultural epicycle on the cognitive ecology. In line with this, the *extended* dimension explores how many cognitive processes involve coupling with other agents (mimesis, social entrainment, participatory sense-making) or with nonbiological objects or cultural artifacts (tools, notebooks, musical instruments; Menary, 2007, 2010a). Last, the *enactive* dimension integrates the other Es, when it describes concerns the self-organizing nature of living systems and the active role organisms play in shaping the environments they inhabit. Such modes of activity are explored over a range of timescales (brief encounters, ontogenesis, evolutionary development), closely aligning with the coevolutionary feedback cycle discussed previously.

It is interesting to note that although Tomlinson (2015) makes no mention of 4E cognition as such, he does appear to move toward this way of thinking when he discusses how cognitive processes emerged and developed in our Paleolithic ancestors through embodied activity that was situated within a milieu that they actively shaped. He also argues that such activity necessarily involved the coordination of multiple agents and the "extension" of individual minds into the sociomaterial environment. In connection with this, it is also important to mention the work of the archaeologist and anthropologist Lambros Malafouris (2013, 2015), who develops 4E principles to better understand how brains, bodies, and objects interacted to form the cognitive ecologies of our prehuman ancestors. Notably, his "material engagement theory" expands the idea of neural plasticity discussed previously to include the extended domain of objects, tools, and the embedded sociocultural environment. This implies a kind of "metaplasticity" that involves a "historical ontology" of different forms of material engagement, in which the objects we make and engage with, as well as the cultural environments we enact, have shaped our minds in different ways and over various timescales. Here we can think again of the musically scaffolded environments discussed in the last chapter and the extended empathic spaces they may have afforded at various points in our evolution. Equally, we can consider again how many of the things we interact with become incorporated by the body as part of our cognitive domain. For example, when a blind person uses a cane to navigate an environment, the cane becomes an extension of their perceptual apparatus; perception extends beyond their hand to the tip of the cane (Merleau-Ponty, 1945). In a similar way, expert cellists will feel an important focal point of contact with the instrument where the bow hair engages with the strings, and they will be able to perceive all the variations in texture, tension, friction, and so on that are required to express themselves through that material extension of their bodily consciousness. In line with the biocultural approach,

Malafouris argues that these forms of "material engagement"—the fabrication and use of objects (tools, instruments, prosthetic devices)—do not simply extend human minds in the here and now but have also shaped the human phenotype (bodies, brains, behavior) over deep evolutionary time through processes of gene-culture coevolution.

Life, Music, and Meaning

An enactivist/biocultural, or 4E, approach paints a compelling picture of the development and meaning of musicality in the human phenotype. It explores the origins of musical behavior in the shared social spaces of our prehuman ancestors—where musicality emerges from the primary forms of (participatory) sense-making that reflect our fundamental condition as situated embodied beings who must coconstruct meaningful personal, social, and cultural realities if we are to survive and flourish. In line with our previous discussion, this view reasserts the remarkable span of musicality in human experience, from primordial bodily, affective, and social dimensions, to complex formal relationships and cultural significances; all of which interact dynamically in the enactment of meaningful musical environments.

This all leads to some important ontological and ethical implications. Indeed, if human music making is rooted in the evolving dynamics of organism-environment adaptivity, joint action, and the forms of empathic-emotional engagement that are central to how we make sense of the sociomaterial realities we enact and live though, then the ways we approach it in practice (e.g., music education, musicology, performance, music therapy) should reflect this fundamental existential reality. In other words, the ways we think about and do music should be guided by a philosophy of music based in a rich understanding of what kinds of beings we are and the role music plays (or could and should play) in our lives. Toward this end, the next two chapters look more closely at how musicality—understood as a form of embodied sense-making—develops in infancy and how this engages fundamental proclivities for exploratory, adaptive-improvisational, and creative activity. The resulting insights will set up the discussion in the concluding chapter, in which we examine the ethical implications of enactive music cognition, with a special focus on education.

8 Teleomusicality

In the previous chapters, we focused mostly on the social aspects of musical life—the various relationships between music, emotion, and empathy—and how the interplay of these factors plays out in the enactment of shared musical environments, including those of infants and our prehuman ancestors. As we have seen, musical sense-making develops within extended sociomaterial environments that are actively shaped by interacting, embodied minds. But although the social aspects we have discussed (e.g., between infants and primary caregivers) can be said to provide the foundations for human musicality, musical development also involves more solitary explorations of the sonic possibilities of the environment one is situated in. These explorations entail bodily engagements with the material affordances for sound making, and this plays a central role in how developing musical minds bring forth meaningful musical ecologies.

In this chapter, we examine how such processes play out in infancy and early childhood, with a special focus on understanding the development of the individual exploratory sense-making activities that ground musical behavior.[1] In doing so, we outline the notion of *teleomusicality* (Schiavio et al., 2017). The Greek term "télos" (τέλoζ) describes the "goal" or "result" that motivates a given behavior. Accordingly, teleomusicality entails the intrinsic goal-directedness of the music-like behaviors that emerge in infancy and that develop in childhood and beyond. This concept will help us to advance a useful distinction between "protomusical" and "teleomusical" activities. We use the former term to refer to music-like utterances and movements that do not entail a primary focus on sound itself (e.g., emotional-affective interactions with the caregiver).[2] The latter term describes the goal-directed actions infants adopt specifically to explore the sonic affordances of their environment and, in doing so, enact recurrent repertoires of musical behavior with and through the sounding things they encounter. As we discuss, "protomusicality" and "teleomusicality" are continuous with each other, and with the exploratory, adaptive, and self-organizing processes associated with biological sense-making more generally. However, we suggest that it is the sonically directed forms of engagement associated with "teleomusicality" that are most necessary for activities to be considered properly "musical"— that is, as opposed to behaviors that are sound involving but that are motivated by other needs or desires.

The Biocognitive Foundations of Infant Musicality

Even before birth, humans display a proclivity to engage with sonic environments (Parncutt, 2006, 2008). Hearing (limited to low frequencies) starts at the twenty-second week of gestation, although both the cochlea and central auditory pathway are still structurally and functionally immature at this time; evidence from scalp-recorded auditory evoked potentials suggests that fetuses are already sensitive to auditory events (Moore & Jeffery, 1994; Smith et al., 2003).[3] Unlike vision, the auditory system is already highly functional at birth, and young infants display great acuity to sounds (Bredberg, 1968; Perani et al., 2010; Trehub, 2009; cf. Keefe et al., 1994). By five to six months of age frequency and temporal perception have matured (Werner, 2002), and infants appear to be able to discriminate differences in frequency of less than one-half step (see Olsho, 1984). Studies suggest that young infants can also distinguish between consonant and dissonant sounds, as well as sound properties such as location, duration, and pitch, as indicated by their prolonged looking (see Trehub, 2003a,b,c,d,e). At eight months of age infants begin to make pitch discriminations based on an awareness of melodic contour (Trehub et al., 1984). Research also indicates that infants are born ready to entrain their respiratory patterns, sucking (both rhythm and intensity), tongue and mouth protrusions, eye opening and closing, limb movements, vocalizations, and more with the rhythmically rich properties of the environments in which they are embedded (e.g., sung lullabies and rhythmic movements) (Adachi & Trehub, 2012; Haslbeck, 2014; Nawrot, 2003). Interestingly, it has been found that, when compared to adults, twelve-month-old infants display greater accuracy in detecting changes in rhythmical patterns after brief exposure to foreign music (Hannon &

Trehub, 2005). However, a preference for rhythmical structures common to the infant's own culture emerges later (Soley & Hannon, 2010). Infants are also found to be better than adults at discriminating between certain typical aspects of Western music (e.g., a pitch variation). Research by music psychologists Laurel Trainor and Sandra Trehub (1992) found that, unlike adults, "infants lack implicit knowledge of key and implied harmony, as reflected in the equal ease with which they detect melodic changes that preserve the key and implied harmony of a tone sequence and those that disrupt those elements" (Trehub & Hannon, 2009, p. 115). While adults are easily able to detect changes that violate expectations associated with Western musical structures (such as nondiatonic variations), they have more difficulties with diatonic changes—that is, with pitch alterations that do not deviate from the tonal structure typical of Western classical music. Infants, on the other hand, appear to detect both changes with equal ease.[4]

This research reveals that infants have the ability to attend to sounds and music in very nuanced ways. But it also implies that over time perceptions of musical events become narrowed through experiential factors such as enculturation. In other words, it appears that the ways we structure sonic-musical phenomena may be driven less by predetermined processing functions and more by how our histories of interactivity with the environment shape recurrent patterns of action and perception. As the psychologist Eleanor Gibson (1988) writes, "A baby is provided by nature with some very helpful equipment to start its long course of learning about and interacting with the world. A baby is provided with an urge to use its perceptual system to explore the world; and it is impelled to direct attention outward toward events, objects and their properties, and the layout of the environment" (p. 7).

This recalls the phenomenological insights considered in chapter 4, in which we discussed how our primordial corporeal-sensory openness to the world—what phenomenologists refer to as "passive synthesis"—provides the basis for perception, consciousness, and intentionality. As we also saw, perceptions and actions become constrained through experience. Thus, our enactment of a world of meaningful things and relationships (the perception of things as things) entails cross-modal processes of "sedimentation" that involve the development of behavioral and perceptual forms. The so-called "natural attitudes" that arise from such engagements reflect the construction of stable organism-environment interactions, which afford the creation and communication of meaning. But while the development of such stability is important for survival and well-being, it can also limit an organism's ability to adapt to contingent changes across bodily and environmental conditions and thereby enact new relationships and perceptions. It follows, then, that an agent's musical growth, and its psychophysical development more generally, strongly depend on the ability to continually strike a balance between stability and change, between sedimentation and adaptivity. In our discussion of the Necker cube and Ghanaian polyrhythm in chapter 4, we saw how sedimented attitudes and perceptions can be identified and loosened through musical engagements, including forms of learning and practice—engagements that involve active, multimodal, and goal-directed processes that reconfigure our possibilities for perception and action. Let's now consider how these kinds of processes emerge and develop early in life through infants' active explorations of the sounding environments they are embedded within.

Perception in Action

An enactive approach to musical development places less emphasis on internal pregiven (or genetic) programs and more on understanding how agent-environment relationships self-organize and develop over time. As such, this perspective examines musical development in terms of the deep continuity between action and perception as the foundation for sense-making. In other words, musical growth—like all biological movements—does not blindly adhere to some predetermined developmental agenda or trajectory, nor is it best understood as a cause-and-response process. Instead, musicality emerges and unfolds as a synergistic process that involves the reciprocal interaction of a range of bodily, environmental, genetic, biological, sociocultural, and technological factors. To better understand this, we can think about how the origins of meaning making arise from an infant's fundamental proclivity for action and perception.

From birth (and before) infants' bodies are constantly moving. Their hands and arms, for example, initially display a wide range of seemingly random movements. But as infants encounter objects in the postnatal environment, they rapidly develop the coordinative ability to engage in goal-directed actions. At around six months of age infants begin to employ controlled grasping and appropriate preconfigurations of the fingers (von Hofsten,

1982), and this quickly develops into intentional reaching and grasping behavior. Before six months of age, when infants see objects that constitute goals for given actions (e.g., something to be grasped), they often do not pay attention to the physical properties of the object; instead, infants prefer to concentrate on the action itself (e.g., the grasping). However, by ten months infants can focus both on actions and their goals—for example, the grasping and the properties of the object to be grasped (Perone et al., 2006, 2009). As the developmental psychologist Sammy Perone and colleagues (2011) discuss:

> Studies examining infants' representation of events—such as a woman brushing her hair or a hand manipulating an object to produce a squeaking sound—have revealed that early in infancy, actions are more salient than the physical properties of objects acted upon.... For example, after habituation with an event in which an action is performed on an object and a sound is produced (e.g., squeezing a purple sphere produces squeaking), 6- to 7-month-old infants dishabituated to a novel *action* (e.g., rolling the same purple object) but not to a novel *object* (e.g., squeezing a pink oblong object). By 10 months of age, however, infants dishabituated to events involving novel actions, objects, or sounds produced.... Together, these findings highlight a developmental shift from selectively encoding actions to encoding actions, the objects acted on, and the sound outcome produced when acting on an object. (p. 2)

The development of these more sophisticated integrative behaviors is not only caused by the environment, nor is it strictly determined by inner genetic programming or brain mechanisms. Rather, the appearance of these actions involves the kinds of synergistic, self-organizing, and cross-modal processes we discussed earlier in connection with emotion and empathy, in which patterns of action and feeling arise through the dynamical interaction of neural, corporeal, and environmental trajectories.

Also interesting is that the emergence of such goal-directed behavior appears to coincide with the onset of the infant's ability to predict the goals of the actions of other people involved in similar activities (i.e., reaching for objects; see Woodward, 1998). Again, this helps reveal the bidirectional way action and perception influence each other (Gerson et al., 2015a; Thelen, 1989, 1994; Thelen et al., 2001; Woodward & Gerson, 2014).[5] As Kanakogi and Itakura (2011) suggest, "the developmental onset of infants' ability to understand an action, reflected by the ability to predict the goal of others' action, is synchronized with the developmental onset of their own ability to perform that action, and ... there is developmental correspondence relationship between the ability to predict the goal of an action and the ability to perform that same motor action" (see also Cannon et al., 2012, 2016; Daum et al., 2011; Gergely et al., 1995; Robson & Kuhlmeier, 2016; van der Meer et al., 1995; Woodward, 1998).

Here we can think again about the mirror and canonical neurons discussed previously. As we mentioned, canonical neurons discharge during the execution of a motor act and in response to the presentation of an object in the observer's peripersonal space. Typically, they display congruence between the action (e.g., grasping) and the physical properties of the observed object (e.g., a rattle). Mirror neurons, instead, fire both during the performance of an object-directed action and when that same action is observed in others. They are elicited not by the precise movements performed but by the goal of the given action (Rizzolatti & Sinigaglia, 2008). In other words, what really matters for mirror neurons is not the kinematics (e.g., contractions of single groups of muscles) but the goal-directedness involved in grasping as such (see Fogassi et al., 2005; Keysers, 2007). Importantly, activations in both sets of neurons are elicited by those actions observers know how to perform—actions that are in their "motor repertoire," which can evolve through learning and growth. In line with our discussion in previous chapters, this points to how the development and functioning of these neurons are shaped through experience—they require a context, a lived history of interactivity and adaptivity within a contingent sociomaterial environment.[6]

The observation that motor and perceptual processes are dynamically determined by each other has allowed scholars to go beyond traditional approaches to the study of human development. The latter were mostly interested in discovering the "motor programs" of the central nervous system and in understanding how such programs may generate behavioral outcomes. In contrast to this, the work by the developmental psychologists Esther Thelen, Linda B. Smith, and their collaborators (see, e.g., Smith & Thelen, 2003; Thelen, 1989, 1994, 2000; Thelen & Smith, 1994; Thelen et al., 2001; see also Spencer et al., 2006) offers an impressive collection of contributions aimed at describing motor development in terms of the "complex and ongoing interplay of arousal, attention, motivation, biomechanics, neuro-motor control, muscle performance, head-arm-trunk posture, and experience" (Galloway, 2005, p. 105). Consider, for example, the developmental transition from one mode of exploratory action, such as crawling, to another, such as walking (see He et al., 2015; Oudgenoeg-Paz et al., 2012). This involves a "shift" in behavior, where the pattern of crawling—which was stable for months—is now destabilized by the pattern of standing and walking (see also Clearfield, 2011). As Smith and Thelen (2003) comment, "There is no "program" for crawling assembled in the genes or wired in the nervous system. It self-organizes as a solution to a problem (move across the room), later to be replaced by a more efficient solution. Development is a series of evolving and dissolving patterns of varying dynamical stability rather than an inevitable march toward maturity" (p. 344).

Similarly, we can think about how patterns of musical behavior emerge through adaptive exploratory processes that involve the deep continuity between an infant's perceptual modalities, a basic proclivity to engage with the world, and the corporeal-environmental affordances, the patterns of action and interaction, that arise and evolve through growth, experience, and learning. Indeed, research has shown that explorative activity is a very common feature in young infants, as they are often seen manually interacting with different objects of their environment (Ruff, 1984; Ruff et al., 1984).[7] In the process of doing this, infants can become interested in the sonic properties of the various objects at hand, stimulating further engagement. For example, by exploring (i.e., squeezing, and shaking) a sound-making toy, a sonic discovery can be made, which may help infants select and specify those patterns of action that more effectively lead to the reproduction and elaboration of the sounds that capture their interest. These developments in infants' interactions with the sonic-material environment can then result in the development of various sonic and musical goals (see Delalande, 2009).

Toward Teleomusicality

In previous chapters, we explored the connection between human musicality and the kinds of emotional-empathic forms of communication that characterize interactions between infants and primary caregivers. These interactions involve rhythmic movement, prosodic speech, and so on, and they play a crucial developmental role across a range of areas, including musicality. Here, we refer to such interactions as "protomusical." This is because these behaviors are arguably not necessarily driven by musical or sonic goals as such, but by other needs and desires associated with caregiver attention and communication, when infants use sound to seek for or get the attention of their carer because they need nutrition, affection, or some other form care or interaction (e.g., to elicit play).

At this point, it is important to briefly address the key notion of "musical goals." This can be broadly defined as the (more or less predictable) outcome of a musically motivated action. That is, the action involves engaging with sounds in a way that is not focused on other primary interests (e.g., seeking attention from a caregiver) but is chiefly concerned with the manipulation of sound.[8] This allows us to distinguish between sound-related activities that appear to be motivated by a genuine intention to play with, explore, and, act upon sounds, and the "protomusical" sound-related activities that are driven by other goals, and therefore may be described as musical only a posteriori. Good examples of the former involve an infant hitting ("drumming") a surface to play with the sounds and waiting for a caregiver to do it too (preferably in novel ways that produce different sounds). This is indicative of how early on sonic-musical goals are often shared and how they can build on the kind of social interactions associated with protomusicality (i.e., where the focus is on sound making, but where the goal may be different). An example of the latter situation, in which no clear musical goal is addressed, may be found in occurrences in which the infant uses the very same action (i.e., hitting a surface) for a different scope, such as calling for attention or communicating a particular need (e.g., hunger or distress). In these cases, the kinematics are the same, but the underlying goals are different. Musical goals, here, are rooted in an intrinsic motivation to play with sound for its own sake. As such, they may involve self-directed actions, such as manipulating a sound-making object to get a desired result and autonomously playing with it. Indeed, as infants gain increased muscular control and engage in more sophisticated goal-directed behaviors with the things they encounter, they are also freer to make more solitary explorations of the world around them. Such activity allows them to grow their motor repertoire and to enact pragmatic understandings (possibilities for action and perception, affordances) within their contingent milieu.

So, what happens when the goal of an action becomes a sound? That is, what happens when the goals of one's actions are not simply directed toward reaching for and moving an object but are primarily focused on the objects' sound properties? And what strategies might infants employ to explore and manipulate the sonic environment created through their interactions with a sounding object? As we saw, sound-producing actions can serve multiple goals, such as making music or communicating with others. The music psychologist Michel Imberty (1995) offers a useful starting point for thinking about this distinction. He suggests that our musical engagements include two key features. The first involves the ability to enact and recognize musical behavior. The second entails the engagement and play with the variable nuances of musical action. Arguably, however, both elements can emerge only once an "attentive shift" has occurred. This shift aligns with the transition discussed earlier, in which infants between six and ten months of age gain the ability to attend to both the actions and the (sonic) properties associated with the objects they encounter. This attentive shift, then, concerns the infant's increased ability and interest in focusing on the *musical goals* of the actions, rather than only on their kinematic and visual dimensions. To be clear, this does not mean that infants are unable to perceive the sounds associated with musical events or other sonic-involving situations before this shift occurs. We have already discussed their considerable perceptual aptitude for sound. Rather, before this time it appears that they do not focus on the sound as the primary object of their action-as-perception. In other words, the "attentive shift" permits the constitution of a first musical context, where the infant's goal can become intrinsically "musical." We refer to this basic form of musical activity as *teleomusicality*. Likewise, we define the basic motor actions directed toward a musical goal as "teleomusical acts."

Varieties of Teleomusical Acts

Teleomusical acts can be understood in two ways. We refer to the first as *original teleomusical acts* (OTAs). These originate in patterns of motor behavior that emerge in early infancy, such as grasping and shaking (e.g., a rattle placed in the hand). These actions are executed more or less spontaneously, enabling the basic forms of action-perception looping required to explore and engage with the sonic environment in which the infant is embedded—where, again, the initial focus is on controlling movement (e.g., grasping the object) rather than on the (sonic) properties of the object itself. It is only by around six months of age that such activities may begin to be understood as properly "teleomusical." It is at this age that infants can also focus specifically on the sonic properties of the objects that are acted upon (as opposed to the kinematic dimensions that seem to occupy them earlier on) and develop repertoires of controlled actions to achieve sonic goals.

We refer to the second variety as *constituted teleomusical acts* (CTAs). These are not "primary" acts because they are built through the unification of sets of OTAs. Yet they rapidly develop kinematic fluidity, allowing the infant to execute them as unitary goal-directed actions. Interestingly, the continuity between OTAs and CTAs can be discerned in musical learning throughout the life span. Consider, for example, the adult experience of learning to play a chord on the piano. This activity requires temporal coordination; sensibility of the fingers, wrists, arms, and back; expressivity; and so on. A beginner must explore these actions, as well as the sounds they produce, to achieve the goal of playing the chord. A skilled pianist, however, would achieve the goal (playing this chord in a certain way) through a fluid, nonreducible (holistic) execution, in which the primary focus remains on the musical sound being produced, rather than on the kinematics behind it.

OTAs begin as spontaneous proclivities for movement—they emerge in infancy seemingly with little or no help from the caregiver. As such, they may be understood as *self-organizing* behaviors. They are also *plastic* and can be easily refined and performed in different contexts (through different strategies and motivations). Indeed, they are ecologically relevant with regard to how sonic affordances develop in a given subject-object relationship. Because of this, they also quickly become goal-directed toward the properties of the sound and the enactment of protomusical structures. Every healthy infant possesses the skills necessary to perform the basic acts of hitting, plucking, or scratching (Delalande, 2009; see also Godøy, 1997, 2003). Accordingly, musical development in infancy could be conceived of as a shift from the basic and spontaneous emergence of OTAs to the acquisition and development of CTAs. It is reasonable to argue that once a repertoire of CTAs is established, the young "musician" may then begin to integrate other qualities into her musical life (such as extramusical values). She may now engage in musical activities that involve more complex forms of social coordination (such as performing together) and develop the affordances of sounding objects in new ways through composition and improvisation. In line with this, we may now consider the three types of musical "conduct" offered by Delalande (2009), each of which is an

Teleomusicality 165

extrapolation on the developmental framework Piaget (1964) refers to as "the phases of play":

1. *The explorative conduct* is based on the discovery of sounds and noises. It corresponds to *sensorimotor play*, which, according to Piaget, dominates the first two years of life. After six months from birth, as we discussed, infants become more sensitive to the *auditory/action possibilities* of the surrounding environment. At this stage we suggest that OTAs are being developed into CTAs.

2. *The expressive conduct* corresponds to the phase Piaget defined as *symbolic play* and characterizes the years of kindergarten. Delalande suggests that during this period the child begins to attribute extramusical values to sounds in association with certain situations, places, social roles, expectations, and so forth. This enriches the primary form of sensorimotor understanding with a broader domain of meaning attribution. This phase, then, may be understood to further strengthen CTAs into musical actions and understandings that are more relevant in cultural contexts.

3. *The organizational conduct* emerges when children discover the enjoyment of applying rules to their own musical games (this corresponds to Piaget's *game of rules*). These rules also play a crucial role in practices such as musical analysis and composition, in which the agent develops and employs a particular and/or personal strategy to achieve the desired goal.[9] This phase thus may be thought of as moving beyond given musical cultural norms, enhancing creativity and curiosity to further understand—and explore—theoretical and analytical musical possibilities.

Another important idea of Delalande's concerns the continuous employment and development of exploratory, expressive, and organizational dynamics throughout one's musical life. That is, modes of engagement that characterize earlier stages of development and conduct are not progressively replaced by those that emerge later on; they continue to be crucial to creative musical engagements across the life span (and could play an important role in therapeutic contexts). Thus, improvisation and compositional practices can be seen as continuous with all of the ontogenetic processes of exploration just described, in which the sounding object explored is the relevant musical instrument(s) (including the voice), and its harmonic, melodic, and timbral possibilities, which may then be organized in new ways. Accordingly, even in older children and adults, the creative enactment of new patterns and relationships necessarily involves reengagement with fundamental musical processes, including the primary body movements associated with the emergence of OTAs.

While many approaches to early musical development rightly stress the importance of the infant's relationship with primary caregivers, we attempt to give equal weight to the (more-or-less) independent exploratory activities of the infant. This may help to make a clearer distinction between actions and relationships that are explicitly musical and those that may best be referred to as "protomusical." To be clear, this distinction need not undermine claims that the social and emotional aspects of our musicality can be traced in large part to infant-caregiver interactions; it is not intended to impose a discontinuity. Instead, this perspective aims to refine our understanding of early musical development by considering how an explicitly *musical* behavior emerges as infants explore their environment and begin to focus on the production of sound. As we have seen, before the attentive shift, the infant's environment contains a number of possibilities for sound making, which may be facilitated by, but not wholly dependent on, the caregiver. The latter may place a rattle in the infant's hand, for example, or the infant may engage with the object directly. However, as we also considered, in the first months the infant's attention is spread across the modalities, often with more focus on the kinematic aspects of the engagement (movement). Importantly, the attentive shift that occurs after six months involves a new kind of musical activity, which may now be directed toward the *sonic* possibilities of the object at hand. This occurs first through the exploration of the relationship between spontaneous movements and sound and then via increasingly controlled goal-directed manipulations resulting in new patterns of behavior (which we refer to as OTAs). It is here, we suggest, that *teleomusicality* as such begins (see figure 8.1). Again, the fact that teleomusicality often involves solitary explorations does not mean that caregivers do not participate in it. Moreover, other social interactions with teachers and peers play an important role later on in fostering more complex and cooperative musical activities associated with the development of CTAs and modes of conduct like those outlined by Delalande.

The ideas of protomusicality and teleomusicality may help explain the different, but interrelated, ontogenetic trajectories by which basic musical

Teleomusicality

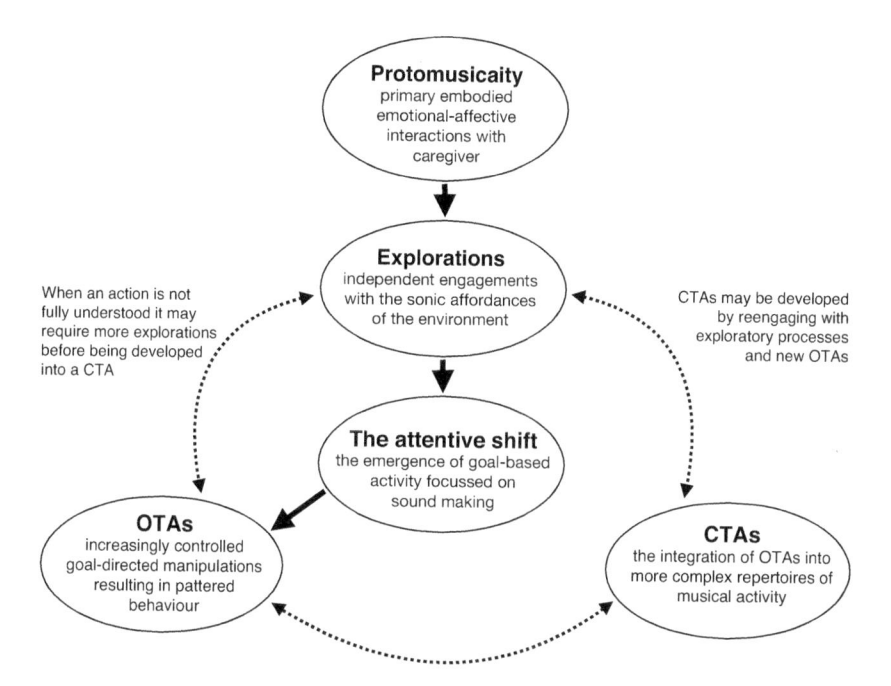

Figure 8.1

This model captures the constitution of teleomusicality through the development of OTAs into CTAs. The *dotted lines* show how, once the attentive shift has occurred, infants may reengage in explorations or OTAs before developing new CTAs. (Adapted from Schiavio et al., 2017)

skills emerge in infancy: for example, the ability to synchronize with music, manipulate sound-making objects toward musical goals, be sensitive to and participate in the different nuances of musical events, and understand the musical actions of others. This framework may also aid in understanding the processes by which musical agents develop such skills beyond infancy (i.e., by constantly implementing and improving their repertoire of CTAs in new, creative ways). For example, it could also be argued that infants who do not transform OTAs in CTAs (perhaps because their musical environment is not affordative enough or simply because their attention is captured more by the visual nuances of an object) will still possess the basic music-related actions they developed during their exploratory sonic behaviors. Indeed, the ability to shake a rattle and listen to its sounds, to push the keys of a piano in various patterns, to move to a rhythm or sound, or to beat a stick against a toy drum all afford the development of basic forms of rhythmic, melodic, and other sonic patterns. Likewise, elementary forms of dancing, singing along, humming, gesturing, and a general sense of "feeling" the music all have roots in early sensorimotor development and exploratory behaviors. These are examples of music-like activities that make up a substrate of goal-directed musical actions that are shared among all human beings but that also develop in various ways, depending on individual histories of interactivity with a given milieu.

9 Creative Musical Bodies

In the previous chapter we explored how early musical development may be conceived of as a dynamical and self-organizing phenomenon, one that cannot be fully captured as genetically determined developmental programs acting in response to environmental stimuli. Instead, we have suggested that the emergence of musicality in infants is better understood in terms of ongoing loops of perceptually guided action by which new affordances arise and evolve. Such sense-making processes are grounded in a primordial proclivity for movement and exploration, which over time result in the enactment of action understandings driven by the interaction and evolution of corporeal, neural, and environmental factors. Interestingly, this synergistic point of view can also offer new insights into the nature of musical creativity.[1] This is important because much research and theory have neglected bodily and interactive aspects, preferring to view creativity in terms of products (ideas, outcomes) that are categorized according to certain criteria and as a cognitive process that is confined to the personal domain (usually the brain) of an individual agent.

For example, creativity is often discussed in terms of categories such as "big-C" and "little-c," in which the former refers to eminent, domain-changing outputs, and the latter to creativity in everyday problem-solving situations and creative expressions, which include the forms of wishful, imaginative, or counterfactual thinking that occur in everyday life (Byrne, 2005). This approach has been developed in different ways (Kozbelt et al., 2010; Runco, 2014). For example, in addition to "big-C" and "little-c," Kaufman and Beghetto (2009) add "mini-c" and "pro-c." The former describes the novel abilities and understanding that stem from an agent's learning processes (e.g., a music student), while the latter concerns the types of products exhibited by professional creators (e.g., a music composer) who have not achieved eminent domain-changing (or big-C) accomplishments in society. Similarly, as Kirton (2003) argues, creativity may also be understood in terms of a spectrum between adaptation and innovation. Adapters develop and improve pre-existing methods and conditions, while innovators initiate more radical changes that may revolutionize the way things are done in a given domain. Another important approach is offered by the psychologist Margaret Boden (1998, 2004), who posits three subtypes of creativity: combinatorial creativity, which entails the ability to combine pre-existing concepts and items in novel ways; exploratory creativity, which arises from the exploration of a given conceptual background; and transformational creativity, which involves a redefinition of a given theoretical or cultural framework. Boden's three subtypes of creativity are enhanced by an overall personal-psychological creativity novel to the cognizer who generates it and a historical creativity recognized by the cultural norms of the society in which the agent is situated.

These categories can be applied to musical contexts in interesting ways. Consider, for example, the new approach to composition developed by Arnold Schoenberg. Schoenberg's work, often referred to as "12-tone" music or "dodecaphony," departs from traditional harmony by using all 12 tones of the chromatic scale equally. This shift resulted in new ways of organizing musical structures that do not depend on the usual hierarchies—there is no tonic, nor dominant; no major or minor keys; and no strong distinction of consonance and dissonance in the relationships between individual tones. The major influence of Schoenberg's contributions on Western composed music implies that he should be placed in the "big-C" category and situated toward the "innovator" side of Kirton's spectrum and that he is representative of transformational and historical creativity in Boden's model. By contrast, the output of film composer John Williams—who juxtaposes pre-existing styles and devices in highly effective ways[2]—might be situated as "adaptive," categorized as "pro-c," and reflect the combinatorial creativity discussed by Boden.

Others have placed more emphasis on the *processes* involved in creative thinking. To take an early example, the pioneering work of Wallas (1926) describes creativity in terms of both conscious (explicit) and subconscious (intuitive) mental processes. These involve stages of (a) preparation, the acquisition of knowledge; (b) incubation, the subconscious restructuring of knowledge; (c) illumination, the flash of insight; and (d) verification, the evaluation and application of the new idea. Others have examined creativity in terms of an agent's capacity to develop multiple possibilities for solving a given problem. Such thought processes are often referred to as "divergent" or "lateral" thinking (see Gardner, 1993, p. 20) and are understood to play a central role in both everyday creativities and the development of domain-changing outcomes. Put simply, this perspective examines how people develop possibilities for action and thought and how they become fluent at adapting such possibilities to the contingencies of a given situation through "convergent" thinking. Along these lines, Koestler (1964) introduces the term "bisociation" to highlight the combinatorial nature of creativity. According to Koestler, creativity arises from the integration (and not merely association) of two frames of thought that at first may seem completely disconnected. In his terminology, two "orthogonal matrices blend together," giving rise to a creative outcome. While this approach does not develop the contextual aspects of creativity in detail, it may help describe how new musical ideas, pieces, or styles result from a process of blending between two or more different concepts, genres, or cultural traditions. In addition to the four processes discussed by Wallas (1926), Hélie and Sun (2010) include creative problem-solving processes that entail the interaction of implicit and explicit forms of knowledge. In doing so they offer subcategories that allow for more nuanced theoretical frameworks and thus greater precision in developing (connectionist) models. Likewise, Finke, Ward, and Smith (1992) describe an approach that involves a stage for generation and one for exploration. The former involves the development of representations, or "pre-inventive structures," as they are referred to. The latter describes the process by which such structures are deployed in novel ways, resulting in creative ideas and action (see also Ward, 1995). These are further divided into additional subprocesses that describe how agents move back and forth between stages depending on their creative needs.[3]

We suggest that, although these perspectives do offer useful insights into certain aspects of creativity, they do not fully account for the types of developmental processes discussed earlier, which necessarily involve adaptive, embodied interactions with the extended sociomaterial environment one is embedded within (Glăveanu, 2014). Throughout this book, we have been examining how our corporeal, perceptual, and exploratory engagements with the environment constitute the ground from which we come to know and understand the world. It is through bodily action that we bring our relationships with the world—and therefore our minds—into being. As we have also seen, these relationships change over time through development and experience: new repertoires of action and interaction are enacted via adaptive patterns of reciprocal causation with the social and physical environment we are situated in. On basic levels, such developments can occur as an organism adapts to perturbations in the environment (to maintain its continued existence and well-being). However, in more complex creatures, like humans, new agent-world configurations can also be guided by a natural inclination to learn, solve problems, create new meanings, and optimize skills and social configurations.

Here we can think again about the experience of learning the Ghanaian polyrhythms discussed in chapter 4. As with many cases of learning and development, this example involved the enactment of new muscular, neural, and environmental (sonic-instrumental-social) relationships. And as we saw, developing these new forms of coordinated action often involves periods of awkwardness, discomfort, and frustration as sedimented perceptions and proclivities for action are loosened and new ones are formed (see also Benedek et al., 2017). This reflects the developmental dynamics discussed in the previous chapter, where the learner explores actions required to produce the patterned sounds (OTAs), which then affords the enactment of more complex musical engagements (CTAs). As we saw, OTAs and CTAs do not simply stand in a linear causal relationship to each other but develop in a circular way as musical agents often have to return to explore the kinematics associated with the creation of new CTAs (e.g., as environments change and new instrumental and social aspects are incorporated). This resonates with a description offered by Eric Clarke (2005), "On first encountering a xylophone, the child's more-or-less unregulated experiments with hands or sticks will result in all kinds of accidental sounds. With unsupervised investigation, the child may discover that different kinds of actions . . . give rise to differentiated results . . . , and even that these distinctions can themselves be used to achieve other goals" (p. 23).

An important point we would like to reassert here is that the exploratory, creative, and improvisational processes associated with musical learning are continuous with the kinds of adaptivity exhibited by all forms of life. This recalls Varela, Thompson, and Rosch's (1991) insight that the situated, embodied mind is like a "path that is laid down in walking," where "cognition as embodied action both poses the problems and specifies those paths that must be tread or laid down for their solution" (p. 205).

To engage in such adaptive behaviors, living systems must constantly work to maintain a balance between constraint and freedom, stability and instability, across corporeal and environmental domains. Again, this reflects an animal's ability to develop consistent patterns of action and relationships with an environment and how it can and must adapt to perturbations in the environment and self-organize new relationships and patterns of action as it moves through life (Schiavio & Benedek, 2020). Interestingly, research using DST modeling can aid in understanding the dynamics involved in the self-organizing behaviors associated with the development of living systems, which can be extended to musical contexts (as well as to the study of emotion and empathy, as we saw in chapters 5 and 6). Let's consider, then, how an approach that draws on insights from enactivism and dynamical systems theory (DST) may help us better understand the continuity between musical development, creativity, and the primary forms of embodied sense-making associated with biological flourishing more generally.

The Corporeal Foundations of Creativity

As we discussed in chapter 2 and elsewhere, DST describes how complex, self-organizing systems emerge and develop over time. In doing so, DST reveals aspects of the system that tend to converge and diverge as patterns of relative stability and instability. These are described as "attractors" and "repellors," respectively, and are shown as a topographical space referred to as a "phase portrait." The latter term describes how the possible states of a system evolve over various timescales.[4] While this approach has provided mathematical descriptions of a range of nonorganic self-organizing systems (Clark, 2001; Strogatz, 1994, 2001), it has also been explored in biological contexts associated with coordinated movement, communication, problem solving, and cognition more generally. Here, important early research involves the famous "finger wagging" experiments by Kelso (1984; see also Haken et al., 1985). These studies involved participants "wagging" the index fingers of both hands at various speeds in two different patterns of relative phase: (a) in-phase patterns, in which the same muscles of each hand and arm were working at the same time, and (b) out-of-phase patterns, in which the same muscles of each hand and arm moved in alternation.

At slower tempos participants maintained coordinated (stable) wagging in both in- and out-of-phase patterns. However, at faster speeds they could only maintain stability with in-phase wagging; once a critical speed was achieved, out-of-phase finger wagging exhibited a brief period of instability before falling into in-phase wagging.

From the DST perspective, the possible states of the finger-wagging system may be understood as tending toward two basins of attraction: a "deep" one at the in-phase state (most stable) and a "shallow" one at the out-of-phase state (less stable). However, at more rapid speeds only the in-phase pattern is stable. Here the shallow basin dissolves and the system organizes itself around the deep attractor only. Haken, Kelso, and Bunz (1985) describe this system with the following potential function, which is often referred to as the "HKB model" after its creators.

$$V(\Phi) = -A \cos \Phi - B \cos 2\Phi$$

The stability of the system is measured by $V(\Phi)$, with the two oscillating components (the index fingers of each hand) described at relative phase Φ. The HKB equation has proven to be very adaptable, and it has been enhanced to describe a range of oscillating systems (Kelso, 1995; Schöner & Kelso, 1988b; see Chemero, 2009), resulting in frameworks that describe a much wider range of behavioral, perceptual, and cognitive phenomena (Kelso, 1995; Schöner & Kelso, 1988b).[5]

This research has revealed that aspects of self-organizing behavior in living systems can be described using differential equations that express the magnitude of variability between pairs of nonlinearly coupled components. For example, DST has been used to explore learning as a type of phase transition—whereby cognitive agents change the attractor layout of the system they are immersed in, creating new phase portraits and a range of new possible phase relations (Schöner & Kelso, 1988a). A relatively simple manifestation of this can be seen in the work of Amazeen and colleagues (1996), who asked participants to learn to wag their fingers in more complex polyrhythmic groupings (e.g., five left for every four right). Initially, the out-of-phase movements could only be maintained at slow tempi (as predicted by the studies just discussed). However, after sustained practice, participants developed the ability to maintain complex out-of-phase patterns at fast tempos. In DST terms, this involves a phase-space deformation, in which, in this case, the basin at the out-of-phase attractor becomes deeper

Creative Musical Bodies

(that is, attains greater magnitude) than the in-phase one through the sustained goal-directed activity of the participant.[6]

Here it is also interesting to note that a number of researchers (e.g., Stephen & Dixon, 2009; Stephen et al., 2009) have documented distinct spikes in system entropy just before the "Ah-ha!" experience associated with the moment of understanding (i.e., the enactment of a new, or "deeper," attractor basin) involved with sustained investigation of a given problem. They also noted that a rapid decrease in entropy follows once the new state is achieved, indicating the "self-organization of a new cognitive structure" (Dixon et al., 2010). This was seen in the finger-wagging experiments, where, again, the finger movements became more entropic just before they switched from the out-of-phase coordination to the in-phase coordination. Similar dynamics have also been noted in more complex problem-solving contexts. Psychologists Damian Stephen, James Dixon, and Robert Isenhower (2009) presented participants with a stationary image that depicted several interconnected gears. Participants were told the rotational direction of the initial "drive gear" and were asked to give the direction of a target gear later in the series. Interestingly, almost all of the participants began by tracing the directions of the gears with their fingers. However, once they realized (through this tracing activity) that each gear moves in an opposite direction to the one it is connected to (1, 3, 5, etc., all move in the same direction; 2, 4, 6 move in the opposite direction) they were then able to use this pattern to predict the direction of a given gear much more quickly. Like the finger-wagging studies, the moment this insight was reached was preceded by increased (entropic) fluctuations in eye and finger movement (as observed in eye- and finger-tracking data). This study, again, highlights the centrality of direct bodily interactions with environmental features for learning, problem solving, and the generation of knowledge. It also shows how these interactions involve the introduction of instability into the system so that new stable configurations, affordances, and understandings can emerge.

The introduction of instability to the system can be the result of the willful activity of the agent and/or involve perturbations in the environment the agent must deal with, perhaps brought about by other agents, social and cultural developments, or other thermodynamic factors. Here the concept of "strange" or "chaotic" attractor can help us to understand such phenomena. These attractors are characterized by varying degrees of entropy and thus evolve over time within a set of constraints. They describe aspects of a system that are stable but that nevertheless exhibit degrees of freedom within such stability and can therefore help to model behavioral flexibility under certain parameters. Under some circumstances, however, the evolution of strange attractors can also lead to bifurcations whereby new patterns of activity arise—that is, new attractors and attractor layouts (Abraham et al., 2012). Such flexibility is crucial for the survival of living systems, which must have the freedom to move, interact, and develop coherent and repeatable patterns of behavior in relatively stable environments but also be able to enact new forms of behavior in changing conditions. Accordingly, we can think of a living agent's phase portrait as divided into a series of strange attractors that permit divergence in various directions within local basins (Schuldberg, 1999): this affords the biocognitive flexibility required to enact new attractor layouts through sustained adaptive behavior. Such alterations of an agent's phase space may involve the embodied types of learning associated with acquiring the ability to walk, speak, and dance, or the process of learning to play an instrument (see Thelen & Smith, 1994; Sudnow, 1978). These developmental processes can involve challenging periods in which the trajectories between neural, emotional, muscular, and environmental (material, social) factors exhibit a high degree of instability, resulting in moments of physical, emotional, and social discomfort (e.g., falling, chaotic crying, frustration). Once new patterns of behavior are stabilized, however, a range of new affordances may emerge: developing new repertoires of action and interaction afford new possibilities for thought, which inspire new action, and so on.

These synergetic dynamics can also describe the development of OTAs and CTAs we introduced in the previous chapter, which involve the discovery and enactment of new patterns of action-as-perception. Likewise, anyone who has attempted to learn a new multilimb pattern on a drum kit, to take one example, will have experienced firsthand a phenomenon like the one described by the polymetric finger-wagging experiment just discussed. As we mentioned previously, the phenomenology of developing new ways of engaging bodily with an instrument (or a musical practice) can involve uncomfortable periods in which one must depart from established patterns of activity, resulting in the system instability described earlier. However, this is necessary to create new and richer possibilities for thought and action (i.e., new basins of attraction and attractor layouts in DST speak), which

Creative Musical Bodies 177

then become "naturalized" as part of the agent's musical motor repertoire.[7] Similarly, a musical ensemble needs to co-realize the patterns of interactivity that allow them to work together effectively (Høffding, 2019; Salice et al., 2017; Seddon & Biasutti, 2009). And here too, sometimes a group must experience moments of instability, frustration, and discomfort where these balances break down. However, by working through these periods the ensemble may develop more flexibility and be able to push the boundaries of their performances in new and sometimes unexpected ways. We suggest that the ability to enact these kinds of dynamical shifts—to initiate, develop, stabilize, and destabilize goal-directed patterns of action and perception—plays a key role in laying the developmental foundations of musical learning and creativity.

The Es of Musical Creativity

The DST perspective offers tools for describing how agents develop new repertoires of meaningful action through bodily action and through interactions with the extended sociomaterial environments they are embedded within. In turn, the 4E approach can provide a useful framework for exploring DST data in phenomenological terms. Here, "embodied" and "embedded" aspects can be thought of as dimensions of experience whereby agents may describe how the interactivity between their corporeal states (muscular, emotional, affective, bodily feeling, and so on) and the dynamics of the environment develop over time (e.g., the feeling of creating a new attractor layout associated with the development of new skills and understandings). Additionally, musical learning and creativity may be facilitated by goal-directed shifts in a musician's relationship with the extended environment—for example, by the desire to learn and develop new musical possibilities—and often in collaboration with other musicians or teachers (Küpers et al., 2014; McPherson et al., 2012; see also Sawyer, 2003). This echoes again the developmental dynamics associated with OTAs and CTAs discussed previously. In line with this, the "extended" dimension can help to illuminate how the material (e.g., musical instruments, sound) and social aspects (e.g., other musicians, teachers) may coconstitute the mental and creative lives of individual agents and how negotiating periods of stability and perturbation can transform different patterns of cognitive offloading, which results in new relationships with instruments and different forms of joint action.

In collaborative music making, participants coenact repertoires of shared action and perception that allow them to collectively scaffold the extended musical environment by taking on and offloading various tasks and meanings to and from each other. As we saw in chapter 6, this could involve entraining with a beat provided by a drummer, the coenactment of phrasing patterns, leading or following harmonic and dynamical shifts, and so on (Keller, 2001; Linson & Clarke, 2017). But just as musicians strive to maintain stability in their interactions, they also sometimes willfully introduce instability into the system. For example, classical musicians performing well-known pieces together initiate and negotiate subtle fluctuations across different music parameters (e.g., dynamics and agogics), which can give rise to unpredicted outcomes. These outcomes, in turn, determine small-scale changes in musical interpretation and expressivity (see also Bishop, 2018; Schiavio & Høffding, 2015). In other forms of music making, such dynamics may be pushed to extremes. Anyone who has engaged in free improvisation, for example, should be able to relate to this from experience. This activity often involves the intentional introduction of instability by the players as a way of keeping the music interesting and vital; master improvisers become highly adept at playing with the limits of interactive musical dynamics (Borgo, 2005; see also Iyer, 2002).

In all, musical behavior involves negotiating moments of stability and instability across bodily and environmental domains. Musical creativity in action, therefore, entails a constant push and pull between constraint and entropy (see also Haught-Tromp, 2017; Medeiros et al., 2014). Too much constraint and the music and musician stagnate; no new relationships are enacted. Too much entropy, however, and the musical environment becomes incoherent and falls to pieces. These in-the-moment aspects of musical creativity are examined in interesting empirical work by Ashely Walton and colleagues (2015, 2017). Their studies used the mathematical tools of DST to analyze how changes in the structure of a musical environment impact the experience of creativity for pairs of interacting keyboard improvisers and how this is reflected in their bodily movements and the ways they adapt to the sonic environment they cocreate. Here, musicians performed improvised duets along with different backing tracks while their head, left arm, and right arm movements were recorded, as was the music they made. Interesting variations in the musicians' movements were found that correlated with periods of stability and instability in the audio documents, as well as with the musicians' reports that they felt most creative in environments that afforded a balance of freedom and constraint, where varying degrees of instability could be introduced and resolved cooperatively.[8]

Studies like these that measure changes in musicians' behaviors in connection with their reported experiences could help us to better identify how certain environmental constraints give rise to dynamics that provide the "right" kind of tensions (the right kind of pushes and pulls between a system's components) for the enactment of new musical possibilities.[9] Among other areas, this has implications for music education in terms of understanding what kinds of environments may foster creativity. It may also have a great deal to offer for researchers in music performance studies. For example, the framework we introduced here may shed light on how a musical ensemble develops the relevant shared patterns of action and perception required to perform a difficult piece of music and how they enact unique ways of communicating as enmeshed components of a communal musical environment by adapting to and/or instigating moments of instability and nudging the system into new shared basins of attraction.

Musical Worldviews

The "enactive" dimension examines what such developments mean for the life-world that is continually brought forth by an agent and what new understandings and relationships have emerged. In connection with this, enactive and DST concepts can offer interesting heuristics for thinking about musical creativity and the negotiation of meaning in historical and cultural domains. One way to do this involves the idea of a musical worldview[10] (Velardo, 2016; see also Gabora, 2017). Ideally, this describes the development (the transforming phase portrait) of a musical agent (x), showing how their musical values, ideas, and beliefs change over time (t) (Gimenes & Miranda, 2011). For example, a person who has listened only to Western-based music may find it difficult to understand and appreciate Indian classical music because this form of music is outside of their worldview. However, this person's worldview could be expanded, among other ways, through sustained engagement with Indian music. This could involve more focused listening, whereby the person may begin to perceive relationships in (and develop a personal relationship with) the music. It may also entail learning to play this kind of music and developing a deeper engagement with the culture it is associated with. These engagements may lead to novel understandings and possibilities for perception and action, the enactment of new attractor basins in DST-speak.

Along similar lines, one may explore the subset of an agent's worldview associated with musical style.[11] In this case, the DST concept of a "strange attractor" we considered earlier could be used to represent a set of stylistic configurations that are similar and coherent; once the state of an agent is inside one of these attractors, they can freely move within its parameters. For example, one could think of an attractor for Baroque music and another for late Romantic music. Here the music of Vivaldi may be understood as confined to (or "embedded in") the Baroque attractor. His musical evolution explores different flavors of Baroque music that correspond to different points within the Baroque attractor. The evolution of a composer like Schoenberg, however, highlights a different possibility as his development is characterized by a more radical change in compositional style from late Romantic music to dodecaphonic serialism. This could be thought of as a change (or addition) of attractors in the composer's musical worldview—that is, from the late Romantic attractor toward the "dodecaphonic basin." In line with our discussion of improvisers earlier, these two examples suggest the possibility of (at least) two general types of (stylistic) evolution: namely, gradual evolution and punctuated evolution.[12] Agents following gradual evolution remain inside (or create within) a given attractor (e.g., Vivaldi). By contrast, in punctuated evolution an agent goes through a phase of increasing system entropy before a bifurcation occurs, and, as a result, the agent exits the current attractor and enters, or indeed, "enacts" a new one (e.g., Schoenberg).

It is also important to note that, from the 4E perspective, such dynamical evolution is inextricably linked with an agent's history of coupling with the embedded environment.[13] This could involve macrolevel shifts whereby an agent's society comes in contact with or becomes more open to other cultures. Changes in worldview can also be associated with the emergence of new technologies (e.g., the pianoforte, electronics, the computer), sociocultural transformations (e.g., free jazz and the civil rights movement; commodification and mass production) or when artists push against the boundaries of existing macrolevel (cultural) attractors associated with aesthetics and practice. For example, consider how Beethoven's late string quartets were highly criticized by his contemporaries for their bold use of quasi-nontonal harmonic structures (Knittel, 1998), owing, perhaps, to a visionary shift in the musical worldview of the composer (brought about by decades of engagement with music) that many of his contemporaries found incoherent with their own. Today, these works are generally considered to be masterpieces and arguably paved the way for later developments in Romantic and posttonal music. One may trace the development of John Coltrane or Jimi Hendrix in similar ways. Both artists were embedded deeply within the musical society of the mid-twentieth century United States, and both played a key role in transforming that society. In the process, they expanded the sets of relationships within established genres of musical practice and enacted new forms of music making. They developed new basins of activity and understanding involving novel embodied patterns of musical action and perception. This resulted in new sonic affordances for the saxophone and guitar, respectively—new extended musician–instrument–ensemble relationships that, in turn, contributed to the evolution of the broader cultural portrait.[14] And so, while creative musical agents do enact musical actions that imply simpler attractors (e.g., playing a repeating rhythm) and move within more established basins of activity associated with a given style, they can also willfully destabilize such attractors. In doing so, they can (a) create music that approaches completely chaotic dynamics or involves stochastic or aleatoric processes (e.g., Iannis Xenakis, John Cage); (b) create music that involves the "collision" of multiple (e.g., stylistic, rhythmic, harmonic) attractors (e.g., Charles Ives and Ornette Coleman); or (c) coenact new basins of attraction characterized by new sets of bodily, sonic, and social relationships (e.g., through free improvisation; Borgo, 2005).

Transferring the properties of complex systems to music agents (which are complex, *living* systems) allows us to think of them as being able to function at the edge of chaos—a term that refers to their ability to initiate episodes of instability and to self-organize new relationships through adaptive goal-directed activity, resulting in coherent "outcomes" that are not completely predictable (Capra, 1996; Schuldberg, 1999; Strogatz, 1994). Moreover, such "outcomes" need not be understood as a "finished product." Rather, they alter the constraints of the environment, leading to new developments in the system. By this light, creative activity—on both historical and in-the-moment timescales—may be said to involve the enactment of ongoing recursive "feedback" and "feedforward" loops between embodied agents and the extended material, social, and cultural environments they are embedded in and actively shape.

Musical Learning as Embodied, Exploratory, and Creative

The view of creativity outlined previously extends the discussion of musical development and "teleomusicality," which was examined as a creative process of exploration that begins with the sensing body's perceptual openness to the world, its basic proclivity to reach out to and make sense of the world (passive synthesis). Accordingly, this approach also has strong implications for what musical learning involves as it highlights the central role played by a situated, engaged, and adaptive body in the emergence of new musical skills and knowledge. Likewise, these insights connect with the discussion in chapter 4, in which we considered the phenomenology of learning Ghanaian polyrhythms and how it required learners to explore and develop new bodily, instrumental, sonic, and social configurations. The idea of the creative musical body also resonates with the discussion of the "metaphorical mind" (see chapter 4), which, again, sees human development in terms of the enactment of meaningful patterns of action, cross-modal perception, and feeling that unfold through the dynamics of embodied exploration (the construction of "perceptual-behavioral forms" mentioned in chapter 3). A crucial point that emerges here is that, by an enactive light, cognitive development cannot be reduced to the reproduction of pregiven procedures, knowledge, and skills, or the accumulation of representations of a fixed "world out there." Indeed, this would deny the fundamentally creative self- and world-making nature of living systems, which is central to the enactive conception of autonomy. As Evan Thompson (2007) reminds us, "[a]n autonomous system becomes informed by virtue of the meaning formation in which it participates, and this meaning formation depends on the way its endogenous dynamics specifies things that make a difference to it" (p. 5). Importantly for our discussion, this means that musical learning is not best understood as a process that begins with the individuation and reproduction of facts and procedures imposed from the "outside." Rather, skills and understandings are acquired and developed in the sense that they are self-constituted by the entire living organism in its embodied relationship with the environment (which often includes collaboration

Creative Musical Bodies 183

with teachers and/or peers). Musical learning, in other words, is an ongoing process in which the learner maintains and develops a condition of self-realization in relation with their contingent milieu (instruments, teachers and peers, social and cultural dynamics, and so on). Here it may be useful to consider a couple of examples to help clarify what's at stake.

Let's begin by thinking of a beginner piano player named Leyla. Leyla loves the music of Bach and Mozart and wishes to become an expert interpreter of classical piano repertoire. She has a basic understanding of how the piano works and the sounds she can get out of it. She has also been exposed to a variety of music, including the genre she wishes to develop. Additionally, Leyla works with a teacher who offers advice on technique and interpretation. She has peers she practices with and accompanies, and audiences she performs for. Leyla's experiences of musical learning are constantly influenced by these personal, social, cultural, and material aspects over various timescales. And the skills she develops become meaningful through the constant integration of various overlapping aspects of this body-world interactivity. The metabolic, neural, muscular, emotional, and social engagement configurations she enacts form the basis for new possibilities for perception, action, and thought. In other words, Leyla's biological organization enacts a self-generated ecology for learning, whereby different components configure and calibrate themselves in relation to each other and to the musical environment in which she participates. Learning processes, then, are essentially determined by ongoing relational dynamics, for it is only through the adaptive unification of these components into various self-organized configurations that Leyla can generate new musical understandings and possibilities (i.e., skills) that may be reused, adapted, and further developed in creative ways.

While musical agents may engage in the conscious construction of representations to help them imagine possibilities, remember musical forms, and learn musical skills (Gruhn, 2006; Hallam & Bautista, 2018; Klahr & Wallace, 1976), these processes are first dependent on the development of the kinds of dynamical body-environment configurations discussed previously. In other words, the factors that contribute to the development of musical skill cannot be limited to behavioral responses nor to inner mental content. A developing musician's ability to make increasingly fluid and accurate movements, to become more proficient at adapting to unforeseen perturbances in the musical environment, and to push their music making in new directions begins with (and returns to) the body's capacity to configure itself in relation to the instrument (which itself becomes an extension of the body).

The dynamical processes involved in embodied learning connect with the idea of "skilled coping" (Dreyfus, 2013; Høffding, 2014). This term refers to an agent's ability to be engaged in contextually situated activities in which the actions and equipment involved become transparent with regard to the moment-to-moment goals and contingencies of the event. Musical behavior involves the development of repertoires of action and perception that can be pushed and pulled in various directions across modalities. These "perceptual-behavioral gestalts" (see chapter 3) are adaptively combined in different ways across individual and collective domains; they play out in the moment and depend deeply on the knowledge possessed by the body (Gallagher, 2005; Johnson, 2007; Sutton et al., 2011). As an experienced piano player performs, she does not need to think about each movement she produces. Nor does she need to objectively account for all the properties of her environment; doing so would impede her performance (Geeves et al., 2014; Montero, 2010, 2016). Instead, she must be able to enact a "flow," a seamless integration of bodily knowledge and thought-in-action (Csikszentmihalyi, 1988, 1990, 1996), in which the equipment and other agents she interacts with become dynamically integrated into her cognitive ecology. Developing high levels of adaptive skilled coping and flow also allows musical performers to take risks by introducing new elements into the system or by placing their musical body in new contexts and configurations that are resolved in the moment.[15]

Recognizing the centrality of the situated body for skilled coping does not mean that there is no thinking or reflecting involved (Geeves et al., 2014; Høffding, 2019; Kimmel et al., 2018; Montero, 2016; Sutton et al., 2011).[16] However, the previous discussion suggests that musical learning and performance contexts require a more dynamical and embodied conception of thought-in-action, in which more deliberative decision-making and problem-solving processes are enmeshed synergistically with movement, context, and feeling. In other words, the bodily knowledge associated with skilled coping and flow does not first involve abstract concepts nor does it represent or reproduce features of an objective outer reality that can be easily measured or categorized using direct description. Rather, it entails the direct interaction between the sensing body and world and the perceptual

Creative Musical Bodies

and behavioral dispositions that arise from such interactions. The development of these dynamics is explored in David Sudnow's *The Ways of the Hand* (1978, 2001), a self-reflective phenomenological description of how he learned to improvise at the piano. Sudnow's account focuses on describing his bodily engagement with the instrument, with a special focus on the emergence of coordinative fluidity in his hands. He examines how these corporeal developments, which often involve awkward and uncomfortable periods of instability, give rise to new musical possibilities and understandings (affordances), which cannot be abstracted away from the direct bodily interaction with the instrument and the emerging musical world he creates each time he plays.

Another example that highlights the primacy of the situated body for musical learning comes from a behavioral study (Schiavio & Timmers, 2016) that compared memory tasks across three groups of participants: pianists, musicians who do not play the piano, and people who do not consider themselves as musicians (they do not play an instrument or engage in musical practice). Participants from the three groups familiarized themselves with three relatively simple but tonally ambiguous melodies in one of the following learning modalities: (1) actually playing the keyboard, (2) playing the keyboard with no auditory feedback, (3) watching a video showing a hand performing the melodies, and (4) listening to the melodies. The ability to learn and memorize these melodies was measured by testing the participants' capacity to recognize them among slightly varied versions. Interestingly, the difference across groups was less significant than expected: pianists, musicians who did not play the piano, and nonmusicians scored equally well. What really mattered, instead, was the learning modality: regardless of their knowledge about music, participants who learned the melodies by actually playing the keyboard (with or without audio) performed better than the others in their task. By physically exploring the sonic environment and by manipulating its properties in ways that are musically meaningful (i.e., playing the keyboard), nonmusicians and nonpianists discovered how to make sense of (and memorize) the musical material in their own way. Pianists who had deeper knowledge of music for piano and who were familiar with its set of action possibilities did not appear to rely on this preexisting musical knowledge to achieve the task. Otherwise, they would have performed better than the other groups. Accordingly, it could be argued that, starting with their first musical experiences, musical learners do not first require sets of organized rules, nor internal models subserving decision making. Instead, they tend to systematically explore the resources of their bodily engagement with the musical environment to optimize their possibilities and to develop meaningful experiences and understandings (including those related to more "high-level" processes, such as memory or imagination). This arguably points to a learning mechanism common to both expert and nonexperts that, again, highlights the primacy of sensorimotor exploration. Both novices and expert musicians need to be motorically engaged with the music to facilitate learning.

Put simply, then, we argue that musical skills are acquired and developed in the sense that they are self-constituted by the entire living organism in its embodied relationship with the environment. This view recalls the dynamically relational model of creativity just discussed, which explores musical sense-making in terms of the interaction of neural, corporeal, and environmental trajectories as the agent self-organizes patterns of action and perception across these domains. Here, skill emerges through the significances and valences that are enacted in each musical action. Accordingly, learning can be advantageously understood as a *creative process* in which the entire brain-body-world network dynamically changes itself though the physical enactment of new relationships and configurations—which, again, involves the goal-directed process of introducing and resolving periods of instability in the system so that new possibilities may emerge. Therefore, musical learning reflects the biological principles of autopoiesis (self-making), autonomy, and sense-making associated with the enactivist approach.

Likewise, studying the development of musical skill could also draw on a 4E framework. Thinking again of Leyla, the dynamics that guide her learning may be explored in terms of her (*embedded*) social and cultural background; her developing relationships with peers, teachers, and instruments (*extended*); her (*embodied*) behavioral dispositions for movement (e.g., muscular linkages); her emotional life (e.g., valenced/affective motivations for action, perception, and meaning making); and how all of these dimensions and more dynamically interact in the ongoing *enactment* of Leyla's musical abilities and worldview.

We conclude this chapter by briefly considering one last example that highlights the creative and exploratory nature of musical learning in a social context. This concerns a recently developed community-based project at the University of Music and Performing Arts of Graz (Austria) led by doctoral student Andrea Gande and her supervisor, music educator Silke Kruse-Weber (Gande & Kruse-Weber, 2017). Meet4Music (M4M) is an informal pedagogical setting developed around the figure of a facilitator (an expert musician) who guides participants in sessions involving musical and dance improvisation. Importantly, the facilitator does not propose a specific set of rules but enables different coordinated possibilities for action and interaction that are explored by each participant in modalities that are meaningful to their personal and cultural backgrounds. Here, the specific boundaries between the participants become fluid and flexible, as they reciprocally adapt to the contingent demands of the musical moments. In the process, the participants collectively enact or, indeed, self-organize their own musical relationships and meanings. The sessions are open to everyone in the community, including recent immigrants and refugees. Qualitative research (Kruse-Weber & Gande, 2017; Schiavio et al., 2019) suggests that the open-ended and improvisational nature of this program provides a way for established residents and newcomers to interact and thereby develop shared musical goals, trust, and friendships even when communicating in spoken language is difficult or impossible—that is, the enactment of new, shared musical-social worldviews. This example is important because it brings together the social dynamics discussed in chapter 6, in which we examined the idea of empathy as dynamically interactive and unfolding process of participatory sense-making. It also highlights again the creative, dynamical, improvisational, and world-making nature of embodied human minds, and the positive role creative musical environments can play for human flourishing and well-being.

The M4M program also offers a positive response to critiques emerging from recent scholarship in philosophy of music education, which warn of the tendency to downplay musical knowledge and activities other than those associated with the Western canon and how this colonizing bias may promote forms of exclusion. Notably, this involves a shift away from traditional "work-based" pedagogical approaches, in which the focus was on developing the technical skills required for the reproduction of musical compositions, and where "creativity" was a domain (tacitly) reserved for the composer (Elliott & Silverman, 2015). Among other things, work in this area explores "non-formal learning" processes (Green, 2001, 2008), which involve the types of self-directed and collaborative development that occurs outside of the (formalized) school environment. In these contexts, learners develop a range of skills and understanding through engagement with improvisational, creative-exploratory activities. For these activities, the processes and outcomes are not strictly defined, and shared worlds of musical understanding are "enacted" cooperatively[17] and in ways that are relevant to their lives and the extended sociocultural environments they participate in (Burnard, 2012; Burnard & Dragovic, 2014). As we explore, these practical and ethical perspectives on musical learning both reflect, and can be refined by, connections with enactive cognitive science.

10 Praxis

In this book we have examined the various ways in which the life-mind continuity associated with the enactivist approach to cognition can help us understand the emergence and meaning of human musicality. In chapter 2, we considered how the living body is always engaged in the process of maintaining itself within a contingent environment. Living systems play an active role in their own cognitive development—they *enact* their own worlds of significance—and musicking is an important way that human animals do this. In chapter 3, we discussed how musical experience is rooted in a primordial openness to the world, a drive to move, perceive, communicate, and synthesize experience cross-modally. However, as we also noted, our early musical actions soon take on more focused and intentional characteristics as we build the perceptual and behavioral forms that guide meaning making. Such insights are developed in the phenomenological discussion of learning the Ghanaian polyrhythms in chapter 4; in chapters 5 and 6, in which we examined the extended and socially interactive nature of the musical emotion, communication, and meaning making; and in the last three chapters in which we explored the biocultural emergence of musicality in the human phenotype, infant musical development, and musical creativity in terms of the (co)enactment of new patterns of action, cross-modal perception, and feeling.

In the process, we explored how music is central to (the movements of) human life: to how we reach out to and make sense of the world. From the perspective of enactive cognitive science, our musical behaviors and experiences are a form of embodied sense-making: by engaging in musical activity, human beings self-organize patterns of action, form relationships with other people and things, contribute to the flourishing of their identity, and thereby bring forth worlds of meaning. As we have seen, this perspective challenges approaches to musical aesthetics that place the locus of musical experience and meaning within the formal elements of the musical work, as well as related conceptions of music as a "decorative art" or a "pleasure technology" with little or no meaning for human survival and well-being. Additionally, an enactive perspective reveals that musical development is not fundamentally based on the retrieval and processing of pregiven external knowledge, methods, and facts, nor is it wholly dependent on hardwired cognitive "programming" or genetic coding. Rather, musical life emerges and evolves through an ongoing engagement of the feeling body in transformative, "4E" interactions with the social and material environment.

In this chapter, we explore some practical and ethical implications of an enactivist approach to the musical mind with a special focus on education.[1] We begin by outlining the influence that Enlightenment-inspired conceptions of musical aesthetics have had on music pedagogy, and we consider how an enactive perspective can support and enhance the current shift toward more engaged, collaborative, and creative approaches to musical learning. To do this, we return to the Aristotelian perspective introduced in chapter 1 to discuss educational approaches informed by his idea of *praxis* and how this so-called "praxial" orientation (and music pedagogy more generally) is enhanced by enactive thinking. We conclude by offering a few final thoughts on how the "life-based" orientation toward education, music, and mind provided by enactive cognitive science can help to reveal new conceptions of human being-in-the-world that are relevant to some of the global challenges we face in the twenty-first century.

Music Education and the "Aesthetic View"

If musicality is continuous with our primary status as self-making, autonomous beings, then the ways we conceive of, and do, music education should reflect this existential reality. However, conceptions of musical development and learning that embrace the adaptive, creative, and improvisational nature of human being-in-the-world have often been overshadowed by Western assumptions that privilege written (score-based), composed works. Accordingly, Western academic music education (e.g., in both classical and jazz contexts) has often adopted a stance in which the main focus has been on imparting the technical skills required to produce faithful reproductions of works and styles through codified performance practice. Keith Sawyer (2007) comments that this perspective reflects an assumed hierarchy whereby "the creation of new music is almost exclusively associated with composers; and where a performer's primary role is to execute those compositions. In this division of labor, instrumentalists do not need to be capable of creating new music, nor do they need the correspondingly deeper conceptual understanding of music that underlies composition.... [I]n our culture's stereotypical view, we do not think their creativity is of the same order as the composer who generates the score" (p. 3).

These assumptions align with the reifying trends associated with the commodification of music in contemporary society: where music is often assumed to be an objective "thing" to be consumed or reproduced (rather than an embodied, interactive, relational, multimodal, sense-making activity). To be clear, the argument here is not that the study and performance of composed works should be eliminated from music education. The issue is that the often-tacit acceptance of this hierarchical conception of music tends to obscure a range of possibilities and meanings that can enrich musical experience of all kinds.

This hierarchical, "work-based" orientation tends to downplay the importance of situated and creative activities common in many forms of musicking around the world, activities that are currently given little or no attention in much contemporary classical music performance training (e.g., playing by ear, improvisation, collaboration, and the creation of original music; Creech et al., 2008; Elliott, 1995; Rodriguez, 2004; Small, 1998). As such, the adoption of a reproduction-based perspective in educational contexts may not encourage the development of unique ensembles and approaches to music making that reflect the day-to-day lives of individuals, social groups, and indigenous or marginalized cultures. It is thus claimed that the standard focus on Western musical works tends to leave little space for the development of personhood as well as the expression and exploration of difference and diversity (Elliott & Silverman, 2015; O'Neill, 2009). Other writers argue that the global dominance of the Western perspective amounts to a form of cultural and epistemological colonialism in music education, it maintains a kind of hegemonic status over local and indigenous musical traditions (Bradley, 2012; Imada, 2012), and it marginalizes certain musical practices (e.g., improvisation; Bailey, 1993; Nettl, 1974).

The growing prevalence of this hierarchical orientation over the late nineteenth and twentieth centuries also bolstered the "formalist" approaches to musical aesthetics mentioned earlier—where, as we noted, a proper aesthetic understanding of a musical work involves the ability to correctly perceive, understand, and reproduce the supposedly objective structural relationships encoded into the score by the composer. While this rather depersonalized orientation toward musical meaning, beauty, and form is now in decline in musicological circles, it has nevertheless exerted a major impact on music education, especially at the secondary and postsecondary levels, where music education was often seen as a form of "aesthetic education" (Reimer, 1989). The idea of "music education as aesthetic education" (MEAE) was presaged (e.g., Leonhard & House, 1959) and promulgated by a number of scholars (e.g., Knieter, 1979; Schwadron, 1966; Swanwick, 1979). However, its chief advocate is Bennett Reimer (1989, 2003), who was the first to proffer a detailed explanation of the foundations and practical implications of MEAE. Because Reimer was highly regarded in the field, his book quickly became the authoritative source on MEAE and the philosophical "bible" for many music education scholars and teachers for several decades.

MEAE rests partly on Leonard Meyer's (1956) concept of absolute expressionism, but much more so on Susanne Langer's (1953, 1957, 1958) version of this theory. MEAE remained essentially unchallenged until Bowman (1991) and Elliott (1991, 1995) unpacked and questioned many of Langer's theoretical claims and Reimer's adaptations of them. However, many music philosophers before and after Elliott and Bowman (e.g., Alperson, 1979; Berleant, 1986; Budd, 1985, 1989; Danto, 1986; Higgins, 2011; Kivy, 1980; Nagel, 1956; Sparshott, 1982) also analyzed and rebutted Langer's arguments. Nevertheless, possibly because of a long tradition of scholarly insularity in music education, all but a few music education scholars remained either unaware of or ignored the literature of music philosophy, not to mention music psychology. This was still largely the case until the early twenty-first century.

Given the importance of Langer's work for music education, a few of her main claims need review. Langer (1958) argues music education is "the education of feeling" (p. 8). Reimer (1970) rewords Langer's argument: "The major function of education in the arts is to help people gain access to the experiences of feeling contained in the aesthetic qualities of things" (p. 53). He continues, in a later work (1989), "Creating art and experiencing art do precisely and exactly for feeling what writing and reading do for reasoning" (p. 33). These claims depend on Langer's theoretical assertion that the constant flux of our feelings is not knowable until ordered or stabilized and humans do this by imposing symbolic forms on their experiences. She argues that music is suited to symbolize feelings because musical structures and emotional patterns have the same logical shape; they are isomorphic. "The tonal structures we call music bear a close logical similarity to the forms of human feeling . . . music is a tonal analogue of emotive life" (Langer, 1953, p. 27). By this light, music provides listeners with a cognitive representation of emotions, moods, mental tensions and resolutions, or a "logical picture" of sentient, responsive life, a source of insight. She concludes that "music is not the cause or the cure of feelings, but their logical expression" (1957, p. 218).

A major reason Reimer (1970, 1989) had a huge impact on the profession was his claim that "[m]usic education is the education of human feeling" (Reimer, 1989, p. 53). This one phrase provided music education philosophers and school music teachers with what they had craved for decades: an apparently unassailable justification for the inclusion of music in the school curriculum and a seemingly sophisticated academic vindication of the worth of their life's work. In this context, few music education scholars were able to (or wanted to) challenge Langer's assertion that music is neither the cause nor the cure of feelings or that music does not arouse actual emotions, feelings, or moods. Of course, these claims defy experience and deny what most people value about music making and listening: deep, direct, and immediate experiences of felt emotions but also a range of other experiences associated with our situated corporality (e.g., movement, empathy). Moreover, listeners arguably do not need "cognitive symbols" to gain an abstract "wordless" knowledge of how they feel because, as we have detailed, the life-mind continuity affords us a moment-to-moment experience of our bodies, emotions, and so forth, without any intervening cognitive symbols. And incidentally, we may ask, what are "educated feelings"? And how do people demonstrate or recognize such things? The latter are crucial questions for music educators because without persuasive answers, they cannot determine whether or not their teaching efforts have made any practical difference in students' lives.

Langer's theory, it may be argued, offers only a limited picture of musical experience when it ignores how musical consciousness involves an active engagement with the living situations in which music is made, enjoyed, and celebrated (see Davies, 1994, p. 134). Indeed, this rather detached conception of music as symbolic or representational of emotive life suggests that such aspects are to be found in the music itself—a perspective that flies against the embodied, socioculturally embedded, and relationally extended approach to musical experience we have been arguing for throughout this book. Likewise, philosopher Kathleen Marie Higgins (2011) observes that Langer's emphasis on music's alleged cognitive-symbolic significance reflects her approval of Hanslick's (Kantian) formalism, which Langer affirms in her *Philosophy in a New Key* (1957, p. 100).

Similarly, Sally Markowitz (1983) argues that the aesthetic paradigm directs listeners to act like musical microbiologists by placing the musical object against a blank background before concentrating on its structural properties alone (p. 34). In terms of classroom instruction (e.g., in the United States and Canada), MEAE meant that "general" (nonperformance) music educators focused on teaching students to "perceive and react" to the "aesthetic qualities" (structural elements and forms) of Western European classical masterworks by having students listen to recordings and follow numbered items listed on "perception charts" that matched cues on the recordings. Music students were taught to avoid being "distracted" by any relationships between pieces of music and the personal, social, cultural, historical, political, ideological, gendered, or other meanings or contexts in which performances, improvisations, musical-social rituals, and so forth, take place and are experienced. Consequently, the MEAE orientation provided little room for listening to, let alone making, the musics of non-Western cultures and popular music.

Rediscovering the Musical Body

Concerns over MEAE have been raised by a range of thinkers in musicology and music education who reassert the primacy of the situated, adaptive, and creative body for musical experience. Music educator Thomas Regelski (1998) argues that "[i]n contrast to the separation between mind and body that serves as the governing condition of disinterested aesthetic contemplation, we need to understand that music listening and all forms of music making depend on the human body" (p. 22). Tia DeNora (2000) agrees that the aesthetic concept of music omits a view of the body as a socialized entity that focuses "on what the body may become as it is situated within different contexts" (p. 75). Musicologist and critical theorist Susan McClary (2000) asserts that "the power of music . . . resides in its ability to shape the ways we experience our bodies, emotions, subjectivities, desires, and social relations. And to study such effects demands that we recognize . . . cultural constructedness" (p. 7). Thomas Turino (2008) argues that "music is not a unitary art form . . . this term [music] refers to fundamentally distinct types of activities that fulfill different needs and ways of being human" (p. 1). Ian Cross (2012) adds that if music is a worldwide activity, then we should not only replace the word *music* with *musics*, we also need to understand that pieces and styles of music make sense in relation to their cultural contexts: the "histories, values, conventions, institutions, and technologies that enfold them; musics can only be approached through culturally situated acts of interpretation" (p. 17).

Ethnomusicologist Bruno Nettl (2005) argues that the assumptions, claims, and criteria of the aesthetic view are deeply flawed because there are millions of musical styles in the world for which it makes "no sense at all" (p. 16). This is because the aesthetic view arguably downplays the numerous ways in which the social situatedness of music (including classical music) influence people's experiences of all types of music listening, music making, musical interpretation, expression, creativity, musical-emotional experiences, musical preferences, and musical-human values; it tricks us into assuming that music is all-one-thing, a "universal language." Similarly, musicologist Richard Taruskin (1995) explains how the aesthetic view fails to provide a full and accurate understanding of the "classical music" it was meant to explain, not to mention most other musics (p. 17). He writes that this orientation not only narrows dramatically our musical attitudes and perceptions but it also narrows and strictly "regulates" our "social [musical] practices" (p. 10). In short, Taruskin argues emphatically that the aesthetic concept and its corollaries are a "utopian lie" (p. 17).

In a nutshell, it is argued that the aesthetic view neglects agency and the situated, enactive, nature of musical experience. It therefore tends to reinforce an essentially anonymous, depersonalized status for the musician, listener, music educator, and student, reducing musical learning to a process of acquiring technical and theoretical knowledge, in which the body is little more than a tool that is trained to perceive and perform "correctly." Music educator Wayne Bowman (2004) warns that this tendency to abstract musical experience from the contingencies of life limits it to "a psychologistic affair, purged of things like muscle, blood, bone, struggle, power, politics—in fact, most of the things that make it momentous. . . .

This leaves the body in an awkward place, if any place at all, and neglects music's status as cultural action. Foremost among the reasons music truly matters educationally is its participatory, enactive, embodied character—and its consequent capacity to highlight the co-origination of body, mind, and culture" (p. 46).

While we can certainly use music to symbolize or represent aspects of life and culture, its significance for human existence goes much deeper than this. As we have seen, music is not simply a product for production and consumption; it is not merely an "analogue" for emotional life (an "emoticon"; Bertinetto, 2020), nor is it an object of aesthetic analysis. Music is a cognitive domain we live through, a human sense-making system with deep ontogenetic and phylogenetic roots; one that integrates action, feeling, and thought across personal, social, cultural, and technological dimensions.

It is interesting to note how the contrast between the aesthetic view and the enactivist orientation echoes age-old debates over the nature of musical experience. As we saw in chapter 1, the Aristotelian perspective adopted by Aristoxenus counters the abstract Platonic-Pythagorean orientation by placing the locus of musical experience in the active, sensing body. Likewise, Mark Johnson (2007) notes that the embodied, cross-modal, and "metaphorical" approach to mind and meaning we discussed in chapter 4 demands an understanding of "aesthetic experience" that goes much deeper than the detached, rationalizing, and analytical approach inherited from Enlightenment thinking. This brings us closer to the original Greek notion of *aesthesis*, which is grounded in the senses (see also Dissanayake, 1995; Shusterman, 2008). Additionally, the way we construct meaning through *aesthesis* involves active histories of (4E) sense-making within the sociomaterial environment: "Meaning includes patterns of embodied experience and pre-conceptual structures of our sensibility (i.e., our mode of perception, or orienting ourselves and of interacting with other objects events, or persons). These embodied patterns do not remain private or peculiar to the person who experiences them. Our community helps us interpret and codify many of our felt patterns. They become shared cultural modes of experience and help to determine the nature of our meaningful, coherent understanding of our 'world'" (Johnson 1987, p. 14).

These insights into the socially embodied nature of human sense-making resonate closely with another ancient idea that is useful for thinking about what an enactivist approach to music education involves: Aristotle's conception of *praxis*. *Praxis* has been discussed and developed by a range of thinkers.[2] However, for our purposes, we focus here on interpretations and uses of the idea by philosophers of music education (Bowman, 2007; Dunne, 1993; Elliott, 1995; Elliott & Silverman, 2015; Regelski, 1998, 2002, 2012) and especially those who make close connections between *praxis* and enactivism.

Praxis, Enaction, and Music Education

Praxis is generally understood to comprise four modes of knowing. These modes are referred to as: (1) *techné* (the technical or procedural knowledge associated with production), (2) *theoria* (theoretical knowledge), (3) *phronēsis* (situated-embodied or practical-ethical knowledge), and (4) *poiesis*, which refers to the activity of production itself and is therefore intimately involved with *techné* (Elliott & Silverman, 2015). These elements constitute the key epistemological components of *praxis* as they make up the forms of knowledge required to initiate positive transformations in the world. Importantly, *praxis* entails not only technique, production, and theory but also the forms of right action that lead to human well-being (*eudaimonia*). In addition to being an epistemological framework, *praxis* has profound ethical implications.

The ethical dimension of *praxis* is expressed most clearly in the idea of *phronēsis*, which concerns our direct involvement with life and the fundamentally caring way we orient ourselves in the world. The knowledge and action associated with *phronēsis* are inherently embodied, affective, and social; these forms of understanding include the ongoing development of pragmatic "knowing-how" that takes relevant circumstances into account. As such, *phronēsis* embraces the deep continuity between embodied action, imagination, and thought—between movement, empathy, affectivity, feeling and motivation, and how we frame the world in rational and ethical terms. It spans and connects the dimensions of human being-in-the-world, and it connects with the primary self-organizing and improvisational capacities of the embodied mind—those that enable "knowing how to negotiate our way through a world that is not fixed and pre-given but that is continually shaped by the types of actions in which we engage" (Varela et al., 1991, p. 144). *Phronēsis* is therefore central to how we enact meaningful relationships within the contingencies of life, which is crucial for well-being and the authentic bringing forth of the self (autopoiesis).

Phronēsis gives living, contextual meaning to technical and theoretical knowledge and guides the various forms of production associated with *poiesis*; it refers to the ways we open up to and make sense of the world and may therefore be cultivated into forms of reflection that reveal the richness of being in a given situation.[3] Importantly, when a focus on *techné* obscures the other three elements, all connection to *praxis* is lost. In this case *techné* is stripped of its ethical responsibility, as "technical skills are not, by themselves, individuating, self-actualizing, creative, or personal growth experiences" (Elliott & Silverman, 2015, p. 46). Put simply, then, a praxial approach to music education seeks to foster a healthy integration of *techné, theoria, poiesis,* and *phronēsis*. In doing so, it embraces a much richer set of possibilities for what music can mean in human life.

Enactivist thinking offers a useful biological grounding for praxis (and specifically *phronēsis*) when it describes the origins of mind and selfhood in terms of the self-generating activity of living creatures as they continually strive to enact sustainable and flourishing relationships with the contingent environments in which they are embedded.[4] Likewise, the idea of *praxis* can help to integrate enactivist biocognitive principles within the domain of thought, production, and the kinds of right action that lead to human flourishing. In light of this, scholars are now drawing on enactivist principles to refine praxis-based pedagogical models (Elliott & Silverman, 2015). Notably, philosopher and music educator Marissa Silverman (2012, 2020) develops important connections between praxial and enactive/4E perspectives, with a special emphasis on how open-ended and collaborative music education environments can afford opportunities to explore relational, empathic sense-making dynamics in ways that encourage human flourishing or *eudaimonia* (see also Smith & Silverman, 2020).

A central aspect of Silverman's work involves articulating the positive ways musical praxis can highlight the emergence of personhood within a sociomaterial network: "[P]ersons emerge—and understand themselves as persons—and are enacted because of the dynamical syntheses of our many embodied processes that are in/of our worlds . . . instrumental musicing is something worth doing for the sake of the self and others" (Silverman, 2020, pp. 5–8). These insights draw on the relational conceptions of selfhood, autonomy, and agency central to enactivism. We developed this in chapter 6 and in the previous chapter, in which we explored the embodied and "cooperative" ways musical agents bring forth meaning and identity as they reach out to their social world, and how, beginning in infancy, our musicality provides numerous interpersonal affordances for such forms of sense-making. As we also considered, the development and well-being of intersubjective musical ecologies is not based in some facile sense of consensus or conformity. Instead, it involves negotiating the dynamics of difference inherent to complex systems comprising directly interacting autonomous agents and how this negotiation unfolds over time, resulting in shared forms of action-as-perception (e.g., the improvising keyboard players in chapter 9; or the discussion in chapter 6 of how ensembles develop extended musical worlds through the enactment of musically scaffolded empathic spaces).[5] From a 4E perspective cooperation emerges from the self-organizing embodied-affective interactions of the individuals involved, and the shared needs, desires, and actions that result from such contingent intersubjectivity (*phronēsis*). In this way, a social group (e.g., a music class) may be understood to enact their own goals and ways of coordinating action through extended adaptive processes.[6]

The praxial approach to music education aims to encourage these creative social and ethical sense-making potentials and the diverse ways they play out in the lives of individuals and social groups. Not surprisingly, the kinds of learning environments praxial educators seek to foster do not rely on traditional musical hierarchies, authoritarian models of education, and teaching through a detached "aesthetic view." Rather, praxially minded teachers aim to help students reveal possibilities for musical development that are relevant to their lives and their shared concerns for being and becoming musical—technique, theory, and production are guided by ethical sense-making (*phronēsis*).

Relationality and the Ethics of Care

Accordingly, Silverman (2012, 2020) and other thinkers (Gilligan, 1982; Held, 1993; Noddings, 2012) draw on "care ethics" to help develop relational approaches to education: "In care ethics, relation is ontologically basic, and the caring relation is ethically (morally) basic. Every human life starts in relation, and it is through relations that a human individual emerges. . . . Care ethics emphasizes the difference between assumed needs and expressed needs. From this perspective, it is important not to confuse what the cared-for wants with that which we think he should want. We must listen, not just 'tell', assuming that we know what the other needs" (Noddings 2012, pp. 771–773).

A care ethics approach builds on the fundamentally coenacted, bidirectional nature of empathic communication we examined in chapter 6. Accordingly, this perspective eschews pedagogical models in which what is taught (and its significance) is externally imposed on the pedagogical system and where the "how" of teaching involves "blind faith in and devotion to a technicist method" (Regelski, 2002, p. 111) or some kind of curriculum for all students everywhere (Noddings, 1995). In light of this, educator and philosopher Nel Noddings (1984) argues that the teacher has two major tasks: "to stretch the student's world by presenting an effective selection of that world with which she is in contact, and to work cooperatively with the student in his struggle toward competence in that world" (p. 177). As she conceives of it, a caring relationship emerges when teachers take responsibility for and attend to the needs of the other, when we perceive students in purely individualistic terms, but as relational and interdependent beings. Noddings argues that the carer seeks out the needs of the cared-for. Both roles (carer, cared-for) are equally important. Without recognition by the cared-for, the caring relation is incomplete: "How good I can be is partly a function of how you—the other—receive and respond to me. Whatever virtue I exercise is completed, fulfilled, in you. The primary aim of all education must be nurturance of the ethical ideal" (Noddings, 1984, p. 178).

So, in addition to helping develop the technical and theoretical knowledge by which students can realize the personal and collective forms of musicking that distinguish the praxes of different agents (e.g., between different jazz, or hip-hop, musicians), a praxial approach requires that teachers are mindful of the personal and collective impact their teaching decisions have on students. A major reason for emphasizing this point is because teachers' and students' musical actions are not automatically "right actions," or musically expressive and creative. They depend mainly on teachers' ongoing commitment to thinking reflectively about why and for whom they are teaching music. Philosopher Hannah Arendt (1958) emphasizes that because praxis is inherently social—because it always occurs within a complex web of human relationships and individuals' needs and desires—the outcomes of praxis can never be predicted with certainty. Thus, musical and educational praxes require teachers to make highly informed diagnoses of students' needs and goals and to determine how to plan, act, and achieve musical and educational outcomes in/for learning situations (Carr & Kemmis, 1986, p. 190). Praxis is of such a nature, however, that even these "oughts" are provisional, contingent and interim "theories" that need to be constantly reconsidered in light of experiences and the changing particulars of the individuals, groups, and contexts involved (Regelski, 1998, p. 29).[7]

A teacher's knowledge (e.g., why-to, what-to, how-to, when-to, where-to, how much-to, and so on) is guided by a "feel" or a "practical sense" (*phronēsis*): such understandings of how a musical situation goes, should go, or will go as it unfolds, is informed by the history of embodied experience that the teacher develops over time and in repeated engagements with a chosen endeavor (Regelski, 2004). The development of this kind of situated knowledge requires ongoing reflection on the phenomenology of teaching to understand what activities, environments, and relationships afford positive growth experiences for learners. In connection with this, a 4E framework can offer a useful way of examining musical development and teaching practice. For example, educators and students may be asked to reflect on questions such as the following:

Embodied: What new instrumental challenges have emerged and what new body-instrument relationships and understandings have developed in the process of meeting them?

Embedded: How have our musical activities explored and developed our understandings of the broader physical, sonic, historical, social, cultural, and gendered world(s) we live through as individuals and social groups? What roles does the sociocultural environment play in shaping the ways we compose, improvise, and collaborate?

Enactive: What new meanings have we opened up through our music making? What new relationships have emerged? And how have they transformed the ways we engage with the world musically, sonically, socially, emotionally, and so on?

Extended: In what ways have creative possibilities been enhanced or made possible through my interactions with other people, technologies, and other nonorganic ecological factors? And how have I helped to facilitate the creative development of others? (van der Schyff, 2019, p. 333)

Of course, these kinds of questions can be refined and reformulated and in various ways. For example, a 4E model may also be used to examine situations in which students are struggling or to identify social and/or material factors that may be negatively impacting the pedagogical environment.

Implications for Curriculum

The introduction of praxial and enactivist thinking opens a pedagogical perspective in which people become socially and ethically responsible for their own learning and teaching processes, which requires the development of an engaged, critically reflective attitude in teachers and students. As Ezequiel Di Paolo (2020) writes, "Critical education is a mutual transformation of teachers and students via concrete problem solving, the breaking down and re-constitution of given wholes, and sustained dialogues that create the conditions to change concrete realities and open new possibilities for action in our own personal and community processes of becoming" (p. 6).

With this in mind, praxial and enactivist inspired learning environments may vitalize more codified forms of musical practice (e.g., the performance of classical music) by highlighting the value of critical reflection, risk taking, and relevant, situated interpretations that connect directly with the lives of creator-performers and the communities they participate in. However, a praxial-enactivist approach also implies the exploration of other forms of creative musicking that engage the possibilities of individuals and group in different ways. These forms involve improvisation and experimentation, the exploration of wider cultural perspectives and practices, the creation of original music, as well as the development of creative collaborative projects that *decenter* a Western academic orientation (e.g., Korsyn, 2003; Powell, 2005). Advocates of ethnomusicological pedagogy have shown that such elements may be introduced early on to foster open, culturally aware, and creative attitudes in children (Campbell, 2009). Likewise, a number of highly promising possibilities for new curricula exist in marginalized music practices associated with free improvisation and experimental music (see van der Schyff, 2019).

To take one example of the latter, James Gutierrez (2019) posits a novel approach to music theory pedagogy that draws on a 4E framework and the conduction techniques developed by improvisor Lawrence "Butch"

Morris, which involve a flexible lexicon of gestures that are used to guide an ensemble improvising in real time. As Gutierrez explains, "[S]tudents bring their instruments to class, form an ensemble, and take turns using signs and gestures to conduct their peers" (p. 1). These gestures indicate the use of various music theory concepts "aligned with learning objectives (e.g., harmonic minor scales, Neapolitan chords, or polytonality)" but also allow students to collaborate and "experiment with musical structure in situ, with minimal or no reliance upon notation" (Gutierrez, p. 1). In this way, students can creatively and collaboratively engage with theoretical ideas in action by drawing on the resources of the extended musical environment they are embedded within. This perspective provides a useful shift from the rather abstract and disembodied orientation that characterizes much music theory education as students are able to develop theoretical concepts within an interactive praxis-based context.

Gutierrez's approach connects with the ways that some forms of early childhood and community music teaching and learning environments engage *praxis*, embodiment, and enaction as they incorporate direct corporeal engagement with the sociomaterial environment. Musical development, understanding, communication, expression, and meaning making are manifest first in the self-organizing forms of action-as-perception enacted by students' environmentally situated musical bodies, including, of course, their singing voices, their listening, moving, dancing, and interacting with musical objects. Taking these possibilities further, educator-researchers (e.g., Greenhead, 2017; Juntunen, 2016; Juntunen & Hyvönen, 2004; Wentink, 2017) are using phenomenological and 4E approaches to examine how activities and concepts associated with Dalcroze Eurhythmics (Jaques-Dalcroze, 1914/1967) can be developed within and beyond early childhood music education. As Rosalind Ridout and John Habron (2020) explain,

> Dalcroze Eurhythmics is an approach to teaching, learning, and understanding music through exploring various music-movement relationships in social, creative, and rigorous ways . . . Dalcroze Eurhythmics is a practical and holistic approach to music education. It is practical in that it combines sensing, action, feeling and thinking. . . . It is holistic in at least two senses. First, as a philosophy of music education, Dalcroze Eurhythmics works to overcome the dualism of mind and body . . . with practitioners choosing instead to emphasize and work with the entire body of the living system. . . . Second, as a pedagogy, it is experienced not only as music education, but also a means to social integration, personal transformation and well-being. (p. 2)

Ridout and Habron employ the lenses of phenomenology and 4E cognitive science to develop a Dalcroze-inspired approach in the context of learning, rehearsing, and performing contemporary music:

> The radically cross-modal nature of Dalcroze pedagogy finds theoretical support in current developments in cognitive science. . . . With the so-called 4E model of music cognition, we can consider how Dalcroze pedagogy affords (i) embodied experiences of music involving the body, as participants move, feel, listen, think, and make contact with each other. . . . These experiences are (ii) embedded within the "agent's ecological niche" . . . which consists of elements such as space, the acoustic, other people, and the cultural context. Thus embedded, Dalcroze students develop their knowing through an (iii) extended cognition, interacting with other agents (students and teacher) and objects that become part of the musician's "cognitive ecology." . . . These objects can include the equipment mentioned above, used to explore the world and receive feedback, and afford knowledge that would not be achievable solely by mental processes. Finally, the (iv) enactive dimension concerns how living systems and their environments mutually shape each other and is most easily seen in improvised settings. (p. 2)

According to Ridout and Habron's cross-case analysis, this approach afforded a number of positive insights for developing performers, which are categorized in eight main themes: body and breath; the body as a "way in"; learning through the body overcomes specific technical difficulties; an embodied relationship with the score; deeper knowledge and connection to music; clarifying one's own interpretations; communication with the audience; a bigger (sociocultural/ethical) picture beyond the instrument.

Embodied, interactive, and collaborative forms of musical *praxis* also characterize many community music environments, such as the Meet4Music program introduced in the previous chapter. As we discussed, this program provides positive social experiences for all involved, including recent immigrants and refugees, through forms of improvisational practice—in which participants communicate their own musical voices and develop shared forms of music making that are unique to the people involved. Schiavio and colleagues (2019) used an enactive model to better understand the interpersonal dynamics involved in contexts like this, exploring qualitative coding categories such as "collaboration," "non-verbal communication," and "sense of togetherness." The study showed that environments that embrace music's potentials as a participatory sense-making capacity can afford embodied and emotional forms of communicating (including expressive movement) and can thereby foster friendships and new communities of practice. Importantly, this research suggests the tremendous possibilities that improvisational forms of musical *praxis* have for fostering an openness to difference and for initiating new shared cultural perspectives.

As Silverman (2014) notes, real-time musical collaboration depends on the remarkable fact that "self and other are able to enact each other because a human being's first-person sense of him/herself has the capacity to recognize him/herself relationally . . . as another" (p. 2). And indeed, safe environments that encourage the self-organizing potentials of musical agents (and extended musical systems) through improvisation and creative collaboration can offer transformational experiences that begin to dissolve acquired fears and dehumanizing assumptions associated with social and cultural "othering." As such, musical praxis can help in developing intersectional "meeting places" for individuals to come together from different stylistic, gendered, and cultural backgrounds to experience each other, imagine together, enact empathic spaces, develop shared understandings, and form new "hybrid" musical cultures (Bhabha, 2004; Greene, 1995; O'Neill, 2009; Powell, 2005; Powell & Lajevic, 2011).

Critical Ontology

The ideas and research outlined previously indicate only a few of the possibilities enactivist thinking can have for music education and community music. Theorists and practitioners are developing enactive and 4E concepts in the domains of music technology (e.g., Tuuri & Koskela, 2020), music composition (e.g., Pohjannoro, 2020), music therapy (e.g., Schiavio & Altenmüller, 2015), and more. But despite this growing recognition of the importance of music for development, socialization, health, and well-being more generally, music education finds itself in an increasingly precarious position in today's globalized free-market economy. In response to this situation, music's role in education has often been conceived of, and advocated for, in terms of its putative value for other nonmusical aspects of life. Claims have been made, for example, that training in music can positively impact test scores in math and science and that it can provide models for teamwork that result in more efficient corporate work environments. While such claims have been supported and critiqued in various ways, a number of thinkers (see Elliott & Silverman, 2015) have argued that regardless of how well intentioned these forms of advocacy may be, they are problematic when they simply reflect the values of a neoliberal corporate worldview and miss the deeper meaning of musicality for human life.

The shift in perspective associated with enactive cognitive science and praxial music education has implications for the meaning of music that go far beyond its value simply as a means to an end (or the assumption that music is something once removed from our lives, an aesthetic representation of emotional life). By an enactive light, the embodied, relational, adaptive, and self-organizing nature of autonomy and agency in living systems becomes the essential ontological guide for an ethical music education. As such, teachers, students, and embodied musical agents more generally are not conceived of as mere responders and reproducers of information but as active (participatory) sense-makers—as self- and world-making beings. This is to say that recognizing the life-mind continuity and the relational nature of (participatory) sense-making that is so well exemplified in human music making has profound existential implications for how we may reconceive of human being and knowing, as well as our relationship to the "natural" world that sustains us. This connects with the work of thinkers across a range of domains who have examined how the focus on technological and economic progress, colonization and globalization, the bureaucratization of work and culture, and the broader commodification of life have resulted in a reduced ontology—one in which the meaning of people, animals, trees, rivers, and oceans, and "nature," more generally, is prescribed by a corporate resource mentality (Evernden, 1993; Heidegger, 1982). Put simply, it is argued that this orientation has imposed a set of ontological assumptions that have led to a worldview in which people, and "nature" more generally, are viewed as "resources" to be exploited for economic gain.[8]

This impoverished ontology extends to all areas of life, including music (and its commodification) and education (Lines, 2005, 2015). As a number of pedagogical theorists have noted, education in the industrial and postindustrial eras has increasingly focused on training people to work within the neoliberal free-market economy and therefore tended to view learners as receivers, processors, and reproducers of information. Keith Sawyer (2003, 2007) refers to this as an "industrial" approach that plays out in a decontextualized and compartmentalized pedagogical culture in which students are trained to memorize and reproduce existing knowledge so that they may standardized testing criteria. This orientation, he argues, essentially involves a "production line" intended to produce workers ready for the corporate techno culture—a prescriptive and "mechanical" approach that affords little practice in developing "the deeper conceptual understanding and adaptive expertise that allow [for the generation] of new knowledge" (2007).

This "industrial attitude" tends to strip living agents of their ontological status as autopoietic and autonomous beings; it reduces the living to the artificial and thereby imposes a kind of "false consciousness" (Eagleton, 1991). Interestingly, this insight also recalls the Aristotelian ontological distinction between the "living" and the "made" we introduced in chapter 1, in which the former is self-making and intrinsically meaningful and the latter is meaningful only instrumentally (in terms of its uses for other agents). Again, made entities (e.g., computers, ships, houses) have their ontological footing outside of themselves, they are not self-organizing and have no "care" for their existence or for that of others. As we saw in chapter 2, a similar distinction is also at the heart of the enactive conceptions of "operational closure," which describes the self-generating bounded metabolism and the forms of affectivity and agency that characterize the relationally autonomous nature of living systems. By contrast, artificial entities are "operationally open" since their coming into being and operations are not self-initiated nor self-relevant.

The important point here is that these kinds of distinctions bring the self- and world-making nature of living cognition to the foreground as an ontoethical first principle. And indeed, the life-based perspective associated with enactivism—and "life philosophy" more generally—has influenced a number of critical thinkers (Bai, 2003; Bateson, 1972; Miller, 1997; Nakagawa, 2000) who seek to identify and dismantle the assumptions and power structures that coerce and constrain human life. Notably, the late pedagogical theorist Joe Kincheloe (2003) argues that we require a reevaluation of what human being and knowing entails—a "critical ontology": "[A] critical ontology positions the body in relation to cognition and the process of life itself. The body is a corporeal reflection of the evolutionary concept of autopoiesis, self-organizing or self-making of life. [I]f life is self-organized, then there are profound ontological, cognitive, and pedagogical implications. By recognizing new patterns and developing new processes, humans exercise much more input into their own evolution than previously imagined. In such a context human agency and possibility is enhanced" (p. 50).

Interestingly, Kincheloe (2008) relates how his thinking was transformed by his early encounters with a community of improvising jazz and blues musicians. Here, he discovered a way of learning and "becoming together" centered around a shared activity, a musical *praxis* that was shaped by the participants themselves, who learned with and from each other.

A central aspect of critical ontology as conceived of by Kincheloe and other thinkers involves breaking free from the "cognitive enclosure" of a colonized mind, a mind that is limited or confined by sedimented ontological, epistemological, and moral assumptions imposed by the dominant power structure (Giroux, 2011; Ng, 2020; Werner, 2020). A relational, 4E conception of mind and self can help to open new possibilities for experience and action that embrace the dynamics of difference and diversity in the extended sociocultural ecology and the positive transformations for perception and thought this can initiate. This fluid and extended conception of human being and knowing can help to loosen categorical, rigid, and hierarchical assumptions imposed by a colonizing mentality.

Breaking up such sedimented attitudes invites opening up to indigenous and other marginalized ways of knowing (Batacharya & Wong, 2018), which can help us to move beyond alienating notions of autonomy and the "disenchanting" machine metaphors that dominate conceptions of mind and life in the modern world (Thompson, 1998; Wexler, 2000). In line with this, an enactive perspective can also help us to reconceive of the cultural and educational institutions we have constructed for music and other domains (see De Jaegher, 2013). Such institutions are characterized by certain engrained modes of behavior and thought that are often thought to exist "in a special normative realm independently of the actual lives of people" (Torrance & Froese, 2011, p. 46). And this reification makes it difficult to see how our institutions and received ways of thinking and acting could be criticized or changed. Enactive perspectives, by contrast, explore how normative contexts "are embedded in the ways people conduct [their] lives—their continued existence requires that they be continually (inter-) enacted, in either word or deed" (ibid). Indeed, the enactive idea of relational autonomy and social cognition highlights the origins and potential *fluidity* of normativity in the complex embodied, contextual, and cooperative processes associated with participatory sense-making (De Jaegher, 2013; De Jaegher & Di Paolo, 2007). Opening up to these possibilities can help in critically examining our institutions and in initiating positive changes that reflect the lives and needs of the people who use them. In a similar way, the relational conception of selfhood associated with enactivism, care ethics, and musical praxis can also help us rethink democratic ideals such as "freedom," which, in this light, becomes not something "I possess" or "I deserve" but is something we do together and for each other.

Expanding on these ideas, Kincheloe (2003) offers 23 basic ideas that underpin the development of critical ontology. These ideas are framed in terms of specific needs related to "conceptualizing new, more just, and more complex ways of being human" (p. 1). They include the need:

- to move beyond mechanistic metaphors of selfhood.
- to appreciate the autopoietic (self-producing) aspect of the "self" to gain a more sophisticated capacity to reshape our lives.
- to understand the importance of sociohistorical consciousness concerning the production of self.
- to recognize dominant power's complicity in self-production vis-à-vis ideologies, discourses, and linguistics.
- to conceptualize new ways of analyzing experience and apply it to the reconstruction of selfhood.
- to become cognizant of the cognitive act as the basic activity of living systems—the process of establishing relationships and new modes of being.
- to grasp the notion that this ontological process of cognition constructs the world rather than reflecting an external world already in existence.
- to realize that the nature of this world, the meanings we make about it, and our relationships with it are never final—thus, humans are always in process.
- to see that the self is not preformed as it enters the world—that it emerges in its relationships to other selves and other things in the world.
- to realize that the nature of the interactions in which the self engages actually changes the structure of the mind. (pp. 1–2)

Since music spans such a wide range of what human sense-making entails, it offers unparalleled possibilities for exploring these ideas and for negotiating (and celebrating) the difference and diversity that characterize our lives inside and outside the institutions we enact. As such, critically aware forms of musical *praxis* have enormous potential to contribute to an "ontological education" when musical learning is situated within the wider ecological, sociopolitical, and economic concerns related to the meaning and future of human being-in-the-world.

Indeed, musical sense-making can provide new perspectives on life, mind, and culture when performers, composers, listeners, students, and teachers reach out to the multiple "voices" in urban and natural worlds through forms of empathic, relational, and critical listening. For example, Dylan Robinson (2020) discusses the important critical insights offered by the forms of "hungry listening" associated with indigenous sound studies and how these can result in richer conceptions of identity and new understandings of the sociocultural dynamics both within and between indigenous and settler communities. Along these lines, philosopher Freya Mathews (2008, p. 53) points out how embodied, empathic, and synergistic activities such as musicking may open us up to a deeper compassionate relationship with all life and thus help "induce in humans a moral point of view with respect to other-than-human life forms" (see also Nollman, 1990, 1999; Rothenberg, 2005, 2014; Rothenberg & Ulvaeus, 2009). She notes that in music and arts pedagogy contexts, students and teachers may be "encouraged to identify imaginatively with wider and wider circles of the [sonic] landscape, until, hopefully, the students acquire an expanded sense of identity, described in the deep ecology literature as the 'ecological self'" (Mathews, 2008, p. 54; see also Mathews, 1991). In this way, our musicality can also help reawaken us to the deep continuity between natural and human worlds (Shevock, 2018).[9] Exploring the diverse ways musical *praxis* is enacted within the extended material, social, cultural, and biological environments we are embedded within can offer important contributions to these kinds of projects by rerevealing the relational interconnectedness of our bodies, ourselves, and the world we inhabit and shape. As such, an enactive approach to the musical mind is also very well positioned to contribute to the transformations in culture and consciousness required for the survival and flourishing of our species and the realization of a more compassionate and ethical world.

Parting Thoughts

In this book we have outlined a conception of musicality as a human sense-making capacity that is continuous with the fundamental self-organizing processes of life. We have examined how musical sense-making extends into different of human existence. And we have argued that the enactive perspective allows for richer, more holistic perspectives on the nature of musical experience, in which body, brain, and environment are integrated as an evolving extended system. Importantly, the lenses of DST and the 4E framework have enabled us to offer accounts of musical creativity, musical development in ontogenesis and evolution, musical emotion and empathy, musical perception and aesthetics, and musical learning that challenge and/or extend previous models. And as we have just discussed, an enactive approach to the musical mind also brings to light a number of critical and ethical concerns; highlights the fundamentally relational nature of human autonomy and sense-making; asserts the self- and world-making status of living agents; and grounds these aspects in bioethical principles associated with human flourishing and well-being.

In all, we have attempted to provide a wide-ranging account of the enactive musical mind, one that is as inclusive as possible with regard to the interdisciplinary trends that concern current music and biocognitive studies. This said, the view on human musicality offered in this book is not intended to be an exhaustive account of what an enactive/4E approach should or could involve. For one thing, many of the models and insights we have employed originated in nonmusical domains (e.g., affective science and social cognition). And while these perspectives are highly relevant to understanding musical cognition, it will be very interesting to see how insights derived from musical research will, in turn, impact enactivist theorizing.

This shift is arguably beginning to occur. For example, cognitive scientists Steve Torrance and Frank Schumann (2019) have recently considered how the interactive and adaptive dynamics of musical improvisation may inform the understanding of cognition from a 4E perspective. Similarly, we suggest that collaborative forms of musicking could have a great deal to offer to second- and third-wave perspectives on the extended mind thesis. As we have touched on many times in our discussion, joint music making involves a range of adaptive sociomaterial agent-environment integrations whereby the extended mind can be explored in relation to technology (e.g., instruments, digital devices for recording and sound manipulation, telecommunications), social dynamics, empathy and emotion, learning and developmental environments, and more. Indeed, an important (and defining) concern of third-wave theorizing involves examining more closely how extended engagements unfold in sociocultural contexts and how such engagements initiate changes to neural structures and body-world configurations (Kirchhoff, 2012). Accordingly, it would be very interesting, for example, to see how research in cross-cultural musical environments (e.g., the M4M program) could inform extended mind theory—for example, how third-wave approaches may connect with critical and social perspectives in current music research.

In line with this, we would also like to reiterate that the field of music therapy is especially well suited to engage with the 4E perspective we have introduced in this book (Maiese, 2020). As we touched on in chapter 6, music therapy interventions often highlight the capacity of neural structures to reshape themselves and develop in novel ways through guided interactions with extended, musically scaffolded environments. Music therapy also offers numerous examples of how such environments can foster improved social dynamics and how they can promote (corporeal and emotional) self-regulation and well-being more generally. Therefore, while music therapy has not yet made strong connections with 4E cognitive science, it nevertheless shows great potential to become an important source for enactivist and 4E theorizing. In turn, the 4E approach may provide useful frameworks for research and analysis in music therapy—for example, for constructing questionnaires, developing data coding categories, and as a theoretical lens for interpreting results.

With such possibilities in mind, we hope that the account we have provided here will inspire new connections between diverse domains of research and practice. Indeed, the merging of perspectives from interdisciplinary musicology and the sciences of life and mind stands to illuminate issues that extend beyond the concerns specific to these areas of inquiry. As we have considered, an enactive approach to the musical mind can help to spark new critical perspectives that pose challenging questions with regard to what it means to be and become a human being and, in doing so, prompt new conceptions of our relationships to the diverse social, cultural, and natural worlds that sustain us. It is with great anticipation, then, that we await the new arguments and insights that advance, challenge, and refine the ideas presented in this book.

Notes

1 Getting Situated

1. We use the term "musicality" to refer to the "set of traits based on and constrained by our cognitive and biological system" (Honing et al., 2015, p. 2) that underlie our capacity to do things musically.

2. For relevant critiques and defenses see Adams & Aizawa, 2009; Aizawa, 2014; Chemero, 2009; Gallagher, 2017; Hasanoglu, 2018; Hutto & Myin, 2012; Matthen, 2014; Menary, 2010a,b,d.

3. Plato's Socrates argues, for example, that "for they must beware of change to a strange form of music, taking it to be a danger to the whole. For never are the ways of music moved without the greatest political laws being moved, as Damon says, and I am persuaded" (*Republic* line 424c; see Bloom, 1968).

4. As Heidegger (1998) writes in his analysis of Aristotle's conception of *phusis* (nature), "[Plants and animals] are beings only insofar as they have their essential abode and ontological footing in movement. However, their being-moved is such that the *archê*, the origin and ordering of their movedness, rules from within those beings themselves" (p. 190).

5. This insight prefigures current research (Babikova et al., 2013; Barto et al., 2011; Gagliano, 2015; Maher, 2017; Song et al., 2010), which reveals that plants, slime moulds (see Jabr, 2012), and other organisms have much more sophisticated perceptual and communicative capacities than previously thought. Many species exhibit selective, learning, and even and "social" behaviors that were once reserved for animal intelligence.

6. For discussions of the impact of this orientation on music education, see Elliott, 1995; Elliott & Silverman, 2015.

7. This point of view also shows itself in the broader consumer culture of the twenty-first century, in which musical conventions, the score (in both the physical and the abstract sense), and, more recently, the recorded product become what is

signified by the word "music." Along these lines, critical theorists have pointed out how this attitude is in line with the continuing process of reification associated with the development of mechanical reproduction and commodity fetishism in Western capitalist society (Adorno, 1973; Adorno & Horkheimer, 2002; Benjamin, 2008; Lukács, 1971).

8. This is evident in the pioneering work by psychologist Carl Seashore (1938), who famously conceived of human musicality as "a simple construct discrete from other aptitudes and abilities, represented and indexed by a person's aptitude for and ability in auditory discrimination" (p. 69).

9. Indeed, the work of Helmholtz in psychoacoustics was central to the early development of research in music cognition as he demonstrated how the input/output functions of psychological mechanisms could be represented mathematically.

10. In the mid-twentieth century, researchers introduced the idea that mental processes could be understood in terms of computations carried out by machines (Gardner, 1985; McCulloch, 1965). Such machines (a brain or computer) would consist of many simple threshold devices (e.g., neurons, silicon chips, or tubes that function in a binary on/off or active/inactive capacity) connected to able to perform logical operations (Heims, 1980; for a thorough account see Pinker, 1997; and Bechtel et al., 1998).

11. Fodor's (1983) initial theory proposed that "functionally specified cognitive systems" (i.e., modules) exist only in localized lower levels of processing. Two of the principal features of such modules, according to Fodor, are "domain-specificity" and "information encapsulation," meaning that each module works on a specific type of information and that the processing in each module cannot be affected by information in other areas of the brain not directly associated with its input/output path. However, the modular approach has been greatly expanded by the field of evolutionary psychology, which, unlike Fodor, understands the brain to be massively modular across all levels of functioning (Pinker, 1997; Tooby & Cosmides, 1989, 1992). In brief, this approach attempts to explain the diversity of human thought, behavior, and culture in terms of the evolution of a large array of such modules, each adapted by natural selection to serve a specific function related to survival.

12. Some current thinkers have posited revised dualistic perspectives. See, for example, David Chalmers' (2017) discussion of "naturalistic dualism."

13. Put simply, Schenkerian analysis is a hierarchical method for examining voice leading and musical form. Here, the coherence of a musical composition is supposedly based in a deep, underlying structure that is thought to be essentially the same across all works of tonal music (see Brown, 2005; Cook, 2007). Schenkerian analysis aims reveal how the unique characteristics of a given work emerge from this common primary structure, hence its perceived affinity with Chomsky's (1975, 1980) universal grammar theory.

14. Leonard Bernstein's (1976) famous lecture series on music entitled "The Unanswered Question" makes significant use of Chomsky's theories on language to discuss musical expressivity and structure in terms of phonology, syntax, and semantics.

15. For an enactive approach to musical composition, see Nagy, 2017.

16. Similar perspectives are offered by researchers in cognitive science and philosophy of mind who maintain that while cognition involves corporeal and environmental factors, the "mark of the cognitive" is restricted to the brain (Adams & Aizawa, 2009).

17. A variety of perspectives have been offered on interpretations of connectionism, neural networks, and the role of representations: see Chemero, 2009; Clark, 1989, 1993, 1995, 1997, 2001; Clarke, 2005; Horgan & Tienson, 1989.

18. We discuss the mirror neuron mechanism in more detail in chapter 2 and elsewhere.

19. For example, see the discussions of Hendrix's famous Woodstock performance of "The Star-Spangled Banner" in DeNora (2011) and Clarke (2005). See also the discussion of the "Biker Boys" in Willis (1978).

2 Basic Principles of Enactive Cognitive Science

1. The section entitled "Sense Making" develops material that originally appears in Schiavio and van der Schyff, 2018.

2. While various interpretations of embodiment and enactivism have been put forward, our perspective is more in line with the "classical," "autopoietic," or "biological" proposal that originates in the work of Varela, Thompson, and Rosch (1991) and that has been developed by Thompson (2007) and Colombetti (2014) among others. This said, our goal here is not contrast this framework with similar accounts such as sensorimotor enactivism (O'Regan & Noë, 2001a,b) or radical enactivism (Hutto & Myin, 2012). Rather, we adopt conceptual tools and models (e.g., dynamical systems theory) that are shared among these perspectives in an attempt to develop the common orientation of these points of view in the context of musical cognition.

3. This echoes the Aristotelian distinctions between the living and the made discussed in the last chapter.

4. For example, the mating behavior of a common housefly could be described in terms of a simple input/output light sensitive movement controller (Poggio & Reichardt, 1976; Reichardt & Poggio, 1976; see also Schöner, 2008). Luminance changes detected on the fly's eye (input) initiate torque changes in the flight motor, adjusting the direction of flight (output). While this simple input/output description does account for the mechanics of the fly's movement, and the responses elicited by environmental stimuli, it has little to say about the *meaning* of such activity in terms of the dynamical coupling between the living system and its niche. When

considered as embedded in its "social" and physical environment, the perceptually guided action of the fly is purposeful on a deep biological level (it seeks its mate). That is, its actions are motivated by "concerns" for survival and reproduction special to the autonomous fly-environment system. (Our thanks to an anonymous reviewer for suggesting this example.)

5. For example, see the discussion of "chemotaxis" in van Duijn et al., 2006; Di Paolo et al., 2017; Egbert et al., 2010 .

6. Readers may be interested to consider the approach taken by Cummins and De Jesus (2016), who offer a critical view of enactivist assumptions regarding the "autopoietic cell." Most centrally, they argue that the cell is asocial and ahistorical, and therefore properly social factors like participatory sense making cannot begin here. They thus caution against the possible anthropomorphism in imposing notions of self, identity, or interaction on such entities. In line with this, they prefer to drop the term "autopoiesis" and refer instead to a "bio-enactive framework." They suggest this is more in line with the classic model outlined in Varela, Thompson, and Rosch (1991) (where the term "autopoiesis" is not, in fact, used) (see also De Jesus, 2015, 2016).

7. See also Bourgine and Stewart, 2004; Thompson, 2007.

8. The nature of this "precariousness" is described by Hanne De Jaegher and Ezequiel Di Paolo (2007) in the following way: "By precarious we mean the fact that in the absence of the organization of the system as a network of processes, under otherwise equal physical conditions, isolated component processes would tend to run down or extinguish" (p. 487).

9. The enactive conception of agency is unpacked in detail in Di Paolo et al. 2017.

10. This is not to meant imply that our approach should be reducible to behaviorism.

11. Thompson and Stapleton (2009) write, "Consider motile bacteria swimming uphill in a food gradient of sugar (Thompson, 2007, pp. 74–75, 157–158; Varela, 1991). The cells tumble about until they hit upon an orientation that increases their exposure to sugar, at which point they swim forward, up-gradient, toward the zone of greatest sugar concentration. Sugar is significant to these organisms and more of it is better than less because of the way their metabolism chemically realizes their autonomous organization. The significance and valence of sugar are not intrinsic to the sugar molecules; they are relational features, tied to the bacteria as autonomous unities. Sugar has significance as food, but only in the milieu that the organism itself enacts through its autonomous dynamics" (pp. 2–3).

12. Such systems are involved in everything from household thermostats to advanced robotics.

13. As MIT roboticist Rodney Brooks (1991) argues in an influential paper entitled "Intelligence Without Representation," "[T]he world is its own best model as it is always exactly up to date and complete in every detail."

14. Mirror neurons code the goal of the action (see Fogassi et al., 2005). "Virtually all mirror neurons display this behavior: they respond only if the action is directed at an object. They therefore respond to the action of grasping and not simply the hand movements involved in grasping" (Keysers, 2007, p. 3).

15. Neuroscientist Vittoro Gallese (2009) reports that there is a significant congruence between "the response during the execution of a specific type of grip, and the visual response to objects that, although differing in shape, nevertheless all 'afford' the same type of grip that excites the neuron when executed" (p. 489).

16. Psychiatrist and philosopher Thomas Fuchs (2017) writes that mirror neurons are "part of the motor system but also have a perceptual function, [and] can be well integrated into enactive concepts: the circular connection of perception and movement proves to also be fundamental for intersubjective relations. The understanding as well as the imitation of the action of others are based on a form of perception already containing movement" (p. 191).

17. The phenomenologist Maurice Merleau-Ponty (1945) discusses how, when a blind person uses a uses a cane to navigate an environment, the cane becomes an extension of their perceptual apparatus, that is, how their perception now extends beyond their hand to the tip of the cane. Similarly, pianists extend their corporeal and sensory capacities through a complex piece of technology. With practice, they achieve a remarkable transparency between their expressive goals, actions, and the instrument. The latter is experienced not simply as an object—the (musical) world is enacted and experienced through it as an extension of the body.

3 Music and Consciousness

1. This chapter is based on a previous article by Schiavio and van der Schyff (2016) entitled, "Beyond Musical Qualia," which appeared in *Psychomusicology: Music, Mind, and Brain*.

2. This term refers to hearing sounds without seeing their originating causes, such as one experiences in concerts of music composed for loudspeaker presentation (as opposed to live performances, where one can identify the musicians and instruments producing the sounds).

3. More precisely, the "hard problem of consciousness" refers to the kind of explanation required to capture the relationship between physical phenomena, such as brain processes, and conscious experience (i.e., thoughts, ideas, and subjective feelings).

4. Hutto and Myin (2012) argue that, from a radically enactivist perspective, there is no hard problem at all, as there can only be a hard problem in a disembodied, representation-driven view of cognition.

5. The homunculus problem refers to the implication of some "inner" perceiver, raising the question of the homunculus' inner perceiver, and so on ad infinitum.

6. Here, minds are equated to "program-level" operations—or "software" implemented by the "hardware" of brain. This implies that any number of material substrates could realize the same mental states so long as they are able to implement the same system language and output the same behavioral function (just as different hardware can support and run the same software).

7. For the latter, it is common to refer to "musical qualia" (Dowling, 2010; Goguen, 2004).

8. One position argues that qualia originate in the environment and that the perceived qualities of a certain experience depend on how we represent the objective features of the outer world in our head (see Dretske, 1995; Lycan, 1996; Tye, 1995). A classic argument against this approach proposes that, although two subjects may share the exact same representational content constructed from the same external properties, the experience may also possess different phenomenal characteristics for each subject (Block, 1990, 1994). This approach has also been drawn into question in empirical contexts by Gestalt psychologists, who have shown that many experiences cannot be reduced to objective features of the environment—they depend on the motivations, history, and sensory systems of the situated perceiver in *relation* to the environment (see Käufer & Chemero, 2015). A second approach posits that qualia are not identifiable with what they represent in the world, but rather with their physical vehicle—with the sets of neurons that become active under certain circumstances. Thus, when we are listening to a glissando, for example, our brain responds in certain ways, and such neural activity *is* the experience we are feeling. It has been argued that in adopting such an approach one may end up with a category mistake, where one attributes properties to X (in this case, neurons) that instead belong to Y (in this case, experience). Evan Thompson (2007, p. 350) points this out very clearly, noting that, although experience is in fact intentional (or "about" something), it is also holistic. That is, it is constituted by the constantly working networks of perceptions, intentions, emotions, and actions and can often have an intransitively self-aware (or a nonreflective) character. Neural activity, as standardly described, has none of these features.

9. As the philosopher W. E. Kennick (1961) writes, direct forms of description involve "the sort of [verbal or written] description one might find on a 'wanted' poster in a post office: Height, 5 11; weight, 170 lb.; color of hair, dark brown; eyes, blue; complexion, ruddy; small horizontal scar over the right eye. . . . More frequently we employ a sort of 'indirect' description which includes a description of the circumstances in which the feeling is felt" (pp. 317–318).

10. For example, Paul Churchland (1985) claims that future research on the brain will shed light on such issues and eventually allow us to substitute the term *qualia* with a much more accurate description based entirely on synaptic connections and brain networks.

11. In a brain, this could initially be brought about through electrochemical responses given off by sense receptors in response to inputs coming from the environment.

According to Dennett (1988) this could result in frequencies of (binary, on-off, 1–0) neural activation leading to the activation of repeatable and thus recognizable patterns.

12. For Dennett (1988), "[T]he properties of the thing experienced are not to be confused with the properties of the event that realizes the experience" (p. 71). This is analogous to the relationship between the icon that appears on the screen of a computer and the coding that produces it.

13. This approach also allows Dennett (1988) to do away with the other attributes that are supposed to characterize qualia—suggesting, for example, that the assumed private nature of experience is just the result of the "idiosyncrasy of our discrimination profiles" (p. 69); "the physical difference between someone's imagining a green cow and imagining a purple cow might be nothing more that the presence or absence of a particular zero or one in one of the brain's registers" (p. 71).

14. Readers may be interested to consider the descriptions of timing and interactivity experienced by jazz performers provided by musicologists Ingrid Monson (1996), Paul Berliner (1994), and Mark Doffman (2009). Additionally, the improvising guitarist Derek Bailey (1993) offers many fascinating reports by improvisers across a range of cultural contexts.

15. It has been argued that the strict focus on grammar, structure, and representation often tacitly assumes a reductive "correspondence theory" approach to communication (in both language and music) that ignores the interactive, embodied, emotional, and often highly idiosyncratic relationships formed between the listener and the environment in the construction of meaning (Clarke, 2005; DeNora, 2011). Likewise, speech act theory has shown that in most verbal exchanges the correlation between grammatical structure and meaning may not be so clear as when language is considered in ideal terms. Although language can refer directly to (or represent) concrete reality, in everyday life meaning is often constructed through self-organizing relational processes where creative and contextual interpretation plays an important role (Searle, 1967; Streek, 1980). It has also been demonstrated that communication of all kinds very often employs, and depends upon, physical movement. This includes meaningful and emotive bodily and facial gestures (Davidson, 2005; Johnson, 2007; Runeson & Frykholm, 1983; Smith, 1998). In other words, these observations strongly suggest that "meaning" is not constructed and communicated solely through linguistic abstractions—asserting, again, the central role of action, feeling and the body in living communication (understood as participatory sense making).

16. As the philosopher Taylor Carman (2008) writes, to perceive is "not to have inner mental states but to know and find your way around in an environment ... Perception and movement are not related to one another as causes and effects, but coexist in a complex interconnected whole" (p. 87).

17. For a useful overview see Käufer & Chemero, 2015, pp. 79–91.

18. Because language allows us to communicate abstract propositions and complex concepts, it is not difficult to see why representational logical-linguistic processes have become synonymous not only with rational thought but with the idea of mind itself. That representation is a central aspect of linguistic communication is an uncontroversial and undeniable fact. However, as Varela, Thompson, and Rosch (1991) note, this weak notion of representation takes on much heavier ontological and epistemological commitments when it is generalized to construct an all-encompassing theory of cognition—that is, the strong computational notion that cognition is necessarily and always grounded in the subpersonal outputs of symbolic-syntactic information processing in the brain.

19. Among other things, this means that if we wish to continue to use the term *qualia*, it can no longer refer to some sort of "inner" private experience. Rather, it will need to be reconceived of in a multidimensional or 4E context.

20. This perspective has been developed in musical contexts by a range of authors who have produced compelling and highly nuanced accounts of what musical experience entails (Clifton, 1976, 1983; Ihde, 1976; Krueger, 2011a,b; Roholt, 2014; Schafer, 1994; Sudnow, 1978).

21. As an aside, we can also note that this perspective casts Jackendoff's (1987) insights into the multiplicity of conscious experience in a new light: it negates the need for complex sets of "intermediate-level representations," highlighting the active and situated nature of the conscious mind (Varela et al., 1991).

4 Phenomenology and the Musical Body

1. This chapter is based on an article by van der Schyff (2016) entitled, "From Necker Cubes to Polyrhythms," which appeared in *Phenomenology and Practice*. The section "Some Implications for Future Research and Theory" develops material from a paper entitled, "Beyond Musical Qualia: Reflecting on the Concept of Experience" by Schiavio and van der Schyff (2016).

2. The terms *noesis* and *noema* originate in the Greek word νόημα, which refers to the "aboutness" of a given thought. They were introduced in phenomenology by Husserl (1962 [1913]) to distinguish the basic (and inseparable) elements of conscious thought, or "intentionality." Put very simply, *noesis* refers to the mental act of perceiving something in a certain way, while *noema* concerns the intended object of experience. These terms have been developed by various ways by a range of thinkers (Woodruff Smith, 2007, pp. 304–306; see also Sokolowski, 2000; Solomon, 1977).

3. One might also examine how the perception of the various attributes of musical sound develop in relation to bodily-emotive states (i.e., synaesthetic perceptions of movement, location, space, texture, feeling; see Clifton, 1976; Johnson, 2007;

Merleau-Ponty, 1945). Similarly, as a performer, one could explore how experience is transformed through the embodied agency of those with whom we coenact musical worlds (Krueger, 2014; Reybrouck, 2005a,b). Or one could begin with an examination of *sedimented* conventions and attitudes in the cultural context (e.g., Clayton et al., 2012; Small, 1998), examining one's relationship to them and considering alternative ways of thinking. Phenomenological inquiry can also afford a more nuanced awareness of time perception, where the focus may shift from the narrow (onsets and the trailing off of sounds) to the broad (the evolution of a tone or form). This may also involve an exploration of the relationships between the "just past" to the anticipation of what is to come next, the *retentions* and *protentions* that characterize the temporal nature of intentionality (see Husserl, 1991; Merleau-Ponty, 1945). These activities all involve a process of moving from taken-for-granted ways of listening, perceiving, and thinking toward a more reflective, present, or mindful attitude—one that may foster more pluralistic, relational, and imaginative ways of attending to the world.

4. Musical engagements necessarily involve cross-modal, embodied, and intersubjective forms of awareness and attention sharing (Clifton, 1976; Johnson, 2007; Leman, 2007). For example, as a listener, one may decide to focus on specific instruments in an ensemble—perhaps one that normally plays a background role—and thus develop a new perspective on a well-known piece of music. In doing so one could also actively seek out new aspects (overtones, harmonics, rhythmic and dynamic nuances, and so on) to develop the experience of the sounds one engages with (Roholt, 2014). Additionally, one could attend to the sonic properties of the acoustic space one occupies (acoustics, reflections, reverberation, diffusion; see Ihde, 1976; Blesser & Salter, 2006). Along these lines, a "focus, core, field and fringe" delineation has been taken up in an auditory/musical context by phenomenologists (Ihde, 1976; see also Schafer, 1994), who point out that, unlike visual experience, the auditory field is omnidirectional, explicitly temporal, and always connected to a sense of action or movement. In this way, musical experience arguably surrounds, permeates, and transforms our being in a way that visual experience does not; even in seemingly passive listening contexts it actively engages the body and does not first impose a strict subject-object separation. This observation is also taken up by Clifton (1976) in a listening context, who describes the bidirectionality of musical experience: "I possess the music, and it possesses me" (p. 76).

5. For a relevant discussion of empirical research see Møller et al., 2021.

6. Readers may also wish to consult the excellent book *West African Rhythms for the Drumset* by Royal Hartigan, Abraham Adzenyah, and Freeman Donkor (1995).

7. As we discuss in more detail in upcoming chapters, musical sense-making begins in infancy, through direct interactions with the sociomaterial environment, and before high-level linguistic competency is achieved.

8. In chapter 7 we will suggest that the former may have emerged in the human phenotype thanks to the evolutionary affordances provided by the latter.

9. For further discussion of this idea see Thompson, 2007, pp. 28–36.

10. This follows William James' famous claim that even the most abstract concepts afforded to us by language begin with bodily perceptions and feelings that result from pragmatic interactions with the world, "more and less mean certain sensations" (1979, p. 38).

11. See also Eitan & Granot, 2006; Eitan & Timmers, 2010.

12. Composer John Sessions (1941) captures this insight when he writes: [It] is easy to trace our primary musical responses to the most primitive movements of our being—to those movements which are at the very basis of animate existence. The feeling for tempo . . . [has a] primitive basis in the involuntary movements of the nervous system and the body in the beating of the heart and . . . in breathing, later in walking. . . . [If] an increase in intensity [(pitch, loudness)] of sound intensifies our dynamic response to music . . . it is because we have already in our vocal experiences—the earliest and most primitive as well as later and more complicated ones—lived through exactly the same effects. (pp. 105–109)

13. See also Lawrence Zbikowski's (2005, 2017) musicological applications of conceptual metaphor theory.

14. For example, Kiverstein & Miller (2015) show how given brain functions, being context dependent, can be understood better when included in a large-scale network with other brain areas, the body, and the world (including other agents).

5 Music and Emotion

1. This chapter expands upon a short paper by van der Schyff and Schiavio (2017b). It also develops ideas originally presented by van der Schyff (2013) and by Schiavio and colleagues (2017). With regard to the latter publication, we would like to thank our coauthors, Julian Céspedes-Guevara and Mark Reybrouck, for their contribution to the ideas presented in this chapter.

2. Similar assumptions have also influenced theorizing in in psychological contexts. For example, the expanded lens model (Juslin, 2000; Juslin & Lindström, 2010) posits that composers encode an emotional message into a work; performers manipulate different musical parameters to increase the probability that listeners will identify the "right" emotion intended for them by the composer.

3. Huron's approach develops the core insights of Leonard Meyer's book, *Emotion and Meaning in Music* (1956). Interestingly, Meyer does not rely on information-processing models but develops Gestalt theory and the thought of pragmatist thinkers such as Charles Sanders Peirce and John Dewey to account for emotional responses to music in terms of the satisfaction and thwarting of expectation.

4. Although most research into the perception and expression of musical emotions has taken for granted that music expresses discrete emotions, much of this work

has been carried out without a clear and explicit theoretical basis. For example, many studies select ad hoc lists of discrete emotions to select their stimuli and to categorize the listeners' physiological responses and reported experiences (see Eerola & Vuoskoski, 2013).

5. For a detailed critical review of BET in musical contexts see Céspedes-Guevara & Eerola, 2018.

6. The music psychologists Laura-Lee Balkwill and William Forde Thompson (1999) offer a useful distinction between universal "psycho-physical" (tempo, pitch) and "culture-specific" (i.e., emotional interpretation of intervals and gestures) cues that may partially explain intercultural agreements and differences. Additionally, examining the differences between the perceived and induced aspects of musical emotion could help explain why people may sometimes perceive emotions *in the music* in ways that coincide with basic emotions (e.g., joy, sadness, anger, etc.), while the emotions they *feel aroused in themselves* (induced) may not be organizable into neat categories (see Céspedes-Guevara & Eerola, 2018).

7. The BRECVEMA theory accounts for a range of other emotional musical experiences: basic emotions (in the case of the contagion mechanism); everyday, "garden variety" emotions (in the case of the episodic memory and visual imagery mechanisms); surprise-like emotions such as interest, anxiety, surprise, and chills (in the case of the expectation mechanism); general arousal (in the case of the brainstem and rhythmic entrainment mechanism); aesthetic emotions (in the case of the aesthetic evaluation mechanism).

8. Indeed, CPM draws on the DST approach to help explain the interactivity between the relevant mechanisms associated with emotion (Scherer, 2007).

9. This recalls Dennett's notion of "pips" discussed in chapter 3.

10. This orientation has been bolstered by relatively recent cultural developments in which listeners are often conceived of as essentially passive consumers of music. Music psychologists Lincoln John Colling and William Forde Thompson (2013) note that before the rise of the concert hall and the recording industry, "music was almost always experienced as a multimodal phenomenon containing auditory, visual and kinaesthetic dimensions"; "music was rarely observed passively . . . music-making was often social, collaborative . . . and had weak boundaries between performer and audience" (p. 197).

11. Some readers may note similarities with "constructionist" approaches (Barrett, 2006; Russell, 2003), which argue that emotions cannot be understood in terms of discrete regions in the brain or be reduced to genetically determined affect programs and the "basic emotions" they support. Rather, from a constructionist perspective, emotions involve dynamical interactions between large-scale networks involved with domain-general processing (Barrett & Satpute, 2013). This means that the constructionist perspective aligns with the enactive approach when it highlights the active role perceivers play in (constructing) emotional experience. However, the

constructionist orientation diverges from enactivism when it tends to minimize of the role of "the biological" in episodes that are considered properly emotional. This is because constructionism argues that "primitive" biological components associated with affectivity (valence and "core affect") are only loosely linked to emotions as such. By this light, proper emotions require higher cognitive processes. These processes entail a set of causally sequenced events that ultimately depend on the linguistic abilities that allow us to categorize emotions as fear, anger, shame, and so on. Thus, for us to recognize an emotion in ourselves or in another person depends on the "mental script" for that emotion (for a critical discussion, see Colombetti, 2014, pp. 46–49). The constructionist approach has been applied in musical contexts only very recently by the music psychologists Julian Céspedes-Guevara and Tuomas Eerola (2018). See also Céspedes-Guevara (2016).

12. Situations where a live musical event is shared by everyone involved can create a sense of cohesion or "co-subjectivity" (Clarke et al., 2015).

13. Thanks to Joel Krueger for bringing this research to our attention.

6 The Empathic Connection

1. This chapter contains parts adapted from an article by van der Schyff and Krueger (2019) that appeared in an ebook entitled *Music, Speech, and Mind* (edited by Antenor Ferreira Correa). We thank Joel Krueger for his contribution to the ideas presented here.

2. As the psychologist Simon Baron-Cohen notes, "A theory of mind is impossible without the capacity to form second-order representations" (see Baron-Cohen et al., 1985, p. 38).

3. It should be noted that Coplan (2011) does not simply ignore the importance of lower-level perceptual processes. Instead, she aims to clarify a term that, she suggests, is applied too widely and is currently too vaguely defined—that is, by narrowing empathy proper to a distinctly representational process in the mind of the empathizer.

4. See Krueger, 2009, 2015; Gabrielsson, 2011; Schiavio & Høffding, 2015.

5. The psychologist Edward Titchener (1909), for instance, posited two forms of empathic experience. One involves the more common ability to perceive the "inner states" of others, which is rooted in our need for socialization. The other takes on a more detached "aesthetic" dimension as it entails the ways we project aspects of ourselves into, or somehow identify with an object of contemplation such as a work of art or music.

6. Nonexperts will also have some basic capacity to understand and ascribe meaning to some of these actions because they will have engaged in similar actions since childhood.

7. This is because theory-theory aligns well with other established approaches in social cognition: for example, social learning theory (Bandura, 1977) and the Theory of Mind (ToM) approach in evolutionary psychology (Tomasello, 1999; Tomasello et al., 2005). Recent work by Tomasello (2014) integrates elements from simulation theory (ST) and theory-theory (TT).

8. Again, these neurons activate in relation to actions that are in our motor repertoire: "These neurons respond to the sound of actions and discriminate between the sounds of different actions, but do not respond to other similarly interesting sounds such as arousing noises, or monkeys' and other animals' vocalizations" (Gallese, 2009, p. 521).

9. Recent studies in embodied music cognition (Bamford & Davidson, 2018; Wallmark et al., 2018) have explored emotional and cognitive empathy in attempts to illuminate and measure "trait empathy." The latter refers to empathy as a stable or dispositional aspect of one's personality that appears to modulate empathic reactions to music in certain contexts (Eerola et al., 2016; Garrido & Schubert, 2011; Greenberg et al., 2015; Kawakami & Katahira, 2015; Kreutz et al., 2008).

10. This echoes Wittgenstein (1980): "We *see* emotion. As opposed to what? We do not see facial contortions and *make the inference* that he is feeling joy, grief, boredom. We describe a face as sad, radiant, bored, even when we are unable to give any other description of the features—Grief, one would like to say, is personified in the face" (§570).

11. This phenomenon continues to be investigated by researchers in developmental psychology who have offered further evidence for the primary role of direct embodied interaction for social cognition (Hobson, 1993; Meltzoff, 2005; Meltzoff & Decety, 2003; Reddy, 2008; Stern, 1985).

12. Such insights may offer interesting ways of interpreting the studies of musical development and education mentioned in at the end of the previous chapter—that is, where children appeared to display increased levels of empathy after engaging in cooperative musical activities (see Kirschner & Tomasello, 2010; Rabinowitch et al., 2012, 2015).

13. For similar discussions see Salice et al., 2017; Schiavio & Høffding, 2015; van der Schyff et al., 2018; Walton et al., 2017.

14. Clarke (2013) writes, music is "inextricably bound up with that wider auditory world, since it sounds within it, incorporates environmental sounds into its own material, and (with the development of recording, broadcast, and listening technologies) takes on fluid relationships with the physical spaces that it occupies—from practical and normative to provocative and paradoxical" (p. 90).

15. Emerging evidence also suggests that solitary listening can evoke empathy, specifically in the form of implicit affiliation toward members of a specific cultural group.

For example, Vuoskoski and colleagues (2017) found that listening to a five-minute track of music from a particular culture can increase positive implicit attitudes toward facial images representing members of that culture.

16. Of course, these aspects of participatory sense making also characterize nonmusical contexts. However, musical interactions appear to intensify their empathic effects. For example, Hove & Risen (2009) asked to engage in in a simple tapping task. Participants who achieved higher levels of synchronization with each other tended to rate their feelings of mutual affiliation as more intense. Similarly, Demos and colleagues (2012) found that when listeners were asked to rock their chairs together without music and then in time to a piece of recorded music, they reported feeling more connection with each other in the musically scaffolded space. The music acted as a kind of "social glue."

17. This has connections with research that shows how neural activity associated with musical engagement may be understood in terms of oscillatory dynamics that do not simply represent but actively resonate with the body and the sonic environment—forms of "dynamical attending" that allow for social phenomena such as rhythmic entrainment (see Large & Jones, 1999; McGrath & Kelly, 1986). This could open further possibilities for studying (social) music cognition as an embodied and ecological phenomenon—that is, by examining how patterns of coordinated (oscillatory) activity and associated neural structures emerge, stabilize, and transform through the dynamical interactivity of the various dimensions of the (musical) brain-body-world system (more on this in the next chapter).

18. As we saw previously, these processes are not exclusively music dependent. When we interact with other people, their expressive actions—gestures, facial expressions, postural adjustments, intonation patterns, movements, manipulations of shared space, and so on—directly impact our bodily responses, and vice versa.

19. Mark Reybrouck (2016, 2017a,b) offers compelling accounts of how symbolic, simulative, and embodied-interactive processes may combine in musical sense making. In doing so, he attempts to integrate computational and embodied-ecological perspectives.

7 The Evolution of the Musical Mind

1. An earlier version of this chapter appeared in *Frontiers in Neuroscience* (see van der Schyff & Schiavio, 2017a).

2. The orthodox adaptationist orientation focuses on fitness optimization as the main driver of evolutionary processes. At the core of Darwin's original theory is the idea that biological evolution occurs through a process of modification by descent—that is, through mutation and the recombination of hereditary material through reproduction. The mechanism responsible for this process is known as

natural selection, which "chooses" the phenotypes that function most effectively within a given environment; organisms are selected on how optimally they fit the environment at hand—hence the famous phrase, "survival of the fittest" (see Sober, 1984). The so-called neo-Darwinian movement emerged in the mid-twentieth century with the advent of modern genetics. Here, evolutionary theory began to focus on the fitness of genes as the main quantitative basis for understanding the adaptive traits organisms exhibit in relation to the environments they inhabit—where the fitness of a given gene and its associated phenotypic trait are understood in terms of a calculus of "abundance" (optimization of surplus offspring and the growth of an interbreeding population) and/or "persistence" (optimization of reproductive permanence; long-term survival) (Dawkins, 1976). Thus, the cognitive capacities of the human phenotype are understood to have emerged from adaptive processes associated with fitness optimization that occurred over an evolutionary timescale. This drives evolutionary psychology's central claims that the human mind evolved toward fitness optimization, that is, toward the capacity to create representations that optimally correspond with a stone-age hunter-gatherer environment and that many of the perceptions, thoughts, behaviors, and desires associated with modern life (a life we are supposedly *not* biologically adapted for) are largely "parasitic," "invasive," or otherwise dependent on mental (computational) processes and structures (modules) that developed deep in human prehistory (Sperber, 1996; Sperber & Hirschfield, 2004; Tooby & Cosmides, 1989, 1992). It has been one of the central projects of evolutionary psychology to discern what human activities and thoughts can be understood as properly adaptive from those that are biologically irrelevant (see Pinker, 1997).

3. Various arguments have emerged in support of each position (see Honing et al., 2015; Huron, 2001; Killin, 2016a, 2017; Mithen, 2005; Patel, 2008; Pinker, 1997; Tomlinson, 2015).

4. Varela, Thompson, and Rosch (1991) summarize this approach: "Evolution is often invoked as an explanation for the kind of cognition that we or other animals presently have. This idea refers to the adaptive value of knowledge and is usually framed along neo-Darwinian lines. . . . Evolution is often used as a source of concepts and metaphors in building cognitive theories. This tendency is clearly visible in the proposal of so-called selective theories of brain function and learning" (p. 193).

5. The term "exaptation" refers to changes in the function of a given physiological or behavioral trait in the process of the biological evolution of an organism. The classic example is bird feathers, which originally evolved for thermoregulation but were later co-opted for mating-territorial display, catching insects, and then flight. The developmental systems approach complicates the causal relation of adaptations and exaptations. Here they stand not in a linear sequence but rather in a cyclical relationship, in which the new uses of an adaptation associated with the exaptation may lead to secondary adaptations and so on (see Anderson, 2007; Gould, 2002;

Gould & Vrba, 1982). Referring to the relationship between adaptations and exaptations Tomlinson (2015) writes, "The first are not necessarily prior to the second, since behaviors originating as exaptations might alter selective pressures in ways leading to new adaptations" (p. 36).

6. It should be noted that Patel (2018) has recently updated his approach to include coevolutionary principles.

7. It has been argued that the strict adaptationist approach does not properly consider epigenetic factors and the important role of ontogenetic and environmental processes (Goodwin et al., 1983; Jablonka & Lamb, 2005). For example, one may consider here the discovery of polygenic traits, as well as the phenomenon of epistasis, in which the regulation and expression of a given gene is dependent on, and contributes to, the activity of other genes in the intra-/intercellular environment via epigenetic processes; environmental and biochemical factors (e.g., hormones) play an important role (Lambert et al., 1986; Ridley, 2003).

8. This substantially expands the orthodox Mendelian understanding of genetic inheritance, which posits an additive "one-directional" schema (genes ⇒ cells ⇒ environment ⇒ phenotype). By the classical view, genes trigger protein production, this guides the functioning of cells, which, with some influence from the environment, produce identifiable traits (Moore, 2003). This older approach works well when explaining so-called single-gene disorders like Huntington's disease or certain elementary physical features like eye color, especially as they develop in relatively static and homogeneous environmental contexts (e.g., Mendel's pea plants). And there are also certain basic biological features that may still be understood in terms of a neo-Darwinian comparative fitness scale (e.g., oxygen consumption). However, a growing number of biologists find classical theories of genetics and evolution lacking in the context of more complex physical, behavioral, and psychological attributes such as personality or cognitive, athletic, and musical ability, which increasingly appear to be heavily influenced by environment, motivation, activity, and experience (Bateson, 2003; Bateson & Mameli, 2007; Ericsson et al., 2006; Meaney, 2001; Sternberg, 2005).

9. Figure 7.1 depicts the most general level of description and does not show the more microlevel "cycles within cycles" that occur, for example, within the intraorganism milieu. These include the patterns of muscular, emotional-affective, neural, and metabolic activity that influence the expression of genes and gene groups over various timescales. This, in turn, helps to guide developmental processes and behavior that impacts the environmental niche, which, in turn affects the development of the organism (gene-culture coevolution).

10. Interestingly, the rhythmic patterns that emerged in this experiment display six statistical universals found across different musical cultures and traditions.

11. Readers may be interested to consider studies that examine the "musical" properties of the stone artifacts and acoustical characteristics of Paleolithic environments (see Blake & Cross, 2008).

12. This involves the integration of phonemes and words into grammatical structures and the development of a generative syntax that provides the "rules" for such processes—or, likewise, the organization of discrete sets of sounds, tones, and pitches into rhythmic/formal hierarchies that could be consciously repeated or manipulated (e.g., melodies and drumming patterns).

13. For similar arguments see Malafouris, 2013, 2015.

14. Fodor (2001) himself has argued that applications of the modular-computational theory of mind have been greatly overestimated. Likewise, the neuroscientists Mireille Besson and Daniele Schön (2012) write that, "The theory of modularity did provide a useful framework.... However evidence has accumulated at the micro (genetics and molecular biology ...) and macro-levels (cognitive psychology, neuroscience, cognitive neuroscience ...) that, in our view, point to the limits of modularity ... advancement of knowledge at various levels of biological organization increasingly shows that biological and cognitive processes are largely influenced by environmental factors ... the expression of genes depends on epigenetic factors ... and cognitive processes unfold as a function of context.... It becomes consequently more and more difficult to consider brain and behavior as linear systems ... that can be decomposed into independent modules and functions. Rather, these functions seem to be highly interactive" (pp. 289–290).

15. The neuroscientist and psychobiologist Jaak Panksepp (2009) comments, "Most cortical specializations (including seemingly genetically preordained cortical processes such as vision and hearing) are currently explained better by epigenetic regulatory influences on cortical specializations arising from lower brain functions.... Most neocortical functions seem to emerge during brain development, guided by many poorly understood environment-sensitive processes" (pp. 233–235). He concludes, therefore, that the human neocortex may in fact contain *no* evolutionary determined modules for either music or language; that the neural origins of musicality are largely subcortical; and that the emergence of "emotional protomusical communications" may have led to the development of both music and propositional language. Thus, it may be that the ancient emotional core of the limbic system provides "the actual instinctual energetic engines that still motivate our music making and continue to be the tap-roots that allow the rich foliage of cultural invention that is modern music to assume the impact it does on our minds" (p. 237).

16. For studies on music and brain plasticity (see Gaser & Schlaug, 2003; Jäncke et al., 2001; Lappe et al., 2008; Large & Jones, 1999; Münte et al., 2002; Pantev et al., 2001; Schlaug, 2001). Additionally, clinical studies have demonstrated music's deep effects on the body as well as its capacity to transform or reorganize neural structures (e.g., Bunt, 1994; Jovanov & Maxfield, 2011; Nayak et al., 2000; Standley, 1995; Tomaino, 2009).

17. See Toiviainen (2000) for an interesting discussion of this phenomenon in the context of music and artificial intelligence.

8 Teleomusicality

1. This chapter develops sections that originally appeared in an article entitled, "When the Sound Becomes the Goal: 4E Cognition and Teleomusicality in Early Infancy" (*Frontiers in Psychology*) by Schiavio, van der Schyff, Silke Kruse-Weber, and Renee Timmers (2017). We thank Silke Kruse-Weber and Renee Timmers for their contribution to the ideas developed in these sections. An early version of the teleomusicality theory was developed by Schiavio and Menin (2011).

2. Others use this term to refer to the evolution of music-like expressions in our prehuman ancestors and other animals. For example, Tecumseh Fitch (2019) writes, "Darwin suggested that initial function of vocal learning was in the production of song-like vocalizations, analogous to those produced by birds. These protomusical utterances were, by hypothesis, free of any specific propositional meanings" (see also Fitch, 2005).

3. Here we can also consider the parallel development and anatomical unity of the cochlea (hearing) and vestibule (balance, orientation) and the prenatal physiological-emotional "communication" that appears to occur between mother and fetus through their shared circulatory system (e.g., hormone release), as well as bodily sounds and movements (see Parncutt, 2006). This, again, highlights the origins of the cross-modal nature of human perception and sense making discussed in chapter 4.

4. Trehub and Hannon (2009) write, unlike adults "infants lack implicit knowledge of key and implied harmony, as reflected in the equal ease with which they detect melodic changes that preserve the key and implied harmony of a tone sequence and those that disrupt those elements" (see also Trainor & Trehub, 1992).

5. This echoes J. J. Gibson's (1979) famous assertion that "[w]e must perceive in order to move, but we must also move in order to perceive" (p. 223), pointing to the mutual influence of perception and action (see also Merleau-Ponty, 1945).

6. A study by Ambrosini et al. (2013) compared the anticipatory gaze of 6-, 8-, and 10-month-old infants (and a control group of adults) during goal-directed observational tasks. The researchers found a significant correlation between the ability to perform a given action and the capacity to successfully predict the goal of an observed action. This was noted in particular when 6-month-old infants observed actions involving the precision grasping of objects (i.e., using thumb + index), which they are still not able to perform. Here they tended to focus on the action and not on what was grasped (they were not able to predict the goal of the action). By contrast, full-hand grasping is a well-established skill in the motor repertoire of most infants of this age. And indeed, when observing these types of actions, infants of this age did appear to predict with more precision the appropriate goal of the given grasping action. A related finding involves the relationship between age and the degree of gaze proactivity. Here, advantages for goal prediction were found from

8 months onward, with 10-month-old subjects showing faster gaze proactivity to precision grasping. Taken together, these results suggest that the ability to perceive and/or predict the goal of observed motor behaviors is action-specific; it depends on whether the infant is capable of actually carrying out the action observed.

7. Philippe Rochat (1989) notes: "Newborns show elements of reaching with arms and hands toward an object moving close to them. . . . Young infants track objects moving in their field of view with both eyes and head. . . . From birth, they orient their heads in the direction of a sound source . . . and appear to selectively orient (root) their mouths in the direction of a familiar odor. . . . Contrary to the view that babies are passive spectators bombarded with stimulation, these observations indicate that from the earliest age infants are actors and, in particular, explorers of their environment" (p. 871).

8. With regard to this point, one could argue that some music-related actions do in fact involve other goals that go beyond the sole production of sounds—like *dancing*, for example. However, we could say that an activity like dancing is usually enacted in a musical context, showing that the goal of similar actions remains intrinsically musical. As we will see, this is different from activities that we could define as "musical" only a posteriori, such as clapping hands for approval, for example. In such cases, the motivation behind these actions cannot be described as "musical."

9. In this sense, even using *chance* as a composition methodology (e.g., John Cage) is still a choice, a self-imposed rule.

9 Creative Musical Bodies

1. This chapter is based on a paper that appeared in *Music and Science* entitled "Musical Creativity and the Embodied Mind: Exploring the Possibilities of 4E Cognition and Dynamical Systems Theory" by van der Schyff, Schiavio, Ashley Walton, Valerio Velardo, and Anthony Chemero (2018). We thank our coauthors for their contribution to the ideas presented in this chapter. This chapter also develops passages of the article by Schiavio and van der Schyff (2018) entitled "4E Music Pedagogy and the Principles of Self-Organization" (*Behavioral Sciences*).

2. Williams has drawn on compositional devices introduced by Wagner, Korngold, Stravinsky, Orff, Elgar, Tchaikovsky, Strauss, and Holst (among others) to produce his scores for the *Star Wars* franchise and many other films (Audissino, 2014).

3. Other scholars have noted that products and processes are deemed creative (or not) by communities of practitioners and consumers—how they are received and categorized (e.g., big-C, little-c, and so on) involves a process of "sociocultural validation" (Cropley, 2011). In addition to the various personal factors that underpin creative outputs, researchers have also argued that we need to understand the external

"pressures" of the cultural context, how they impact the creative person, and thereby shape the products a person produces (Rhodes, 1961).

4. Areas where the system's state tends to evolve toward an attractor are shown as basins of attraction.

5. For an excellent overview, see Chemero, 2009.

6. For a related study on musical synchronization, see Large, 2000.

7. Similar insights have also been developed by Sudnow (1978, 2001) in his famous self-reflective phenomenological description of how he learned to improvise at the piano.

8. An interesting aspect of this study is the lack of a significant effect concerning visual information between the participating musicians. In other words, it seems that the degree of coordination may not always be affected by whether musicians are able to observe each other or not. A possible interpretation here is that the emerging musical environment does not necessarily require the integration of visual aspects for it to become self-sustaining. Instead, it may be that as long as certain basic requirements are met (e.g., "having a sense of freedom while performing" or being open to interact) musical creativity may develop in a range of contexts.

9. For related work concerning creativity, the generation of quasi-randomness, patterning and nonpatterning, see Benedek et al., 2012; Zabelina et al., 2012.

10. Thanks to Valerio Velardo for bringing this idea to our attention.

11. It should be noted that practical applications of this approach are currently being developed. Preliminary work is being done employing relevant mathematical tools associated with DST to model virtual creative music agents (see Gimenes, 2013; Gimenes & Miranda, 2008, 2011; Velardo, 2016).

12. Here, readers may note a similarity between the concepts of gradual/punctuated evolution and those of exploratory/transformational creativity as defined by Boden (2004; see previous). Exploratory creativity is a form of gradual evolution, where an agent creates new artefacts within a given conceptual space (i.e., "inside" an attractor). Transformational creativity, by contrast, is a form of punctuated evolution in that it entails the generation of new possibilities for perception and action (and resulting artefacts) that are substantially different from those previously generated (i.e., motion from one attractor to another).

13. For a similar approach in the context of jazz composition see Rosenberg, 2010.

14. From this perspective Schoenberg's move to serialism cannot be properly understood as restricted to the artist himself. Rather, it was facilitated by his interaction with a range of environmental factors that include the emerging critiques of European culture and aesthetics, the composers and other artists of his day, as well as his many brilliant students, who went on to develop their own influential worldviews.

15. For discussions of these kinds of dynamics in martial arts and dance see Kimmel & Rogler, 2018; Kimmel et al., 2018.

16. In connection with this, Simon Høffding (2018) has examined the experiences involved in collective music making through a series of interviews with the Danish String Quartet. Notably, this research develops a richer understanding of what "musical absorption" involves—one that ranges from something close to a total immersion in the shared activity, to "frustrated playing" (associated with overcoming obstacles and challenges), to more reflective experiences while performing. Such insights are helping to revise conceptions of what flow and skilled coping entail to include forms of deliberate thought-in-action.

17. Such concerns have been addressed (among others) by Keith Sawyer (2003, 2006), who draws on recent developments in social and distributed cognition, and real-life experiences in musical and theatrical improvisation. In doing so, he develops a rich model for exploring creativity as a distributed and emergent phenomenon (see also Csikszentmihalyi, 1988, 1990; Gardner, 1993), where creative activity entails (at least) the following four characteristics: (1) The activity has an unpredictable outcome, rather than a scripted, known endpoint. (2) There is moment-to-moment contingency: each person's action depends on the action just before. (3) The interactional effect of any given action can be changed by the subsequent actions of other participants. (4) The process is collaborative (see Sawyer & DeZutter, 2009, p. 82).

10 Praxis

1. This chapter develops passages and ideas drawn from van der Schyff, Schiavio, and Elliott (2016) and van der Schyff (2015).

2. Although the concept of *praxis* originated with Aristotle, numerous scholars since Greek times have appropriated, broadened, or deepened its original meanings and implications. Major contributors to *praxis* studies include Immanuel Kant, G. W. F. Hegel, Martin Heidegger, Hans Georg Gadamer, Karl Marx, Antonio Gramsci, Jürgen Habermas, William James, John Dewey, Hannah Arendt, Alasdair MacIntyre, David Carr, Wilfred Carr, Stephen Kemmis, and Nel Noddings.

3. The primacy of *phronēsis* resonates closely with the fundamental enactivist insight that "[t]he greatest ability of living cognition [. . .] consists in being able to pose, within broad constraints, the relevant issues that need to be addressed at each moment" (Varela et al., 1991, p. 145).

4. Enactivist insights into the self-organizing nature of biological systems also find an antecedent in the ancient conception of *phusis*. This term is often translated as "nature," but this concept has more to do with the animate way in which the world is experienced by creatures who discover themselves enmeshed within it. *Phusis* has phenomenological connotations when it refers to the experience of how the world continually surges up and transforms (weather, emotions, plants, animals). People

and things move in and out of existence, and other entities emerge to take their place; matter and form are enmeshed in an endless process of interactive transformation (Dreyfus, 1992; Dreyfus & Kelly, 2011). With this in mind, *phusis* is not an object or an objective property of the world (something we are over-and-against). It is not, first and foremost, a kind of "knowledge of" but is a mode of "disclosure" in which we are inextricably implicated as living, self-making beings—the primordial way being is revealed by a being who cares about being. Accordingly, the idea of *phusis* also connects with the fundamental forms of affective, embodied, sense making associated with "passive synthesis" (see chapter 4). Additionally, *phusis* underpins *praxis* in the sense that the kinds of transformations the latter initiates in human worlds are predicated on the more primary ontology of the former. This means, for example, that the kinds of human making and doing described by *poiesis* are understood as continuous with the movements of life and the environment—as a form of *ekstasis*. This last term refers to how a thing moves from its standing as one thing to become another—the unfolding of a thing out of itself (a plant emerging from a seed, the evolution of a musical idea, and so on), or the emergence of "thing-ness" from "no-thingness," being from void. By this light, *poiesis* can be viewed as a kind of human nurturing in which things are dealt with as needing to be helped to come forth (Dreyfus & Spinosa, 1997), as in art making, friendship, child rearing, and education.

5. Here the well-being of such intersubjective ecologies is not understood to be based in some facile sense of consensus or conformity, but rather in terms of the ongoing dynamics of difference inherent in larger systems comprising autonomous agents (e.g., a class or ensemble). Bateson (1972) reminds us that difference is "the pattern that connects"; reaching out to difference asserts the existence of a self or a point of view, while at the same time showing that the "self" cannot be extricated from the complex system of organism-environment interactions it emerges from and that sustains it (see also Small, 1998). It follows, then, that the idea of "cooperation" need not be based in "higher" or representational-conceptual modes of knowledge or communication (or empathy-as-simulation). Nor is it necessarily motivated by some external pregiven goal. Instead, cooperation emerges from the self-organizing, embodied-affective interactions of the individuals involved, and the shared needs, desires, and actions that result from such contingent intersubjectivity (*phronēsis*). In this way, a social group may be understood to enact their own goals and ways of coordinating action through dynamical adaptive processes (Fantasia et al., 2014). As Hubley and Trevarthen (1979) write, "[C]ooperation means that each of the subjects is taking account of the other's interests and objectives in some relation to the extra-personal context and is acting to complement the other's response" (p. 58).

6. For instance, data from recent qualitative work examining differences and continuities between individual and collective teaching settings (Schiavio et al., 2020) suggest that music teachers are less "present" when involved in group teaching when compared to one-to-one tuition. This reported lack of presence, it should be noted, does not result in a negative learning environment where students are left by themselves. On the contrary, the written preoccupations of the respondents point

to a form of "distributed" teaching, in which the educator's direct presence is traded for a more flexible environment where students are given responsibilities by means of teaching roles and goals.

7. Of course, "good" and "right" results vary from each teaching and learning situation to another. Our choices of and commitments to the different values of students' engagements in specific musical processes (e.g., improvising, composing, performing, and dancing) and various kinds of musical production and "products" (e.g., technological compositions, music videos, rock concerts, chamber music performances, jazz vocal recordings) and other situated musicking are determined by the general requirement that teachers do not violate their obligations to be welcoming, open-minded, respectful, and caring toward the people they teach. And because the values of music education depend on the varied aims, desires, and locations of teachers and learners, it is illogical and unethical to insist that there is one music education methodology, one set of "standards," or one set of governmental policies that are valid for every teacher and learner in every situation. In contrast to focusing on only questions that still occupy many music educators worldwide—including "What works?" and "How well?" (Bowman, 2007, p. 5)—music education also develops ethically guided "action plans" (or curricula based on "action ideals") and actions for specific groups of music students and teachers related to questions such as "Why?" "For whom?" "In what situations?" and "Under what circumstances?" However, regardless of how comprehensive, new, or "innovative" music education offerings seem or promise to be in schools and universities, there is no reasonable guarantee that students will benefit from curricular "action ideals" and "action plans" plans that lack teachers' primary responsibility to grapple with deep and significant educational questions comprehensively and ethically long before deciding "how-to" issues, not to mention eradicating disembodied input/output and transmission/reception models of education.

8. Martin Heidegger (1982) refers to this as an impoverished technological "enframing" (*Gestell*) that dominates the human understanding of being in the modern world. Here it is important to note that, for Heidegger (1982), "technology" does not first and foremost concern machines, nor is it necessarily a negative aspect of human existence. Rather, technology is a basic human potential, a central aspect of how we reveal the world to ourselves and make it intelligible as rational beings. However, a serious problem arises in the modern world when a fascination with reason, technology, and progress obscures other ways of knowing and being. In connection with this, Neil Evernden (1993) writes, "By describing something as a resource we seem to have cause to protect it. But all we really have is a license to exploit it" (p. 23).

9. These potentials connect with the aims of life philosophy—"connecting human life with the fundamental life" (Nakagawa, 2000, p. 79)—and in particular with the "Eastern" perspectives it draws on. Likewise, early enactivist thinking (Varela et al., 1991) is strongly influenced by Buddhist psychology and the mindful awareness tradition, which highlight the interpenetrative and co-arising nature of mind and world. These ideas are discussed in the context of music education by van der Schyff (2015).

References

Abraham, F. D., Krippner, S., & Richards, R. (2012). Dynamical concepts used in creativity and chaos. *NeuroQuantology, 10*(2), 177–182.

Abraham, R., & Shaw, C. (1985). *Dynamics: The geometry of behavior*. Santa-Cruz, CA: Aerial Press.

Adachi, M., & Trehub, S. E. (2012). Musical lives of infants. In G. McPherson & G. Welch (Eds.), *The Oxford handbook of music education* (pp. 229–247). New York: The Oxford University Press.

Adams, F., & Aizawa, K. (2009). Why the mind is still in the head. In P. Robbins & M. Aydede (Eds.), *The Cambridge handbook of situated cognition* (pp. 78–95). New York: Cambridge University Press.

Adorno, T. (1973). *Negative dialectics*. New York: Seabury Press.

Adorno, T., & Horkheimer, M. (2002). *Dialectic of enlightenment*. Stanford, CA: Stanford University Press.

Aizawa, K. (2014). The enactivist revolution. *Avant, 5*(2), 19–42.

Alperson, P. A. (1979). *The special status of music* [Unpublished doctoral dissertation]. University of Toronto.

Altenmüller, E. O. (2001). How many music centers are in the brain? *Annals of the New York Academy of Sciences, 930*, 273–280. https://doi.org/10.1111/j.1749-6632.2001.tb05738.x

Amazeen, E. L., Sternad, D., & Turvey, M. T. (1996). Predicting the nonlinear shift of stable equilibria in interlimb rhythmic coordination. *Human Movement Science, 15*, 521–542.

Ambrosini, E., Reddy, V., de Looper A., Costantini, M., Lopez, B., & Sinigaglia, C. (2013). Looking ahead: Anticipatory gaze and motor ability in infancy. *PLoS ONE, 8*(7), e6791610. https://doi.org/10.1371/journal.pone.0067916

Anderson, D. E., & Patel, A. D. (2018). Infants born preterm, stress, and neurodevelopment in the neonatal intensive care unit: might music have an impact? *Developmental Medicine & Child Neurology, 60,* 256–266. https://doi.org/10.1111/dmcn.13663

Anderson, M. L. (2007). Massive redeployment, exaptation, and the functional integration of cognitive operations. *Synthese, 159*(3), 329–345.

Anderson, M. L. (2014). *After phrenology: Neural reuse and the interactive brain.* Cambridge, MA: MIT Press.

Anderson, W. D. (1966). *Ethos and education in Greek music.* Cambridge, MA: Harvard University Press.

Anderson, W. D. (1994). *Music and musicians in Ancient Greece.* Ithaca, NY: Cornell University Press.

Arendt, H. (1958). *The human condition.* Chicago: University of Chicago Press.

Attneave, F. (1971). Multistability in perception. *Scientific American, 255*(6), 63–71.

Audissino, E. (2014). *John Williams' film music: Jaws, Star Wars, Raiders of the Lost Ark, and the return of the classical Hollywood music style.* Madison: University of Wisconsin Press.

Babikova, Z., Gilbert, L., Bruce, T. J., et al. (2013). Underground signals carried through common mycelial networks warn neighboring plants of aphid attack. *Ecology Letters, 16*(7), 835–843.

Bai, H. (2003). Learning from Zen arts: A lesson in intrinsic valuation. *Journal of the Canadian Association of Curriculum Studies, 1,* 1–14.

Bailey, D. (1993). *Improvisation: Its nature and practice in music.* New York: Da Capo Press.

Balkwill, L., & Thompson, W. F. (1999). A cross-cultural investigation of the perception of emotion in music: Psychophysical and cultural cues. *Music Perception, 17*(1), 43–64.

Balter, M. (2004). Seeking the key to music. *Science, 306*(5699), 1120–1122.

Bamberger, J. (2006). What develops in musical development? A view of development as learning. In G. McPherson (Ed.), *The child as musician: Musical development from conception to adolescence* (pp. 69–92). New York: Oxford University Press.

Bamford, J. M. S., & Davidson, J. W. (2017). Trait empathy associated with agreeableness and rhythmic entrainment in a spontaneous movement to music task: Preliminary exploratory investigations. *Musicae Scientiae, 23*(1), 5–24.

Bandura, A. (1977). *Social learning theory.* Englewood Cliffs, NJ: Prentice Hall.

References

Bannan, N. (2016). Darwin, music, and evolution: New insights from family correspondence on The Descent of Man. *Musicae Scientiae, 21*, 3–25.

Barker, A. (1989). *Greek musical writings* (Vol. 1). Cambridge: Cambridge University Press.

Barker, A. (2007). *The science of harmonics in classical Greece*. Cambridge: Cambridge University Press.

Baron-Cohen, S. (2011). *Zero degrees of empathy: A new theory of human cruelty and kindness*. London: Allen Lane.

Baron-Cohen, S., Leslie, A. M., & Frith, U. (1985). Does the autistic child have a "theory of mind"? *Cognition, 21*(1), 37–46.

Barrett, H. C., & Kurzban, R. (2006). Modularity in cognition: Framing the debate. *Psychological Review, 113*(1), 628–647.

Barrett, L. (2011). *Beyond the brain: How body and environment shape animal and human minds*. Princeton, NJ: Princeton University Press.

Barrett, L. (2015a). A better kind of continuity. *Southern Journal of Philosophy— Supplement: Alternative Models of the Mind, 53*, 28–49.

Barrett, L. (2015b). Why brains aren't computers, why behaviorism isn't Satanism and why dolphins aren't aquatic apes. *The Behavior Analyst, 39*, 9–23.

Barrett, L. (2018). The evolution of cognition: A 4E perspective. In A. Newen, L. De Bruin, & S. Gallagher (Eds.), *The Oxford handbook of 4E cognition* (pp. 719–34). New York: Oxford University Press.

Barrett, L. F. (2006). Solving the emotion paradox: Categorization and the experience of emotion. *Personality and Social Psychology Review: An Official Journal of the Society for Personality and Social Psychology Inc., 10*(1), 20–46.

Barrett, L. F., & Satpute, A. (2013). Large-scale brain networks in affective and social neuroscience: towards an integrative architecture of the human brain. *Current Opinion in Neurobiology, 23*, 361–372.

Barsalou, L. W. (2005). Continuity of the conceptual system across species. *Trends in Cognitive Science, 9*, 309–311.

Barto, K. E., Hilker, M., Müller, F., Mohney, B. K., Weidenhamer, J. D., & Rillig, M. C. (2011). The fungal fast lane: Common mycorrhizal networks extend bioactive zones of allelochemicals in soils. *PLoS ONE, 6*(11), e27195. https://doi.org/10.1371/journal.pone.0027195

Batacharya, S., & Wong, Y.-L. R. (Eds.). (2018). *Sharing breath: Embodied learning and decolonization*. Edmonton, AB: Athabasca University Press.

Bateson, G. (1972). *Steps to an ecology of mind*. New York: Ballantine Books.

Bateson, M. C. (1975). Mother infant exchanges: The epigenesis of conversational interaction. In D. Aronson & R. W. Reiber (Eds.), *Annals of the New York Academy of Sciences: Developmental psycholinguistics and developmental disorders* (Vol. 263, pp. 101–113). New York: New York Academy of Sciences.

Bateson, P. (2003). Behavioral development and Darwinian evolution. In S. Oyama et al. (Eds.), *Cycles of contingency: Developmental systems and evolution*. Cambridge, MA: MIT Press.

Bateson, P., & Mameli, M. (2007). The innate and the acquired: Useful clusters or a residual distinction from folk biology? *Developmental Psychology, 49*(8), 818–831.

Bechtel, W. (1990). Representations and cognitive explanations: Assessing the dynamicist's challenge in cognitive science. *Cognitive Science, 22*, 295–318.

Bechtel, W. (2008). *Mental mechanisms: Philosophical perspectives on cognitive neuroscience*. London: Routledge.

Bechtel, W., Abrahamsen, A., & Graham, G. (Eds.) (1998). *A companion to cognitive science*. Oxford: Blackwell.

Becker, J. (2010). Exploring the habitus of listening: anthropological perspectives. In P. Juslin & J. Sloboda (Eds.), *Handbook of music and emotion: Theory, research, applications* (pp. 127–157). Oxford: Oxford University Press.

Beer, R. D. (1995a). Computational and dynamical languages for autonomous agents. In R. Port & T. van Gelder (Eds.), *Mind as motion* (pp. 121–147). Cambridge, MA: MIT Press.

Beer, R. D. (1995b). A dynamical systems perspective on agent-environment interaction. *Artificial Intelligence, 72*, 173–215.

Beer, R. D. (2003). The dynamics of active categorical perception in an evolved model agent. *Adaptive Behavior, 11*(4), 209–243.

Beer, R. D. (2005). Dynamical approaches to cognitive science. *Trends in Cognitive Science, 4*, 91–99.

Bekius, A., Cope, T. E., & Grube, M. (2016). The beat to read: A cross-lingual link between rhythmic regularity perception and reading skill. *Frontiers in Human Neuroscience*, 10:425. https://doi.org/10.3389/fnhum.2016.00425

Benedek, M., Kenett, Y., Umdasch, K., Anaki, D., Faust, M., & Neubauer, A. (2017). How semantic memory structure and intelligence contribute to creative thought: A network science approach. *Thinking and Reasoning, 23*, 158–183.

Benedek, M., Könen, T., & Neubauer, A. C. (2012). Associative abilities underlying creativity. *Psychology of Aesthetics, Creativity, and the Arts, 6*(3), 273–281.

References

Benjamin, W. (2008). *The work of art in the age of its technological reproducibility and other writings on media*. Cambridge, MA: Harvard University Press.

Benson, C. (2001). *The cultural psychology of self: Place, morality, and art in human worlds*. London: Routledge.

Benzon, W. L. (2001). *Beethoven's anvil: Music in mind and culture*. New York: Basic Books.

Berleant, A. (1986). Cultivating an urban aesthetic. *Diogenes, 34*(136), 1–18.

Berliner, P. (1994). *Thinking in jazz: The infinite art of improvisation*. Chicago: University of Chicago Press.

Bernstein, L. (1976). *The unanswered question: Six talks at Harvard*. Cambridge, MA: Harvard University Press.

Bertinetto, A. G. (2020). Music is not an emoticon: against a reductionist account of musical expressivity. In F. Scassillo (Ed.), *Resounding spaces. Approaching musical atmospheres* (pp. 47–58). Rome: Mimesis International.

Besson, M., & Schön, D. (2012). What remains of modularity? In P. Rebuschat, M. Rohrmeier, J. Hawkins, & I. Cross (Eds.) *Language and music as cognitive systems* (pp. 283–291). Oxford: Oxford University Press.

Bhabha, H. K. (2004). *The location of culture*. Abingdon: Routledge.

Bickerton, D. (1990). *Language and species*. Chicago: University of Chicago Press.

Bickerton, D. (2002). Foraging versus social intelligence in the evolution of protolanguage. In A. Wray (Ed.), *The transition to language* (pp. 207–225). Oxford: Oxford University Press.

Bishop, L. (2018). Collaborative musical creativity: How ensembles coordinate spontaneity. *Frontiers in Performance Science, 9*, 1285. doi: 10.3389/fpsyg.2018.01285

Bittman, B., Berk, L., Shannon, M., Sharaf, M., Westengard, J., Guegler, K. J., & Ruff, D. W. (2005). Recreational music-making modulates the human stress response: A preliminary individualized gene expression strategy. *Medical Science Monitor, 11*, 31–40.

Bittman, B., Croft, D. T., Brinker, J., van Laar, R., Vernalis, M. N., & Ellsworth, D. L. (2013). Recreational music-making alters gene expression pathways in patients with coronary heart disease. *Medical Science Monitor, 19*, 139–147. https://doi.org/10.12659/MSM.883807

Blacking, J. (1976). *How musical is man?* London: Faber.

Blacking, J. (1995). *Music, culture and experience*. Chicago: University of Chicago Press.

Blake, E. C., & Cross I. (2008). Flint tools as portable sound-producing objects in the Upper Paleolithic context: An experimental study. In P. Cunningham, J. Heeb,

& R. Paardekooper (Eds.), *Experiencing archaeology by experiment* (pp. 1–19). Oxford: Oxbow Books.

Blesser, B., & Salter, L. R. (2006). *Places speak, are you listening? Experiencing aural architecture.* Cambridge, MA: MIT Press.

Block, N. (1980). What is functionalism? In N. Block (Ed.), *Readings in the philosophy of psychology* (Vol. I). Cambridge, MA: Harvard University Press.

Block, N. (1990). Inverted earth. In J. Tomberlin (Ed.), *Philosophical perspectives 4, action theory and philosophy of mind* (pp. 53–79). Atascadero, CA: Ridgeview.

Block, N. (1994). Qualia. In S. Guttenplan (Ed.), *A companion to the philosophy of mind* (pp. 214–220). Oxford: Blackwell.

Bloom, A. (1968). *The Republic of Plato: Translated with an interpretive essay.* New York: Basic Books.

Boden, M. A. (1998). Creativity and artificial intelligence. *Artificial Intelligence, 103,* 347–356.

Boden, M. A. (2004). *The creative mind: Myths and mechanisms.* London: Routledge.

Bohlman, P. (1999). Ontologies of music. In N. Cook & M. Everist (Eds.), *Rethinking music* (pp. 17–34). Oxford: Oxford University Press.

Borgo, D. (2005). *Sync or swarm: Improvising music in a complex age.* New York: Continuum.

Bourgine, P., & Stewart, J. (2004). Autopoiesis and cognition. *Artificial Life, 10*(3), 327–345.

Bowman, W. (1991). An essay review of Bennett Reimer's "A philosophy of music education." *The Quarterly Journal of Music Teaching and Learning, 2*(3), 76–87.

Bowman, W. (2004). Cognition and the body: Perspectives from music education. In L. Bresler (Ed.), *Knowing bodies, moving minds: Toward embodied teaching and learning* (pp. 29–50). Netherlands: Kluwer Academic Press.

Bowman, W. (2007). Who is the "we"? Rethinking professionalism in music education. *Action, Criticism, and Theory for Music Education, 6*(4), 109–131.

Bradley, D. (2012). Good for what, good for whom? Decolonizing music education philosophies. In W. Bowman & L. Frega (Eds.), *The handbook of philosophy in music education* (pp. 409–433). New York: Oxford University Press.

Bredberg, G. (1968). Cellular pattern and nerve supply of the human organ of Corti. *Acta Otolaryngol Suppl* (Stockh) *236,* 1–135.

Brooks, R. (1986). A robust layered control system for a mobile robot. *IEEE Journal of Robotics and Automation, 2*(1), 14–23.

References

Brooks, R. (1989). A robot that walks: Emergent behaviors from a carefully evolved network. *Neural Computation, 1*(2), 253–262.

Brooks, R. (1991). Intelligence without representation. *Artificial Intelligence, 47*, 139–159.

Brown, M. (2005). *Explaining tonality: Schenkerian theory and beyond*. Rochester, NY: University of Rochester Press.

Brown, S. (2000). The "musilanguage" model of human evolution. In N. L. Wallin, B. Merker, & S. Brown (Eds.), *The origins of music* (pp. 271–300). Cambridge, MA: MIT Press.

Buccino, G., Lui, F., Canessa, N., Patteri, I., Lagravinese, G., Benuzzi, F., Porro, C. A., & Rizzolatti, G. (2004). Neural circuits involved in the recognition of actions performed by non-conspecifics: An fMRI study. *Journal of Cognitive Neuroscience, 16*(1), 114–126.

Budd, M. (1985). *Music and the emotions: The philosophical theories*. London: Routledge.

Budd, M. (1989). Music and the communication of emotion. *The Journal of Aesthetics and Art Criticism, 47*(2), 129–138.

Bull, M. (2000). *Sounding out the city: Personal stereos and the management of everyday life*. Oxford: Bloomsbury Academic.

Bull, M. (2008). *Sound moves: iPod culture and urban experience*. New York: Routledge.

Bundrick, S. (2005). *Music and image in classical Athens*. Cambridge: Cambridge University Press.

Bunt, L. (1994). *Music therapy: An art beyond words*. London: Routledge.

Bunt, L. (1997). Clinical and therapeutic uses of music. In D. J. Hargreaves & A. C. North (Eds.), *The social psychology of music* (pp. 249–267). London: Routledge.

Burnard, P. (2012). *Musical creativities in practice*. Oxford: Oxford University Press.

Burnard, P., & Dragovic, T. (2014). Characterizing communal creativity in instrumental group learning. *Departures in Critical Qualitative Research, 3*(3), 336–362.

Byrne, R. M. J. (2005). *The rational imagination: How people create counterfactual alternatives to reality*. Cambridge, MA: MIT Press.

Campbell, P. S. (2009). Learning to improvise music, improvising to learn. In G. Solis & B. Nettl (Eds.), *Musical improvisation: Art, education, and society* (pp. 119–142). Urbana: University of Illinois Press.

Camras, L. A., & Witherington, D. C. (2005). Dynamical system approaches in emotional development. *Developmental Review, 25*, 328–350.

Cannon, E. N., Simpson, E. A., Fox, N. A., Vanderwert, R. E., Woodward, A. L., & Ferrari, P. F. (2016). Relations between infants' emerging reach-grasp competence and event-related desynchronization in EEG. *Developmental Science, 19*, 50–62.

Cannon, E. N., Woodward, A. L., Gredebäck, G., von Hofsten, C., & Turek, C. (2012). Action production influences 12-month-old infants' attention to others' actions. *Developmental Science, 15*(1), 35–42.

Cano, R. L. (2006). What kind of affordances are musical affordances? A semiotic approach. Paper presented at *L'ascolto muicale: condotte, pratiche, grammatiche. Terzo Simposio Internazionale sulle Scienze del Linguaggio Musicale*. Bologna, February 23–25, 2006. http://www.geocities.ws/lopezcano/Articulos/2006.What_Kind.pdf

Cappuccio, M. (2009). Constructing the space of action: From bio-robotics to mirror neurons. *World Futures, 65*, 126–132.

Capra, F. (1996). *The web of life: A new scientific understanding of living systems*. New York: Anchor Books.

Carman, T. (2008). *Merleau-Ponty*. London: Routledge.

Carr, W., & Kemmis, S. (1986). *Becoming critical: Education, knowledge and action research*. Philadelphia: Falmer Press.

Carruthers, P., & Smith, P. K. (1996). *Theories of theories of mind*. Cambridge: Cambridge University Press.

Céspedes-Guevara, J. (2016). *Towards a constructionist theory of musically-induced emotions* [Unpublished doctoral dissertation]. University of Sheffield.

Céspedes-Guevara, J., & Eerola, T. (2018). Music communicates affects, not basic emotions—A constructionist account of attribution of emotional meanings to music. *Frontiers in Psychology, 9*, 215. https://doi.org/10.3389/fpsyg.2018.00215

Chaiken, S., & Trope, Y. (1999). *Dual-process theories in social psychology*. New York: Guilford Press.

Chalmers, D. (1990). Why Fodor and Pylyshyn were wrong: The simplest refutation. *Proceedings of the 12th Annual Conference of the Cognitive Science Society* (pp. 340–347). Hillsdale, NJ: Lawrence Erlbaum.

Chalmers, D. (1996). *The conscious mind: In search of a fundamental theory*. Oxford: Oxford University Press.

Chalmers, D. (2017). Naturalistic dualism. In S. Schneider & M. Velmans (Eds.), *The Blackwell companion to consciousness, second edition* (pp. 363–373). New York: John Wiley & Sons.

Chemero, A. (2003). An outline of a theory of affordances. *Ecological Psychology, 15*(2), 181–195.

References

Chemero, A. (2009). *Radical embodied cognitive science*. Cambridge, MA: MIT Press.

Chemero, A., & Turvey, M. (2007). Hypersets, complexity, and the ecological approach to perception-action. *Biological Theory, 2*(1), 23–36.

Cheney, D., & Seyfarth, R. (2008). *Baboon metaphysics: The evolution of a social mind.* Chicago: University of Chicago Press.

Chomsky, N. (1975). *Reflections on language*. New York: Pantheon.

Chomsky, N. (1980). Rules and representations. *Behavioral and Brain Sciences, 3*(1), 1–61.

Churchland, P. M. (1983). Consciousness: The transmutation of a concept. *Pacific Philosophical Quarterly, 64*, 80–95.

Churchland, P. M. (1985). Reduction, qualia, and the direct introspection of brain states. *The Journal of Philosophy, 82*, 8–28.

Churchland, P. S. (2002). *Brain-wise: Studies in neurophilosophy*. Cambridge, MA: MIT Press.

Clark, A. (1989). *Microcognition: Philosophy, cognitive science, and parallel distributed processing*. Cambridge, MA: MIT Press.

Clark, A. (1990 [1995]). Connectionist minds. *Proceedings of the Aristotelian Society, 90*, 83–102. https://doi.org/10.1093/aristotelian/90.1.83

Clark, A. (1993). *Associative engines: Connectionism, concepts, and representational change*. Cambridge, MA: MIT Press.

Clark, A. (1997). *Being there: Putting brain body and world together*. Cambridge, MA: MIT Press.

Clark, A. (2001). *Mindware: An introduction to the philosophy of cognitive science*. Oxford: Oxford University Press.

Clark, A., & Chalmers, D. (1998). The extended mind. *Analysis, 58*, 7–19.

Clark, J. A. (2010). Relations of homology between higher cognitive emotions and basic emotions. *Biology & Philosophy, 25*(1), 75–94.

Clarke, E. F. (2005). *Ways of listening: An ecological approach to the perception of musical meaning*. Oxford: Oxford University Press.

Clarke, E. F. (2013). Music, space, and subjectivity. In G. Born (Ed.), *Music, sound, and space: Transformations of public and private experience* (pp. 90–110). Cambridge: Cambridge University Press.

Clarke, E. F. (2019). Empathy and the ecology of musical consciousness. In D. Clarke, R. Herbert, & E. F. Clarke (Eds.), *Music and consciousness II*. New York: Oxford University Press.

Clarke, E. F., & Davidson, J. W. (1998). The body in performance. In W. Thomas (Ed.), *Composition—performance—reception* (pp. 74–92). Aldershott: Ashgate.

Clarke, E. F., DeNora, T., & Vuoskoski, J. (2015). Music, empathy and cultural understanding. *Physics of Life Reviews, 15*, 61–88.

Clarke, E. F., Dibben, N., & Pitts, S. E. (Eds.) (2010). *Music and mind in everyday life.* Oxford: Oxford University Press.

Clayton, M., Herbert, T., & Middleton, R. (Eds.) (2012). *The cultural study of music.* London: Routledge.

Clearfield, M. (2011). Learning to walk changes infants' social interactions. *Infant Behavior and Development, 34* (1), 15–25.

Clifton, T. (1976). Music as constituted object. *Music and Man, 2*, 73–98.

Clifton, T. (1983). *Music as heard: A study in applied phenomenology.* New Haven, CT: Yale University Press.

Clore, G. L., & Ortony, A. (2000). Cognition in emotion; always, sometimes or never? In L. Nadel, R. Lane, & G. L. Ahern (Eds.), *Cognitive neuroscience of emotion* (pp. 24–61). Oxford: Oxford University Press.

Cochrane, T. (2010). Using the persona to express complex emotions in music. *Music Analysis, 29*(1–3), 264–275.

Colling, J. L., & Thomson, W. F. (2013). Music, action, affect. In T. Cochrane, B. Fantini, and K. R. Scherer (Eds.), *The emotional power of music: Multidisciplinary perspectives on musical arousal, expression, and social control* (pp. 197–212). Oxford: Oxford University Press.

Collins, S., & Kuck, K. (1990). Music therapy in the neonatal intensive care unit. *Neonatal Network, 9*(6), 23–26.

Colombetti, G. (2014). *The feeling body: Affective science meets the enactive mind.* Cambridge, MA: MIT Press.

Coltheart, M. (1999). Modularity and cognition. *Trends in Cognitive Science, 3*(3), 115–20.

Comotti, G. (1989). *Music in Greek and Roman culture* (R.V. Munson, Trans.). Baltimore, MD: Johns Hopkins University Press.

Cone, E. T. (1974). *The composer's voice.* Berkeley: University of California Press.

Cook, N. (2001). Theorizing musical meaning. *Music Theory Spectrum, 23*(2), 170–195.

Cook, N. (2006). Playing God: Creativity, analysis, and aesthetic inclusion. In I. Deliège, & G. Wiggins (Eds.), *Musical creativity: Multidisciplinary research in theory and practice* (pp. 9–24). New York: Psychology Press.

References

Cook, N. (2007). *The Schenker project: Culture, race, and music theory in Fin-de-siècle Vienna*. New York: Oxford University Press.

Cook, N. (2013). *Beyond the score: Music as performance*. Oxford: Oxford University Press.

Cooke, D. (1960). *The language of music*. Oxford: The Oxford University Press.

Coplan, A. (2011). Understanding empathy: its features and effects. In A. Coplan & P. Goldie (Ed.), *Empathy: Philosophical and psychological perspectives* (pp. 3–18). Oxford: Oxford University Press.

Cox, A. (2016). *Music and embodied cognition: Listening, moving, feeling, and thinking*. Bloomington: Indiana University Press.

Creech, A., Papageorgi, I., Duffy, C., Morton, F., Hadden, E., Potter, J., De Bezenac, C., Whyton, T., Himonides, E., & Welch, G. (2008). Investigating musical performance: Commonality and diversity among classical and non-classical musicians. *Music Education Research, 10*(2), 215–234.

Cropley, A. (2011). Definitions of creativity. In M. A. Runco & S. R. Pritzker (Eds.), *Encyclopedia of creativity* (2nd ed, pp. 358–368). Elsevier.

Cross, I. (1999). Is music the most important thing we ever did? Music, development and evolution. In S. W. Yi (Ed.), *Music, mind, and science* (pp. 10–39). Seoul: Seoul National University Press.

Cross, I. (2001). Music, mind and evolution. *Psychology of Music, 29*, 95–102.

Cross, I. (2003). Music and evolution: consequences and causes. *Contemporary Music Review, 22*(3), 79–89.

Cross, I. (2007). Music and cognitive evolution. In L. Barrett & R. Dunbar (Eds.), *Oxford handbook of evolutionary psychology*. Oxford: Oxford University Press.

Cross, I. (2010). The evolutionary basis of meaning in music: some neurological and neuroscientific implications. In F. Clifford (Ed.), *The neurology of music* (pp. 1–15). London: Imperial College Press.

Cross, I. (2012). Music and bio-cultural evolution. In M. Clayton & H. Herbert (Eds.), *The cultural study of music: A critical introduction* (pp. 17–27). London: Routledge.

Csikszentmihalyi, M. (1988). The flow experience and its significance for human psychology. In M. Csikszentmihalyi & I. S. Csikszentmihalyi (Eds.), *Optimal experience: Psychological studies of flow in consciousness* (pp. 15–35). New York: Cambridge University Press.

Csikszentmihalyi, M. (1990). *Flow: The psychology of optimal experience*. New York: Harper & Row.

Csikszentmihalyi, M. (1996). *Creativity: Flow and the psychology of discovery and invention*. New York: Harper/Collins.

Cuffari, E. C., Di Paolo, E., & De Jaegher, H. (2014). From participatory sense-making to language: there and back again. *Phenomenology and the Cognitive Sciences, 14,* 1089–1125.

Cummins, F. (2015). Rhythm and speech. In M. A. Redford (Ed.), *Blackwell handbooks in linguistics: The handbook of speech production* (1st ed., pp. 158–177). Hoboken, NY: Wiley.

Cummins, F., & De Jesus, P. (2016). The loneliness of the enactive cell: Towards a bio-enactive framework. *Adaptive Behavior, 24*(3), 149–159.

Cummins, F., & Port, R. F. (1996). Rhythmic commonalities between hand gestures and speech. In *Proceedings of the Eighteenth Annual Conference of the Cognitive Science Society* (pp. 415–419). Hillsdale, NJ: Lawrence Erlbaum Associates.

Currie, A., & Killin, A. (2016). Musical pluralism and the science of music. *European Journal for Philosophy of Science, 6,* 9–30.

Damasio, A. (1994). *Descartes' error: Emotion, reason, and the human brain.* New York: G. P. Putnam's Sons.

Damasio, A. (2003). *Looking for Spinoza: Joy, sorrow, and the feeling brain.* New York: Harvest.

Danto, A. (1986). *The philosophical disenfranchisement of art.* New York: Columbia University Press.

Darwin, C. R. (1871). *The descent of man, and selection in relation to sex.* London: John Murray.

Daum, M. M., Prinz, W., & Aschersleben, G. (2011). Perception and production of object-related grasping in 6-month-old infants. *Journal of Experimental Child Psychology, 108,* 810–818.

D'Ausilio, A. (2007). The role of the mirror system in mapping complex sounds into actions. *The Journal of Neuroscience, 27,* 5847–5848.

D'Ausilio, A. (2009). Mirror-like mechanisms and music. *The Scientific World Journal, 9,* 1415–1422.

Davidson, I. (2002). The "finished artefact fallacy": Acheulean hand axes and language origins. In A. Wray (Ed.), *Transitions to language* (pp. 180–203). Oxford: Oxford University Press.

Davidson, J. W. (1993). Visual perception of performance manner in the movements of solo musicians. *Psychology of Music, 21*(2), 103–113.

Davidson, J. W. (2005). Bodily communication in musical performance. In D. Miell, R. MacDonald, & D. J. Hargreaves (Eds.), *Musical communication* (pp. 215–228). Oxford: Oxford University Press.

References

Davidson, J. W. (2012). Bodily movement and facial actions in expressive musical performance by solo and duo instrumentalists: Two distinctive case studies. *Psychology of Music, 40*, 595–633.

Davidson, J. W., & Correia, J. S. (2001). Meaningful musical performance: A bodily experience. *Research Studies in Music Education, 17*(1), 70–83.

Davies, S. (1994). *Musical meaning and expression*. Ithaca, NY: Cornell University Press.

Davies, S. (1997). Contra the hypothetical persona in music. In M. Hjort & S. Laver (Eds.), *Emotion and the arts* (pp. 95–109). Oxford: Oxford University Press.

Davies, S. (2012). *The artful species: Aesthetics, art, and evolution*. Oxford: Oxford University Press.

Dawkins, R. (1976). *The selfish gene*. Oxford: Oxford University Press.

Deacon, T. W. (1997). *The symbolic species: The co-evolution of language and the human brain*. London: Penguin.

Deacon, T. W. (2010). On the human: Rethinking the natural selection of human language. https://nationalhumanitiescenter.org/on-the-human/2010/02/on-the-human-rethinking-the-natural-selection-of-human-language/

Deacon, T. W. (2012). *Incomplete nature: How mind emerged from matter*. New York: Norton.

De Jaegher, H. (2013). Rigid and fluid interactions with institutions. *Cognitive Systems Research, 25–26*(0), 19–25.

De Jaegher, H., & Di Paolo, E. A. (2007). Participatory sense-making: An enactive approach to social cognition. *Phenomenology and the Cognitive Sciences, 6*(4), 485–507.

De Jaegher, H., Di Paolo, E., & Gallagher, S. (2010). Can social interaction constitute social cognition? *Trends in Cognitive Sciences, 14*(10), 441–447.

De Jaegher, H., Pieper, B., & Clénin, D. (2017). Grasping intersubjectivity: An invitation to embody social interaction research. *Phenomenology and the Cognitive Sciences, 16*, 491–523.

De Jesus, P. (2015). Autopoietic enactivism, phenomenology and the deep continuity between life and mind. *Phenomenology and the Cognitive Sciences, 15*, 265–289.

De Jesus, P. (2016). Making sense of (autopoietic) enactive embodiment: A gentle appraisal. *Phainomena, 25*(98–99), 33–56.

Delalande, F. (Ed.) (2009). *La nascita della musica. Esplorazioni sonore nella prima infanzia*. Milano: Franco Angeli.

de Landa, M. (1992). Non-organic life. In J. Crary & S. Kwinter (Eds.), *Incorporations* (pp. 129–167). New York: Zone.

Demos, A. P., Chaffin, R., Begosh, K. T., Daniels, J. R., & Marsh, K. L. (2012). Rocking to the beat: Effects of music and partner's movements on spontaneous interpersonal coordination. *Journal of Experimental Psychology, 141*(1), 49–53.

Dennett, D. (1978). *Brainstorms.* Cambridge, MA: MIT Press.

Dennett, D. (1979). On the absence of phenomenology. In D. F. Gustafson & B. L. Tapscott (Eds.), *Body, mind, and method* (pp. 93–113). Dordrecht: Reidel.

Dennett, D. (1988). Quining qualia. In A. Marcel & E. Bisiach (Eds.), *Consciousness in modern science* (pp. 42–77). Oxford: Oxford University Press.

Dennett, D. (1991). *Consciousness explained.* Boston: Little, Brown and Company.

DeNora, T. (1986). How is extra-musical meaning possible? Music as a place and space for "work." *Sociological Theory, 4*, 84–94.

DeNora, T. (2000). *Music in everyday life.* New York: Cambridge University Press.

DeNora, T. (2011). *Music in action: Selected essays in sonic ecology.* Burlington, VT: Ashgate Publishing Company.

Desain, P., & Honing, H. (1991). Quantization of musical time: a connectionist approach. In P. M. Todd & G. Loy (Eds.), *Music and connectionism* (pp. 150–167). Cambridge, MA: MIT Press.

Desain, P., & Honing, H. (2003). The formation of rhythmic categories and metric priming. *Perception, 32*, 341–365.

De Souza, J. (2017). *Music at hand: Instruments, bodies, and cognition.* New York: Oxford University Press.

Deutsch, D. (1999). The processing of pitch combinations. In D. Deutsch (Ed.), *The psychology of music* (pp. 349–411). New York: Academic Press.

Dewey, J. (1938/1991). Logic: The theory of inquiry. In J. A. Boydston (Ed.), *John Dewey: The later works, 1925—1953* (Vol. 12, pp. 1–527). Carbondale, IL: SIU Press.

Di Paolo, E., Cuffari, E. C., & De Jaegher, H. (2018). *Linguistic bodies: The continuity between life and language.* Cambridge, MA: MIT Press.

Di Paolo, E. A. (2005). Autopoiesis, adaptivity, teleology, agency. *Phenomenology and the Cognitive Sciences, 4*, 97–125.

Di Paolo, E. A. (2009). Extended life. *Topoi, 28*, 9–21.

Di Paolo, E. A. (2020). Enactive becoming. *Phenomenology and the Cognitive Sciences*, https://doi.org/10.1007/s11097-019-09654-1

Di Paolo, E. A., Buhrmann, T., & Barandiaran, X. E. (2017). *Sensorimotor life: An enactive proposal.* New York: Oxford University Press.

di Pellegrino, G., Fadiga, L., Fogassi, L., Gallese, V., & Rizzolatti, G. (1992). Understanding motor events: A neurophysiological study. *Experimental Brain Research, 91*, 176–180.

Dissanayake, E. (1995). *Homo aestheticus: Where art comes from and why*. Seattle: University of Washington Press.

Dissanayake, E. (2000). Antecedents of the temporal arts in early mother-infant interaction. In N. L. Wallin, B. Merker & S. Brown (Eds.), *The origins of music* (pp. 389–410). Cambridge, MA: MIT Press.

Dixon, J. A., Stephen, D. G., Boncoddo, R. A., & Anastas, J. (2010). The self-organization of cognitive structure. In B. Ross (Ed.), *The psychology of learning & motivation* (Vol. 52, pp. 343–384). San Diego: Elsevier.

Doffman, M. (2009). Making it groove! Entrainment, participation and discrepancy in the "conversation" of a jazz trio. *Language and History, 52*(1), 130–147.

Doidge, N. (2007). *The brain that changes itself: Stories of personal triumph from the frontiers of brain science*. New York: Penguin.

Donald, M. (1991). *Origins of the modern mind: Three stages in the evolution of culture and cognition*. Cambridge, MA: Harvard University Press.

Donald, M. (2001). *A mind so rare: The evolution of human consciousness*. New York: Norton.

Dowling, W. J. (2010). Qualia as intervening variables in the understanding of music cognition. *Musica Humana, 2*, 1–20.

Drake, C., Jones, M. R., & Baruch, C. (2000). The development of rhythmic attending in auditory sequences: Attunement, referent period, focal attending. *Cognition, 77*, 251–288.

Dretske, F. (1995). *Naturalizing the mind*. Cambridge, MA: MIT Press.

Dreyfus, H. L. (1979). *What computers can't do*. New York: Harper and Row.

Dreyfus, H. L. (1992). *What computers still can't do: A critique of artificial reason*. Cambridge, MA: MIT Press.

Dreyfus, H. L. (2002). Intelligence without representation: Merleau-Ponty's critique of mental representation. *Phenomenology and the Cognitive Sciences, 1*, 367–383.

Dreyfus, H. L. (2013). The myth of the pervasiveness of the mental. In in J. K. Schear (Ed.), *Mind, reason, and being-in-the-world*. London: Routledge.

Dreyfus, H. L., & Kelly, S. D. (2011). *All things shining: Reading Western classics to find meaning in a secular age*. New York: Simon and Schuster.

Dreyus, H. L., & Spinosa, C. (1997). Highway bridges and feasts: Heidegger and Borgmann on how to affirm technology. *Continental Philosophy Review, 30*, 159–178. https://doi.org/10.1023/A:1004299524653

Dunbar, R. (1996). *Grooming, gossip and the evolution of language*. Cambridge, MA: Harvard University Press.

Dunbar, R. (2003). The origin and subsequent evolution of language. In M. H. Christiansen & S. Kirby (Eds.), *Language evolution* (pp. 219–234). Oxford: Oxford University Press.

Dunbar, R. (2012). On the evolutionary function of song and dance. In N. Bannan (Ed.), *Music, language, and human evolution* (pp. 201–214). Oxford: Oxford University Press.

Dunne, J. (1993). *Back to the rough ground*. London: University of Notre Dame Press.

Eagleton, T. (1991). *Ideology: An introduction*. London: Verso.

Eerola, T., & Vuoskoski, J. K. (2013). A review of music and emotion studies: Approaches, emotion models, and stimuli. *Music Perception, 30*(3), 307–340.

Eerola, T., Vuoskoski, J. K., & Kautiainen, H. (2016). Being moved by unfamiliar sad music is associated with high empathy. *Frontiers in Psychology*, 7:1176. https://doi.org/10.3389/fpsyg.2016.01176

Egbert, M., Barandiaran, X., & Di Paolo, E. (2010). A minimal model of metabolism-based chemotaxis. *PLoS Computational Biology, 6*(12), e1001004. https://doi.org/10.1371/journal.pcbi.1001004

Ehrenfels, M. (1890/1988). On gestalt qualities. In B. Smith (Ed.), *Foundations of Gestalt theory*. Munich, Vienna: Philosophia Verlag.

Eibl-Eibesfeldt, I. (1989). *Human ethology*. New York: Aldine de Gruyter.

Eitan, Z., & Granot, R. Y. (2006). How music moves: Musical parameters and listeners' images of motion. *Music Perception, 23*(3), 221–247.

Eitan, Z., & Timmers, R. (2010). Beethoven's last piano sonata and those who follow crocodiles: Cross-domain mappings of auditory pitch in a musical context. *Cognition, 114*(3), 405–422.

Ekman, P. (1980). Biological and cultural contributions to body and facial movement in the expression of emotions. In A. O. Rorty (Ed.), *Explaining emotions* (pp. 73–102). Berkeley: University of California Press.

Ekman, P. (1992). An argument for basic emotions. *Cognition & Emotion, 6*(3–4), 169–200.

Ekman, P. (2003). *Emotions revealed: Understanding faces and feelings*. London: Weidenfeld & Nicolson.

Ekman, P., & Cordaro, D. (2011). What is meant by calling emotions basic. *Emotion Review, 3*(4), 364–370.

References

Ekman, P., & Friesen, W. V. (1971). Constants across cultures in the face and emotion. *Journal of Personality and Social Psychology*, 17, 124–129.

Ekman, P., & Friesen, W. V. (1986). A new pan-cultural facial expression of emotion. *Motivation and Emotion*, 10, 159–168.

Elliott, D. J. (1989). Key concepts in multicultural music education. *International Journal of Music Education*, 13(1), 11–18.

Elliott, D. J. (1991). Music education as aesthetic education: A critical inquiry. *Quarterly Journal of Music Teaching and Learning*, 2(3), 57.

Elliott, D. J. (1995). *Music matters: A new philosophy of music education*. Oxford: Oxford University Press.

Elliott, D. J., & Silverman, M. (2015). *Music matters: A philosophy of music education* (2nd ed.). New York: Oxford University Press.

Ericsson, K. A., Hoffman, R. R., Kozbelt, A., & Williams, A. M. (Eds.) (2006). *The Cambridge handbook of expertise and expert performance*. Cambridge: Cambridge University Press.

Evernden, N. (1993). *The natural alien*. Toronto: University of Toronto Press.

Falk, D. (2000). Hominid brain evolution and the origins of music. In N. L. Wallin, B. Merker, & S. Brown (Eds.), *The origins of music* (pp. 197–216). Cambridge, MA: MIT Press.

Falk, D. (2004). Prelinguistic evolution in early hominins: Whence motherese? *Behavioral and Brain Sciences*, 27(4), 450–491.

Fan, Y., Duncan, N. W., de Greck, M., & Northoff, G. (2011). Is there a core neural network in empathy? An fMRI based quantitative meta-analysis. *Neuroscience and Biobehavioral Reviews*, 35, 903–911. https://doi.org/10.1016/j.neubiorev.2010.10.009

Fantasia, V., De Jaegher, H., & Fasulo, A. (2014). We can work it out: An enactive look at cooperation. *Frontiers in Psychology*, 5:874. https://doi.org/10.3389/fpsyg.2014.00874

Fantini, B. (2014). Forms of thought between music and science. In T. Cochrane, B. Fantini, K. R Scherer (Eds.), *The emotional power of music: Multidisciplinary perspectives on musical arousal, expression, and social control* (pp. 257–269). Oxford: Oxford University Press.

Finke, R. A., Ward, T. B., & Smith, S. M. (1992). *Creative cognition*. Cambridge, MA: MIT Press.

Finnegan, R. (2012). Music, experience, and the anthropology of emotion. In M. Clayton, T. Herbert, & R. Middleton (Eds.), *The cultural study of music: A critical introduction* (pp. 181–192). New York: Routledge.

Fitch, W. T. (2005). Protomusic and protolanguage as alternatives to protosign (comment on Arbib). *Behavioral & Brain Sciences, 28*(2), 132–133.

Fitch, W. T. (2006). The biology and evolution of music: A comparative perspective. *Cognition, 100*(1), 173–215.

Fitch, W. T. (2010). *The evolution of language*. Cambridge: Cambridge University Press.

Fitch, W. T. (2012). The biology and evolution of rhythm: unraveling a paradox. In P. Rebuschat, M. Rohmeier, J. A. Hawkins, and I. Cross (Eds.), *Language and music as cognitive systems* (pp. 73–95). Oxford: Oxford University Press.

Fitch, W. T. (2017). Cultural evolution: Lab-cultured musical universals. *Nature Human Behaviour, 1,* 1–2.

Fodor, J. A. (1983). *The modularity of mind*. Cambridge, MA: MIT Press.

Fodor, J. A. (2001). *The mind doesn't work that way: The scope and limits of computational psychology*. Cambridge, MA: MIT Press.

Fodor, J., & Piattelli-Palmarini, M. (2010). *What Darwin got wrong*. New York: Farrar, Straus, and Giroux.

Fogassi, L., Ferrari, P. F., Gesierich, B., Rozzi, S., Chersi, F., & Rizzolatti, G. (2005). Parietal lobe: From action organization to intention understanding. *Science, 308,* 662–7.

Fogel, A., Nwokah, E., Dedo, J. Y., Messinger, D., Dickson, K. L., Matusov, E., & Holt, S. A. (1992). Social process theory of emotion: A dynamic systems approach. *Social Development, 1,* 122–142.

Fogel, A., & Thelen, E. (1987). Development of early expressive and communicative action: Reinterpreting the evidence from a dynamical systems perspective. *Developmental Psychology, 23,* 747–761.

Freeman, W. (1999). *How brains make up their minds*. London: Orion Press.

Freeman, W. (2000). A neurobiological role of music in social bonding. In N. Wallin, B. Merker, & S. Brown (Eds.), *The origins of music*. Cambridge, MA: MIT Press.

Friston, K. (2009). The free energy principle: A rough guide to the brain? *Trends in Cognitive Sciences, 13*(1), 293–301.

Fritz, T., Jentschke, S., Gosselin, N., Sammler, D., Peretz, I., Turner, R., Friederici, A. D., & Koelsch, S. (2009). Universal recognition of three basic emotions in music. *Current Biology, 19*(7), 573–576.

Fritz, T. H., Halfpaap, J., Grahl, S., Kirkland, A., & Villringer, A. (2013). Musical feedback during exercise machine workout enhances mood. *Frontiers in Psychology,* 4:921. https://doi.org/10.3389/fpsyg.2013.00921

Froese, T., & Di Paolo, E. A. (2011). The enactive approach: Theoretical sketches from cell to society. *Pragmatics & Cognition, 19*(1), 1–36.

Fuchs, T. (2011). The brain—a mediating organ. *Journal of Consciousness Studies, 18*(7–8), 196–221.

Fuchs, T. (2017). *Ecology of the brain: The phenomenology and biology of the embodied mind*. New York: Oxford University Press.

Gabora, L. (2017). Honing theory: A complex systems framework for creativity. *Nonlinear Dynamics, Psychology, and Life Sciences, 21*(1), 35–88.

Gabrielsson, A. (2011). *Strong experiences with music* (R. Bradbury, Trans.). Oxford: Oxford University Press.

Gagliano, M. (2015). In a green frame of mind: perspectives on the behavioural ecology and cognitive nature of plants. *AoB PLANTS, 7*, plu075. https://doi.org/10.1093/aobpla/plu075

Gallagher, S. (2001). The practice of mind: Theory, simulation, or primary interaction? *Journal of Consciousness Studies, 8* (5–7), 83–107.

Gallagher, S. (2005). *How the body shapes the mind*. Oxford: Oxford University Press.

Gallagher, S. (2008a). Direct perception in the intersubjective context. *Consciousness and Cognition, 17*(2), 535–543.

Gallagher, S. (2008b). Understanding others: embodied social cognition. In P. Calva & T. Gomila (Eds.), *Handbook of cognitive science: An embodied approach* (pp. 441–449). San Diego: Elsevier.

Gallagher, S. (2011). Interpretations of embodied cognition. In W. Tschacher & C. Bergomi (Eds.), *The Implications of embodiment: Cognition and communication* (pp. 59–71). Exeter: Imprint Academic.

Gallagher, S. (2012). *Phenomenology*. New York: Palgrave Macmillan.

Gallagher, S. (2017). *Enactivist interventions: Rethinking the mind*. New York: Oxford University Press:

Gallagher, S. (2020). *Action and interaction*. New York: Oxford University Press.

Gallagher, S., & Crisafi, A. (2009). Mental institutions. *Topoi: An International Review of Philosophy, 28*(1), 45–51.

Gallagher, S., & Hutto, D. (2008). Understanding others through primary interaction and narrative practice. In J. Zlatev, T. Racine, C. Sinha, & E. Itkonen (Eds.), *The shared mind: Perspectives on intersubjectivity. Converging evidence in language and communication research* (Vol. 12, pp. 17–38). New York: John Benjamins Publishing Company.

Gallagher, S., & Varga, S. (2013). Social constraints on the direct perception of emotions and intentions. *Topoi, 33,* 185–199.

Gallagher, S., & Zahavi, D. (2008). *The phenomenological mind: An introduction to philosophy of mind and cognitive science.* New York: Routledge.

Gallese, V. (2001). The "shared manifold" hypothesis: from mirror neurons to empathy. *Journal of Consciousness Studies, 8*(5–7), 33–50.

Gallese, V. (2003a). The manifold nature of interpersonal relations: The quest for a common mechanism. *Philosophical Transactions: Biological Sciences, 358*(1431), 517–528.

Gallese, V. (2003b). The roots of empathy: The shared manifold hypothesis and the neural basis of intersubjectivity. *Psychopathology, 36* (4), 171–180.

Gallese, V. (2005). Embodied simulation: From neurons to phenomenal experience. *Phenomenology and the Cognitive Sciences, 4,* 23–48.

Gallese, V. (2009). Motor abstraction: A neuroscientific account of how action goals and intentions are mapped and understood. *Psychological Research, 73,* 486–498.

Gallese V. (2014). Bodily selves in relation: embodied simulation as second-person perspective on intersubjectivity. *Philosophical transactions of the Royal Society of London. Series B, Biological sciences, 369*(1644), 20130177. https://doi.org/10.1098/rstb.2013.0177

Gallese, V. (2017). Visions of the body: embodied simulation and aesthetic experience. *Aisthesis 1*(1), 41–50. https://doi.org/10.13128/Aisthesis-20902

Gallese, V., Fadiga, L., Fogassi, L., & Rizzolatti, G. (1996). Action recognition in the premotor cortex. *Brain, 119,* 593–609.

Gallese, V., & Goldman, A. (1998). Mirror neurons and the simulation theory of mind-reading. *Trends in Cognitive Science, 2*(12), 493–501.

Gallese, V., & Sinigaglia, C. (2011). How the body in action shapes the self. *Journal of Consciousness Studies, 18*(7–8), 117–143.

Galloway, G. C. (2005). In memoriam: Esther Thelen May 20, 1941–December 29, 2004. *Developmental Psychobiology, 47,* 103–107.

Gamble, C. (1999). *The paleolithic societies of Europe.* Cambridge: Cambridge University Press.

Gande, A., & Kruse-Weber, S. (2017). Addressing new challenges for a community music project in the context of higher music education: A conceptual framework. *London Review of Education, 15*(3), 372–387.

Gardner, H. (1985). *The mind's new science: A history of the cognitive revolution.* New York: Basic Books.

Gardner, H. (1993). *Frames of mind: The theory of multiple intelligences.* New York: Basic Books.

References

Garrido, S., & Schubert, E. (2011). Individual differences in the enjoyment of negative emotion in music: a literature review and experiment. *Music Perception, 28*, 279–296.

Gaser, C., & Schlaug, G. (2003). Brain structures differ between musicians and non-musicians. *Journal of Neuroscience, 23*(27), 9240–9245.

Gazzola, V., Aziz-Zadeh, L., & Keysers, C. (2006). Empathy and the somatotopic auditory mirror system in humans. *Current Biology, 16*, 1824–1829.

Geeves, A., McIlwain, D., Sutton, J., & Christensen, W. (2014). To think or not to think: The apparent paradox of expert skill in music performance. *Educational Philosophy and Theory, 46*(6), 674–691.

Gergely, G., Ndasdy, Z., Csibra, G., & Biro, G. (1995). Taking the intentional stance at 12 months of age. *Cognition, 56*, 165–193. https://doi.org/10.1016/0010-0277(95)00661-H

Gerson, S. A., Bekkering, H., & Hunnius, S. (2015a). Short-term motor training, but not observational training, alters neurocognitive mechanisms of action processing in infancy. *Journal of Cognitive Neuroscience, 27*, 1207–1214. https://doi.org/10.1162/jocn_a_00774

Gerson, S. A., Schiavio, A., Timmers, R., & Hunnius, S. (2015b). Active drumming experience increases infants' sensitivity to audiovisual synchronicity during observed drumming actions. *PLoS ONE, 10*(6), e0130960. https://doi.org/10.1371/journal.pone.0130960

Gibson, E. J. (1988). Exploratory behavior in the development of perceiving, acting, and the acquiring of knowledge. *Annual Review of Psychology, 39*, 1–42.

Gibson, J. J. (1966). *The senses considered as perceptual systems*. Boston: Houghton-Mifflin.

Gibson, J. J. (1979). *The ecological approach to visual perception*. Boston: Houghton-Mifflin.

Gilligan, C. (1982). *In a different voice: Psychological theory and women's development*. Cambridge: Harvard University Press.

Gimenes, M. (2013). Improved believability in agent-based computer musical systems designed to study music evolution. *International Symposium on Computer Music Multidisciplinary Research*, Marseille.

Gimenes, M., & Miranda, E. R. (2008). An A-life approach to machine learning of musical worldviews for improvisation systems. *Proceedings of 5th Sound and Music Computing Conference*, Berlin, Germany.

Gimenes, M., & Miranda, E. (2011). Emergent worldviews: An ontomemetic approach to musical intelligence. In E. R. Miranda (Ed.), *A-life for music: On music and computer models of living systems*. Middleton, WI: A-R Editions.

Giroux, H. (2011). *On critical pedagogy*. New York: Continuum.

Glăveanu, V. P. (2014). *Distributed creativity: Thinking outside the box of the creative individual*. Cham: Springer.

Godøy, R. I. (1997). Knowledge in music theory by shapes of musical objects and sound producing actions. In M. Leman (Ed.), *Music, gestalt and computing. Studies in cognitive and systematic musicology* (pp. 106–110). Berlin-Heidelberg: Springer Verlag.

Godøy, R. I. (2003). Motor-mimetic music cognition. *Leonardo, 36*, 317–319.

Godøy, R. I., Song, M., Nymoen, K., Haugen, M. R., & Jensenius, A. R. (2016). Exploring sound-motion similarity in musical experience. *Journal of New Music Research, 45*, 210–222.

Goehr, L. (1992). *The imaginary museum of musical works: An essay in the philosophy of music*. Oxford: Clarendon Press.

Goguen, J. (2004). Musical qualia, context, time and emotion. *Journal of Consciousness Studies, 11*, 117–147.

Goldman, A. (2006). *Simulating minds: The philosophy, psychology, and neuroscience of mindreading*. New York: Oxford University Press.

Goldman, A., & Gallese, V. (1998). Mirror neurons and the simulation theory of mind-reading. *Trends in Cognitive Sciences, 12*, 493–501.

Goodwin, B., Holder, N. & Wyles, C. (Eds.) (1983). *Development and evolution*. Cambridge: Cambridge University Press.

Gould, S. J. (2002). *The structure of evolutionary theory*. Cambridge, MA: Harvard University Press.

Gould, S. J., & Lewontin, R. (1979). The spandrels of San Marco and the Panglossian paradigm: A critique of the adaptationist programme. *Proceedings of the Royal Society*, B 205, 581–598.

Gould, S. J., & Vrba, E. S. (1982). Exaptation: A missing term in the science of form. *Paleobiology, 8*(1), 4–15.

Grahn, J. (2012). Advances in neuroimaging techniques: Implications for the shared syntactic resource hypothesis. In P. Rebuschat, M. Rohrmeier, J. Hawkins, & I. Cross (Eds.), *Language and music as cognitive systems* (pp. 235–241). Oxford: Oxford University Press.

Green, L. (2001). *How popular musicians learn: A way ahead for music education*. London: Routledge.

Green, L. (2008). *Music, informal learning and the school: A new classroom pedagogy*. London: Ashgate Press.

References

Greenberg, D. M., Rentfrow, P. J., & Baron-Cohen, S. (2015). Can music increase empathy? Interpreting musical experience through the empathizing—systemizing (E-S) theory: Implications for autism. *Empirical Musicology Review, 10*(1), 79–94.

Greene, M. (1995). *Releasing the imagination*. San Francisco: Jossey-Bass.

Greenhead, K. (2017). Applying Dalcroze principles to the rehearsal and performance of musical repertoire: A brief account of dynamic rehearsal and its origins. In S. Del Bianco, S. Morgenegg, & H. Nicolet (Eds.), *Pédagogie, art et science: L'apprentissage par et pour la musique selon la méthode Jaques-Dalcroze—actes de congrés de L'Institut Jaques-Dalcroze* (pp. 153–164). Geneva: HEM—Haute école de musique de Genève.

Gregory, A. H., & Varney, N. (1996). Cross-cultural comparisons in the affective response to music. *Psychology of Music, 24*(1), 47–52.

Griffith, N., & Todd, P. (Eds.) (1999). *Musical networks: Parallel distributed perception and performance*. Cambridge: MIT Press.

Gruhn, W. (2006). Music learning in schools: Perspectives of a new foundation for music teaching and learning. *Action, Criticism, and Theory for Music Education, 5*(2), 1–27.

Grush, R. (1997). The architecture of representation. *Philosophical Psychology, 10*(1), 5–23.

Grush, R. (2004). The emulation theory of representation: Motor control, imagery, and perception. *Behavioral and Brain Sciences, 27*, 377–442.

Gutierrez, J. (2019). An enactive approach to learning music theory? Obstacles and openings. *Frontiers in Education*, 4:133. https://doi.org/10.3389/feduc.2019.00133

Hagoort, P. (2005). On Broca, brain, and binding: A new framework. *Trends in Cognitive Sciences, 9*, 416–423.

Haken, H. (1977). *Synergetics—an introduction: Non-equilibrium phase transitions and self-organization in physics*. Berlin: Springer.

Haken, H., Kelso, J. A. S., & Bunz, H. (1985). A theoretical model of phase transitions in human hand movements. *Biological Cybernetics, 51*, 347–356.

Hallam, S., & Bautista, A. (2018). Processes of instrumental learning: the development of musical expertise. In G. McPherson, & G. F. Welch (Eds.), *Vocal, instrumental, and ensemble learning. An Oxford handbook of music education* (Vol. 3, pp. 108–125). New York: Oxford University Press.

Hannon, E. E., & Trehub, S. E. (2005). Tuning in to musical rhythms: Infants learn more readily than adults. *Proceedings of the National Academy of Sciences USA, 102*, 12639–12643.

Hanslick, E. (1891). *The beautiful in music: A contribution to the revisal of musical aesthetics* (G. Cohen, Trans.). London: Novello.

Hartigan, R., Adzenyah, A., & Donkor, F. (1995). *West African rhythms for the drumset*. New York: Alfred Music.

Hasanoglu, K. (2018). Accounting for the specious present: A defense of enactivism. *Journal of Mind and Behavior, 39*(3), 181–204.

Haslbeck, F. B. (2014). The interactive potential of creative music therapy with premature infants and their parents: A qualitative analysis. *Nordic Journal of Music Therapy, 23*, 36–70.

Haslbeck, F. B., & Bassler, D. (2018). Music from the very beginning: A neuroscience-based framework for music as therapy for preterm infants and their parents. *Frontiers in Behavioral Neuroscience*, 12:112. https://doi.org/10.3389/fnbeh.2018.00112

Haugeland, J. (Ed.) (1981). *Mind design: Philosophy, psychology artificial intelligence*. Cambridge, MA: MIT Press.

Haugeland, J. (1985). *Artificial intelligence: The very idea*. Cambridge, MA: MIT Press.

Haught-Tromp, C. (2017). The green eggs and ham hypothesis: How constraints facilitate creativity. *Psychology of Aesthetics, Creativity, and the Arts, 11*(1), 10–17.

Hauser, M., & McDermott, J. (2003). The evolution of the music faculty: A comparative perspective. *Nature Neuroscience, 6*(7), 663–668.

Hawkins, J. (1868). *A general history of the science and practice of music*. (Vol. 1). London: Alfred Novello.

He, M., Walle, E. A., & Campos, J. J. (2015). A cross-national investigation of the relationship between infant walking and language development. *Infancy, 20*, 283–305.

Hebb, D. O. (1949). *The organization of behavior*. New York: Wiley & Sons.

Heft, H. (2001). *Ecological psychology in context: James Gibson, Roger Barker, and the legacy of William James's radical empiricism*. Mahwah, NJ: Erlbaum.

Heidbreder, E. (1933). *Seven psychologies*. London: Century/Random House.

Heidegger, M. (1982). *The question concerning technology and other essays*. New York: Harper Perennial.

Heidegger, M. (1998). *Pathmarks*. Cambridge: Cambridge University Press.

Heims, S. (1980). *John van Neumann and Norbert Weiner*. Cambridge, MA: MIT Press

Held, V. (1993). *Feminist morality: Transforming culture, society, and politics*. Chicago: University of Chicago Press.

References

Hélie, S., & Sun, R. (2010). Incubation, insight, and creative problem solving: A unified theory and a connectionist model. *Psychological Review, 117*, 994–1024.

Henderson, S. M. (1983). Effects of a music therapy program upon awareness of mood in music, group cohesion, and self-esteem among hospitalized adolescent patients. *Journal of Music Therapy, 20*(1), 14–20.

Hergenhahn, B. R. (2001). *An introduction to the history of psychology* (4th ed.). Wadsworth: Thomson Learning.

Herholz, S. C., Lappe, C., Knief, A., & Pantev, C. (2008). Neural basis of music imagery and the effect of musical expertise. *European Journal of Neuroscience, 28*, 2352–2360.

Higgins, K. M. (2011). *The music of our lives*. New York: Lexington Books.

Higham, T., Basell, L., Jacobi, R., Wood, R., Ramsay, C. B., & Conrad, M. J. (2012). Testing models for the beginnings of the Aurignacian and the advent of figurative art and music: the radiocarbon chronology of Geißenklösterle. *Journal of Human Evolution, 62*, 664–676.

Ho, M., & Saunders, P. (1984). *Beyond neo-Darwinism*. New York: Academic Press.

Hobson, P. R. (1993). The emotional origins of social understanding. *Philosophical Psychology, 6*(3), 227–249.

Høffding, S. (2014). What is skilled coping? Experts on expertise. *Journal of Consciousness Studies, 21*(9–10), 49–73.

Høffding, S. (2019). *A phenomenology of musical absorption*. New York: Palgrave Macmillan.

Høffding, S., & Satne, G. (2019). Interactive expertise in solo and joint musical performance. *Synthese*, 1–19.

Honing, H., ten Cate, C., Peretz, I., & Trehub, S. (2015). Without it no music: Cognition, biology and evolution of musicality. *Philosophical Transactions of the Royal Society of London. Series B, Biological Sciences*, 370:20140088. https://doi.org/10.1098/rstb.2014.0088

Horgan, T. E., & Tienson, J. (1989). Representations without rules. *Philosophical Topics, 17*(1), 147–174.

Hove, M. J., & Risen, J. L. (2009). It's all in the timing: Interpersonal synchrony increases affiliation. *Social Cognition, 27*(6), 949–960.

Hsu, H. C., & A. Fogel. (2003). Stability and transitions in mother-infant face-to-face communication during the first six months: A microhistorical approach. *Developmental Psychology, 39*, 1061–1082.

Hubley, P., & Trevarthen, C. (1979). Sharing a task in infancy. *New Directions in Child and Adolescent Development, 4,* 57–80.

Hurley, S. (1999). *Consciousness in action.* London: Harvard University Press.

Hurley, S. (2001). Perception and action: Alternative views. *Synthese, 129,* 3–40.

Hurley, S. (2010). Varieties of externalism. In R. Menary (Ed.), *The extended mind* (pp. 101–153). Cambridge, MA: MIT Press.

Huron, D. (2001). Is music an evolutionary adaptation? *Annals of the New York Academy of Sciences, 930*(1), 43–61.

Huron, D. (2006). *Sweet anticipation: Music and the psychology of expectation.* Cambridge, MA: MIT Press.

Husserl, E. (1950/1992). *Ideen zu einer reinen phaenomenologie und phaenomenologischen philosophie 1: Allgemeine einfuehrung in die reine phaenomenologie* [Ideas pertaining to a pure phenomenology and to a phenomenological philosophy 1: General introduction to a pure phenomenology]. Text nach Husserliana III/1 und V. In E. Ströker (Ed.), *Gesammelte schriften/ Edmund Husserl [Collected Writings/ Edmund Husserl]* (Vol. 5). Hamburg: Meiner.

Husserl, E. (1960). *Cartesian meditations.* Dordrecht: Kluwer.

Husserl, E. (1962 [1913]). *Ideas: General introduction to pure phenomenology.* New York: Collier Books.

Husserl, E. (1991). *On the phenomenology of the consciousness of internal time (18931917)* (J. B. Brough, Ed. and Trans). Dordrecht: Kluwer.

Husserl, E. (2006). *The basic problems of phenomenology: From the lectures, winter semester, 1910–1911* (I. Farin & J. G. Hart, Trans.). Dordrecht: Springer.

Hutchison, W. D., Davis, K. D., Lozano, A. M., Tasker, R. R., & Dostrovsky, J. O. (1999). Pain-related neurons in the human cingulate cortex. *Nature Neuroscience, 2*(5), 403–405.

Hutto, D., & Myin, E. (2012). *Radicalizing enactivism: Basic minds without content.* Cambridge, MA: MIT Press.

Hutto, D., & Myin, E. (2017). *Evolving enactivism: Basic minds meet content.* Cambridge, MA: MIT Press.

Ihde, D. (1976). *Listening and voice: A phenomenology of sound.* Athens: Ohio University Press.

Ihde, D. (1977). *Experimental phenomenology: An introduction.* New York: Putnam.

Imada, T. (2012). The grain of the music: does music education "mean" something in Japan? In W. Bowman & L. Frega (Eds.) *The Oxford handbook of philosophy in music education* (pp. 147–162). Oxford: Oxford University Press.

References

Imberty, I. (1995). Développement linguistique et musical de l'enfant d'âge préscolaire et scolaire [Lingistic and musical development of the preschool and schoolage child]. In I. Deliege, & J. Sloboda (Eds.), *Naissance et développement du sens musical* [Birth and development of the musical sense] (pp. 223–250). Paris: PUF.

Ingold, T. (1999). Tools for the hand, language for the face: An appreciation of Leroi-Gourhan's gesture and speech. *Studies in History and Philosophy of Science Part C, 30*(4), 411–445.

Innes-Ker, A., & Niedenthal, P. M. (2002). Emotion concepts and emotional states in social judgment and categorization. *Journal of Personality and Social Psychology, 83*(4), 804–816.

Iversen, J. R. (2016). In the beginning was the beat: evolutionary origins of musical rhythm in humans. In R. Hartenberger (Ed.), *The Cambridge companion to percussion* (pp. 281–295). Cambridge: Cambridge University Press.

Iyer, V. (2002). Embodied mind, situated cognition, and expressive microtiming in African-American music. *Music Perception, 19*(3), 387–414.

Iyer, V. (2004). Improvisation, temporality, and embodied experience. *Journal of Consciousness Studies, 11*(3–4), 159–173.

Jablonka, E., & Lamb, M. (2005). *Evolution in four dimensions.* Cambridge, MA: MIT Press.

Jabr, F. (2012). How brainless slime molds redefine intelligence. *Scientific American*, 7 Nov. 2012. https://www.scientificamerican.com/article/brainless-slime-molds/

Jackendoff, R. (1987). *Consciousness and the computational mind.* Cambridge, MA: MIT Press.

Jackson, F. (1982). Epiphenomenal qualia. *The Philosophical Quarterly, 32*, 127–136.

Jackson, P. L., Meltzoff, A. N., & Decety, J. (2005). How do we perceive the pain of others? A window into the neural processes involved in empathy. *Neuroimage, 24*(3), 771–779.

James, W. (1911/1979). Percept and concept. In W. James (Ed.), *Some problems of philosophy* (pp. 21–60). Cambridge, MA: Harvard University Press.

Jäncke, L., Gaab, N., Wüstenberg, T., Scheich, H., & Heinze, H. J. (2001). Short-term functional plasticity in the human auditory cortex: An fMRI study. *Brain Research (Cognitive Brain Research), 12*(3), 479–485.

Jaques-Dalcroze, É. (1914/1967). *Rhythm, music, and education* (H. F. Rubenstein, Trans.). London: The Dalcroze Society.

Jeannerod, M., Arbib, M. A., Rizzolatti, G., & Sakata, H. (1995). Grasping objects: the cortical mechanisms of visuomotor transformation. *Trends in Neuroscience, 18*, 314–320.

Johnson, M. (1987). *The body in the mind: The bodily basis of imagination, reason, and meaning*. Chicago: University of Chicago Press.

Johnson, M. (2007). *The meaning of the body: Aesthetics of human understanding*. Chicago: University of Chicago Press.

Jonas, H. (1966). *The phenomenon of life*. Chicago: University of Chicago Press.

Jones, M. R. (2009). Musical time. In S. Hallam, I. Cross, & M. Thaut (Eds.), *Oxford handbook of music psychology* (1st ed., pp. 81–92). Oxford: Oxford University Press.

Jovanov, E., & Maxfield, M. C. (2011). Entraining the brain and body. In J. Berger & G. Turow (Eds.), *Music, science and the rhythmic brain: Cultural and clinical implications* (pp. 31–48). London: Routledge.

Juntunen, M. L. (2016). The Dalcroze approach: experiencing and knowing music through embodied exploration. In C. R. Abril & B. M. Gault (Eds.), *Approaches to teaching general music: Methods, issues, and viewpoints* (pp. 141–167). New York: Oxford University Press.

Juntunen, M. L., & Hyvönen, L. (2004). Embodiment in musical knowing: How body movement facilitates learning within Dalcroze eurhythmics. *British Journal of Music Education, 21*, 199–214.

Juslin, P. N. (2000). Cue utilization in communication of emotion in music performance: Relating performance to perception. *Journal of Experimental Psychology: Human Perception and Performance, 26*, 1797–1813.

Juslin, P. N. (2013a). From everyday emotions to aesthetic emotions: Towards a unified theory of musical emotions. *Physics of Life Reviews, 10*, 235–266.

Juslin, P. N. (2013b). What does music express? Basic emotions and beyond. *Frontiers in Psychology,* 4:596. https://doi.org/10.3389/fpsyg.2013.00596

Juslin, P. N. (2019). *Musical emotions explained. Unlocking the secrets of musical affect*. New York: Oxford University Press.

Juslin, P. N., & Laukka, P. (2003). Communication of emotions in vocal expression and music performance: Different channels, same code? *Psychological Bulletin, 129*(5), 770–814.

Juslin, P. N., & Laukka, P. (2004). Expression, perception, and induction of musical emotions: A review and a questionnaire study of everyday listening. *Journal of New Music Research, 33*(3), 217–238.

Juslin, P. N., Liljeström, S., & Västfjäll, D. (2010). How does music evoke emotions? Exploring the underlying mechanisms. In P. N. Juslin & J. A. Sloboda (Eds.), *Handbook of music and emotion: Theory, research, applications* (pp. 605–642). Oxford: Oxford University Press.

References

Juslin, P. N., & Lindström, E. (2010). Musical expression of emotions: Modelling listeners' judgements of composed and performed features. *Music Analysis, 29*(1/3), 334–364.

Juslin, P. N., & Sloboda, J. A. (Eds.) (2010). *Oxford handbook of music and emotion*. Oxford: Oxford University Press.

Juslin, P. N., & Västfjäll, D. (2008). Emotional responses to music: The need to consider underlying mechanisms. *Behavioral and Brain Sciences, 31*, 559–575.

Justus, T., & Hutsler, J. J. (2005). Fundamental issues in the evolutionary psychology of music: Assessing innateness and domain-specificity. *Music Perception, 23*(3), 1–27.

Kaminski, J., & Hall, W. (1996). The effect of soothing music on neonatal behavioral states in the hospital newborn nursery. *Neonatal Network, 15*(1), 45–54.

Kanakogi, Y., & Itakura S. (2011). Developmental correspondence between action prediction and motor ability in early infancy. *Nature Communications, 2*, 341.

Kanduri, C., Kuusi, T., Ahvenainen, M., Philips, A. K., Lähdesmäki, H., & Järvelä, I. (2015). The effect of music performance on the transcriptome of professional musicians. *Scientific Reports* 5:9506. https://doi.org/10.1038/srep09506

Kant, I. (2001[1790]). *Kritik der Urteilskraft*. Hamburg: Meiner.

Karmiloff-Smith, A. (1992). *Beyond modularity: A developmental approach to cognitive science*. Cambridge, MA: MIT Press.

Käufer, S., & Chemero, A. (2015). *Phenomenology: An introduction*. Cambridge: Polity Press.

Kaufman, J. C., & Beghetto, R. A. (2009). Beyond big and little: The four C model of creativity. *Review of General Psychology, 13*(1), 1–12.

Kawakami, A., & Katahira, K. (2015). Influence of trait empathy on the emotion evoked by sad music and on the preference for it. *Frontiers in Psychology, 6*:1541. https://doi.org/10.3389/fpsyg.2015.01541

Keefe, D. H., Bulen, J. C., Campbell, S. L., & Burns, E. M. (1994). Pressure transfer function and absorption cross section from the diffuse field to the human infant ear canal. *Journal of the Acoustical Society of America, 95*(1), 355–371.

Keller, P. (2001). Attentional resource allocation in musical ensemble performance. *Psychology of Music, 29*, 20–38.

Kelso, J. A. S. (1984). Phase transitions and critical behavior in human bimanual coordination. *American Journal of Physiology: Regulatory, Integrative, and Comparative, 15*, R1000–R1004.

Kelso, J. A. S. (1995). *Dynamic patterns*. Cambridge, MA: MIT Press.

Kennick, W. E. (1961). Art and the ineffable. *The Journal of Philosophy, 58,* 309–320.

Keysers, C. (2007). Mirror neurons. *New Encyclopaedia of Neuroscience.* Amsterdam: Elsevier Press.

Killin, A. (2013). The arts and human nature: Evolutionary aesthetics and the evolutionary status of art behaviors. *Biology & Philosophy, 28,* 703–718.

Killin, A. (2016a). Rethinking music's status as adaptation versus technology: a niche construction perspective. *Ethnomusicology Forum, 25*(2), 210–233.

Killin, A. (2016b). Musicality and the evolution of mind, mimesis, and entrainment. *Biology & Philosophy, 31,* 421–434.

Killin, A. (2017). Coevolution of cognition, sociality, and music. *Biological Theory, 12,* 222–235.

Kimmel, M., Hristova, D., & Kussmaul, K. (2018). Sources of embodied creativity: Interactivity and ideation in contact improvisation. *Behavioral Sciences, 8*:52. https://doi.org/10.3390/bs8060052

Kimmel, M., & Rogler, C. R. (2018). Affordances in interaction: The case of aikido. *Ecological Psychology, 30*(3), 195–223.

Kincheloe, J. L. (2003). Critical ontology: Visions of selfhood and curriculum. *Journal of Curriculum Theorizing, 19,* 47–64.

Kincheloe, J. L. (2008). *Knowledge and critical pedagogy: An introduction.* London: Springer.

Kirchhoff, M. D. (2012). Extended cognition and fixed properties: Steps to a third-wave version of extended cognition. *Phenomenology and the Cognitive Sciences, 11*(2), 287–308.

Kirschner, S., & Tomasello, M. (2010). Joint music making promotes prosocial behavior in 4-year-old children. *Evolution and Human Behavior, 31*(5), 354–364.

Kirsh, D. (1995). Complementary strategies: Why we use our hands when we think. In *Proceedings of the Seventeenth Annual Conference of the Cognitive Science Society.* Hillsdale, NJ: Lawrence Erlbaum Associates.

Kirton, M. J. (2003). *Adaption-innovation: In the context of diversity and change.* New York: Routledge.

Kiverstein, J., & Miller, M. (2015). The embodied brain: Towards a radical embodied cognitive neuroscience. *Frontiers in Human Neuroscience, 9*:237. https://doi.org/10.3389/fnhum.2015.00237

Kivy, P. (1980). *The corded shell: Reflections on musical expression.* Princeton, NJ: Princeton University Press.

Kivy, P. (1990). *Music alone: Philosophical reflections on the purely musical experience.* Ithaca, NY: Cornell University Press.

Kivy, P. (2002). *Introduction to a philosophy of music.* Oxford: Clarendon Press.

Klahr, D., & Wallace, J. G. (1976). *Cognitive development: An information-processing view.* New York: Lawrence Erlbaum.

Knieter, G. L. (1979). Music as aesthetic education. *NASSP Bulletin, 63*(430), 19–27.

Knight, S., Spiro, N., & Cross, I. (2017). Look, listen and learn: Exploring effects of passive entrainment on social judgements of observed others. *Psychology of Music, 45,* 99–115.

Knittel, K. M. (1998). Wagner, deafness, and the reception of Beethoven's late style. *Journal of the American Musicological Society, 51*(1), 49–82.

Knoblich, G., & Sebanz, N. (2008). Evolving intentions for social interaction: From entrainment to joint action. *Philosophical transactions of the Royal Society of London. Series B, Biological sciences, 363,* 2021–2031.

Koelsch, S. (2005). Neural substrates of processing syntax and semantics in music. *Current Opinion in Neurobiology, 15*(1), 207–212.

Koelsch, S. (2012). Response to target article "Language, music and the brain: A resource-sharing framework." In P. Rebuschat, M. Rohrmeier, J. Hawkins & I. Cross (Eds.), *Language and music as cognitive systems* (pp. 224–234). Oxford: Oxford University Press.

Koelsch, S. (2013). Striking a chord in the brain: The neurophysiological correlates of music-evoked positive emotions. In T. Cochrane, B. Fantini, and K. Scherer (Eds.), *The emotional power of music: Multidisciplinary perspectives on musical expression, arousal, and social control* (pp. 177–196). Oxford: Oxford University Press.

Koelsch, S., Gunter, T. C., van Cramon, D. Y., Zysset S., Lohmann, G., & Friederici, A. D. (2002). A cortical "language network" serves the processing of music. *Neuroimage, 17*(1), 956–966.

Koelsch, S., Kilches, S., Steinbeis, N., & Schelinski, S. (2008). Effects of unexpected chords and of performer's expression on brain responses and electrodermal activity. *PLoS ONE, 3*(7), e2631. https://doi.org/10.1371/journal.pone.0002631

Koestler, A. (1964). *The act of creation.* London: Pan Books.

Koffka, K. (1922). Perception: An introduction to the Gestalt-Theorie. *Psychological Bulletin, 19,* 531–585.

Kohler, E., Keysers, C., Umiltà, M. A., Fogassi, L., Gallese, V., & Rizzolatti, G. (2002). Hearing sounds, understanding actions: Action representation in mirror neurons. *Science, 297*(5582), 846–848.

Köhler, W. (1929). *Gestalt psychology*. New York: Liveright.

Köhler, W. (1959). Gestalt psychology today. *American Psychologist, 14*, 727–734.

Korsyn. K. (2003). *Decentering music: A critique of contemporary musical research*. New York: Oxford University Press.

Kozak, M. (2019). *Enacting musical time: The bodily experience of new music*. New York: Oxford University Press.

Kozbelt, A., Beghetto, R. A., & Runco, M. A. (2010). Theories of creativity. In J. C. Kaufman & R. J. Sternberg (Eds.) *The Cambridge handbook of creativity*. Cambridge: Cambridge University Press.

Kreutz, G., Schubert, E., & Mitchell, L. A. (2008). Cognitive styles of music listening. *Music Perception, 26*, 57–73. https://doi.org/10.1525/mp.2008.26.1.57

Krueger, J. (2009). Enacting musical experience. *Journal of Consciousness Studies, 16*(2–3), 98–123.

Krueger, J. (2011a). Doing things with music. *Phenomenology and the Cognitive Sciences, 10*(1), 1–22.

Krueger, J. (2011b). Enacting musical content. In R. Manzotti (Ed.) *Situated aesthetics: Art beyond the skin* (pp. 63–85). Exeter: Imprint Academic.

Krueger, J. (2011c). Extended cognition and the space of social interaction. *Consciousness and Cognition, 20*(3), 643–657.

Krueger, J. (2013). Empathy, enaction, and shared musical experience: Evidence from infant cognition. In T. Cochrane, B. Fantini, & K. Scherer (Eds.), *The emotional power of music: Multidisciplinary perspectives on musical expression, arousal, and social control* (pp. 177–196). Oxford: Oxford University Press.

Krueger, J. (2014). Affordances and the musically extended mind. *Frontiers in Psychology 4*, 1003. https://doi.org/10.3389/fpsyg.2013.01003

Krueger, J. (2015). Musicing, materiality, and the emotional niche. *Action, Criticism, and Theory for Music Education, 14*(3), 43–62.

Krueger, J. (2018a). Direct social perception. In A. Newen, L. de Bruin, & S. Gallagher (Eds.), *The Oxford handbook of 4e cognition*. New York: Oxford University Press.

Krueger, J. (2018b). Music as affective scaffolding. In D. Clarke, R. Herbert, & E. Clarke (Eds.), *Music and consciousness II*. Oxford: Oxford University Press.

Krueger, J., & Szanto, T. (2016). Extended emotions. *Philosophy Compass, 11*, 863–878. https://doi.org/10.1111/phc3.12390

Krumhansl, C. (1997). An exploratory study of musical emotions and psychophysiology. *Canadian Journal of Experimental Psychology, 51*(4), 336–353.

Krumhansl, C., & Kessler, E. (1982). Tracing the dynamic ranges in perceived tonal organization in a spatial representation of musical keys. *Psychological Review*, *89*(4), 334–368.

Krumhansl, C., & Toiviainen, P. (2001). Tonal cognition. *Annals of the New York Academy of Sciences*, *930*, 77–91.

Küpers, E., van Dijk, M., McPherson, G., & van Geert, P. (2014). A dynamic model that links skill acquisition with self-determination in instrumental music lessons. *Musicae Scientiae*, *18*(1), 17–34.

Kyselo, M. (2014). The body social: An enactive approach to the self. *Frontiers in Psychology*, 5:986. https://doi.org/10.3389/fpsyg.2014.00986

Kyselo, M. (2020). More than our body: Minimal and enactive selfhood in global paralysis. *Neuroethics*, *13*, 203–220.

Labbé, C., & Grandjean, D. (2014). Musical emotions predicted by feelings of entrainment. *Music Perception*, *32*(2), 170–185.

Laible, D., & Thompson, R. (2000). Mother–child discourse, attachment security, shared positive affect, and early conscience development. *Child Development*, *71*, 1424–1440.

Lakoff, G., & Johnson, M. (1980). *Metaphors we live by*. Chicago: University of Chicago Press.

Lakoff, G., & Johnson, M. (1999). *Philosophy in the flesh: The embodied mind and its challenge to Western thought*. New York: Basic Books.

Laland, K. N., Odling-Smee, J., & Myles, S. (2010). How culture shaped the human genome: Bringing genetics and the human sciences together. *Nature Reviews Genetics*, *11*(2), 137–148.

Lambert, D., Millar, C., & Hughes, A. J. (1986). On the classic case of natural selection. *Biology Forum*, *79*(1), 11–49.

Langer, S. K. (1953). *Feeling and form: A theory of art*. New York: Charles Scribner's Sons.

Langer, S. K. (1957). *Philosophy in a new key* (3rd ed.). Cambridge, MA: Harvard University Press.

Langer, S. K. (1958). The cultural importance of the arts. In M. F. Andrews (Ed.), *Aesthetic form and education* (pp. 1–18). Syracuse, NY: Syracuse University Press.

Lappe, C., Herholz, S. C., Trainor, L. J., & Pantev, C. (2008). Cortical plasticity induced by short-term unimodal and multimodal musical training. *Journal of Neuroscience*, *28*, 9632–9639.

Large, E. (2000). On synchronizing movements to music. *Human Movement Science*, *19*(4), 527–566.

Large, E. W., & Gray, P. M. (2015). Spontaneous tempo and rhythmic entrainment in a Bonobo (*Pan Paniscus*). *Journal of Comparative Psychology, 129*, 317–328. https://doi.org/10.1037/com0000011

Large, E. W., Herrera, J. A., & Velasco, M. J. (2015). Neural networks for beat perception in musical rhythm. *Frontiers in Systems Neuroscience, 9*:159. https://doi.org/10.3389/fnsys.2015.00159

Large, E. W., & Jones, M. R. (1999). The dynamics of attending: How people track time-varying events. *Psychological Review, 106*(1), 119–159.

Large, E. W., Kim, J. C., Flaig, N., Bharucha, J., & Krumhansl, C. L. (2016). A neurodynamic account of musical tonality. *Music Perception, 33*(3), 319–331.

Laurence, F. (2007). Music and empathy. In O. Urbain & P. Van den Dungen (Eds.), *Music and conflict transformation: Harmonies and dissonances in geopolitics* (pp. 13–25). London: I. B. Tauris.

Lawson, F. R. S. (2014). Is music an adaptation or a technology? Ethnomusicological perspectives from the analysis of Chinese Shuochang. *Ethnomusicology Forum, 23*, 3–26.

Layman, D. L., Hussey, D. L., & Laing, S. J. (2002). Music therapy assessment for severely emotionally disturbed children: A pilot study. *Journal of Music Therapy, 39*(3), 164–187.

Lazarus, R. S. (1982). Thoughts on the relations between emotion and cognition. *American Psychologist, 37*(9), 1019–1024.

LeDoux, J. E. (1996). *The emotional brain*. New York: Simon & Schuster.

LeDoux, J. E. (2002). *Synaptic self: How our brains become who we are*. New York: Viking.

Lee, C. J., Andrade, E. B., & Palmer, S. E. (2013). Interpersonal relationships and preferences for mood-congruency in aesthetic experiences. *The Journal of Consumer Research, 40*(2), 382–391.

Leman, M. (2007). *Embodied music cognition and mediation technology*. Cambridge, MA: MIT Press.

Leman, M., & Maes, P.-J. (2014). The role of embodiment in the perception of music. *Empirical Musicology Review: EMR, 9*(3–4), 236–246.

Leman, M., Nijs, L., Maes, P.-J., & Van Dyck, E. (2018). What is embodied music cognition? In R. Bader (Ed.), *Springer handbook of systematic musicology*. Berlin: Springer Verlag.

Leonard, J. (1992). Music therapy: Fertile ground for application of research in practice. *Neonatal Network, 12*(2), 47–48.

Leonhard, C., & House, R. W. (1959). *Foundations and principles of music education*. New York: McGraw-Hill.

Lerdahl, F., & Jackendoff, R. (1983). *A generative theory of tonal music*. Cambridge, MA: MIT Press.

Leroi-Gourhan, A. (1993 [1964]). *Gesture and speech* (A. Bostock Berger, Trans.). Cambridge, MA: MIT Press.

Levinson, J. (1996). *The pleasure of aesthetics*. Ithaca, NY: Cornell University Press.

Levitin, D. (2006). *This is your brain on music: The science of a human obsession*. New York: Penguin.

Lewis, G. E. (2007). Improvisation and pedagogy: Background and focus of inquiry. *Critical Studies in Improvisation*, *3*(2), 1–5. http://www.criticalimprov.com/article/view/412/659

Lewis, G. E. (2009). *A power stronger than itself: The AACM and American experimental music*. Chicago: University of Chicago Press.

Lewis, M. D. (2000). Emotional organization at three timescales. In M. D. Lewis & I. Granic (Eds.), *Emotion, development and self-organization* (pp. 37–69). Cambridge: Cambridge University Press.

Lewis, M. D. (2005). Bridging emotion theory and neurobiology through dynamical systems modeling. *Behavioral and Brain Sciences*, *28*, 169–245.

Lewontin, R. (1983). The organism as subject and object of evolution. *Scientia*, *118*(1), 65–82.

Lewontin, R., Rose, S., & Kamin, L. (1984). *Not in our genes: Biology, ideology and human nature*. New York: Pantheon Books.

Lickliter, R., & Honeycutt, H. (2003). Developmental dynamics: Toward a biologically plausible evolutionary psychology. *Psychological Bulletin*, *129*, 819–835.

Likowski, K. U., Mühlberger, A., Gerdes, A. B. M., Wieser, M. J., Pauli, P., & Weyers, P. (2012). Facial mimicry and the mirror neuron system: Simultaneous acquisition of facial electromyography and functional magnetic resonance imaging. *Frontiers in Human Neuroscience* 6:214. https://doi.org/10.3389/fnhum.2012.00214

Lines, D. (2005). Working with music: A Heideggerian perspective of music education. *Educational Philosophy and Theory*, *37*(1), 63–73.

Lines, D. (2015). Ways of revealing: Music education responses to music technology. In F. Pio and Ø. Varkoy (Eds.), *Philosophy of music education challenged: Heideggerian inspirations* (pp. 61–74). Dordrecht: Springer.

Linson, A., & Clarke, E. F. (2017). Distributed cognition, ecological theory, and group improvisation. In E. F. Clarke & M. Doffman (Eds.), *Distributed creativity: Collaboration and improvisation in contemporary music*. New York: Oxford University Press.

Lipps, T. (1907). *Ästhetik* [Aesthetics]. Berlin: B.G. Teubner.

Loaiza, J. (2016). Musicking, embodiment and the participatory enaction of music: Outline and key points. *Connection Science, 28*, 410–422.

Lukács, G. (1971). *History and class consciousness: Studies in Marxist dialectics*. London: Merlin.

Lundqvist, L. O., Carlsson, F., Hilmersson, P., & Juslin, P. N. (2008). Emotional responses to music: Experience, expression, and physiology. *Psychology of Music, 37*(1), 61–90.

Lycan, W. (1996). *Consciousness and experience*. Cambridge, MA: MIT Press.

Maes, P., Leman, M., Palmer, C., & Wanderley, M. M. (2014). Action-based effects on music perception. *Frontiers in Psychology*, 4:1008. https://doi.org/10.3389/fpsyg.2013.01008

Magee, W. L., & Davidson, J. W. (2002). The effect of music therapy on mood states in neurological patients: A pilot study. *Journal of Music Therapy, 39*(1), 20–29.

Maher, C. (2017). *Plant minds: A philosophical defense*. London: Routledge.

Maiese, M. (2020). An enactivist approach to treating depression: Cultivating online intelligence through dance and music. *Phenomenology and the Cognitive Sciences, 19*(3), 523–547.

Malafouris, L. (2013). *How things shape the mind: A theory of material engagement*. Cambridge, MA: MIT Press.

Malafouris, L. (2015). Metaplasticity and the primacy of material engagement. *Time and Mind, 8*(4), 351–371.

Malloch, S., & Trevarthen, C. (Eds.) (2009). *Communicative musicality*. Oxford: Oxford University Press.

Markowitz, S. (1983). *Art and the tyranny of the aesthetic* [Unpublished doctoral dissertation]. University of Michigan.

Marraffa, M. (2011). *Theory of mind*. Retrieved from Internet Encyclopedia of Philosophy (IEP). http://www.iep.utm.edu/theomind/

Martin, P. J. (1995). *Sounds and society: Themes in the sociology of music*. Manchester: Manchester University Press.

Mathews, F. (1991). *The ecological self*. London: Routledge.

Mathews, F. (2008). Thinking from within the calyx of nature. *Environmental Values, 17*, 41–65.

Matravers, D. (1998). *Art and emotion*. Oxford: Clarendon.

References

Matthen, M. (2014). Debunking enactivism: A critical notice of Hutto and Myin's radicalizing enactivism. *Canadian Journal of Philosophy, 44*(1), 118–128.

Maturana, H., & Varela, F. (1980). *Autopoiesis and cognition: The realization of the living*. Boston: Reidel.

Maturana, H., & Varela, F. (1984). *The tree of knowledge: The biological roots of human understanding*. Boston: New Science Library.

McClary, S. (2000). *Conventional wisdom*. Berkeley: University of California Press.

McCulloch, W. S. (1965). *Embodiments of mind*. Cambridge, MA: MIT Press.

McDermott, J., & Hauser, M. D. (2005). The origins of music: Innateness, development, and evolution. *Music Perception, 23*(3), 29–59.

McFerran, K. (2019). Music therapy in adolescent groups. In C. Haen & N. Boyd-Webb (Eds.), *Creative arts-based group therapy with adolescents: Theory and practice*. New York: Routledge.

McFerran, K., Roberts, M., O'Grady, L. (2010). Music therapy with bereaved teenagers: a mixed methods perspective. *Death Studies, 34*(6), 541–565.

McGrath, J. E., & Kelly, J. R. (1986). *Time and human interaction: Toward a social psychology of time*. New York: Guilford Press.

McKeon, R. (Ed.) (2001). *The basic works of Aristotle*. New York: Random House.

McLean, E., McFerran, K., & Thompson, G. A. (2019). Parents' musical engagement with their baby in the neonatal unit to support emerging parental identity: A grounded theory study. *Journal of Neonatal Nursing, 25*(2), 78–85.

McPherson, G. E., Davidson, J. W., & Faulkner, R. (2012). *Music in our lives: Rethinking musical ability, development and identity*. New York: Oxford University Press.

Meaney, M. J. (2001). Nature, nurture, and the disunity of knowledge. *Annals of the New York Academy of Sciences, 935*, 50–61.

Medeiros, K. E., Partlow, P. J., & Mumford, M. D. (2014). Not too much, not too little: The influence of constraints on creative problem solving. *Psychology of Aesthetics, Creativity, and the Arts, 8*(2), 198–210.

Meltzoff, A. N. (2005). Imitation and other minds: The "like me" hypothesis. In S. Hurley (Ed.), *Perspectives on imitation: From neuroscience to social science* (Vol. 2, pp. 55–77). Cambridge, MA: MIT Press.

Meltzoff, A. N., & Decety, J. (2003). What imitation tells us about social cognition: A rapprochement between developmental psychology and cognitive neuroscience. *Philosophical Transactions of the Royal Society of London. Series B, Biological Sciences, 358*(1431), 491–500.

Meltzoff, A. N., & Moore, M. K. (1977). Imitation of facial and manual gestures by human neonates. *Science*, 198(4312), 75–78.

Meltzoff, A. N., & Prinz, W. (Eds.) (2002). *The imitative mind: Development, evolution and brain bases*. Cambridge: Cambridge University Press.

Menary, R. (2006). Attacking the bounds of cognition. *Philosophical Psychology*, 19, 329–344.

Menary, R. (2007). *Cognitive integration: Mind and cognition unbounded*. Basingstoke, UK: Palgrave Macmillan.

Menary, R. (Ed.) (2010a). *The extended mind*. Cambridge, MA: MIT Press.

Menary, R. (2010b). Cognitive integration and the extended mind. In R. Menary (Ed.), *The extended mind* (pp. 227–243). Cambridge, MA: MIT Press.

Menary, R. (2010c). Dimensions of mind. *Phenomenology and the Cognitive Sciences*, 9(4), 561–578.

Menary, R. (2010d). The holy grail of cognitivism: A response to Adams and Aizawa. *Phenomenology and the Cognitive Sciences*, 9, 605–618.

Merchant, H., & Bartolo, R. (2017). Primate beta oscillations and rhythmic behaviors. *Journal of Neural Transmission*, 124(5), 20–31.

Merchant, H., Grahn, J., Trainor, L., Rohrmeier, M., & Fitch, W. T. (2015). Finding the beat: A neural perspective across humans and non-human primates. *Philosophical Transactions of the Royal Society B* 370:20140093. https://doi.org/10.1098/rstb.2014.0093

Merleau-Ponty, M. (1942). *La structure du comportement*. Paris: Presses Universitaires de France.

Merleau-Ponty, M. (1945). *Phénoménologie de la percepton*. Paris: Éditions Gallimard.

Meyer, L. B. (1956). *Emotion and meaning in music*. Chicago: University of Chicago Press.

Michaels, S. (2012, February 27). *The healing power of music*. PBS News Hour, MacNeil/Lehrer Productions [television broadcast]. Washington, DC: PBS. https://www.pbs.org/newshour/show/the-healing-power-of-music

Miller, G. F. (2000). Evolution of human music through sexual selection. In N. L. Wallin, B. Merker, & S. Brown (Eds.), *The origins of music* (pp. 329–360). Cambridge, MA: MIT Press.

Miller, R. (1997). *What are schools for? Holistic education in American culture*. Brandon, VT: Holistic Education Press.

Millikan, R. (1995). "Pushmi-pullu" representations. *Philosophical Perspectives*, 9, 185–200.

References

Mithen, S. (2005). *The singing Neanderthals: The origins of music, language, mind, and body*. London: Weidenfeld & Nicholson.

Møller, C., Stupacher, J., Celma-Miralles, A., & Vuust, P. (2021). Beat perception in polyrhythms: Time is structured in binary units. *bioRxiv*.05.12.443747. https://doi.org/10.1101/2021.05.12.443747

Molnar-Szakacs, I., & Overy, K. (2006). Music and mirror neurons: From motion to 'e'motion. *Social Cognitive and Affective Neuroscience, 1*, 235–241. https://doi.org/10.1093/scan/nsl029

Monson, I. (1996). *Saying something: Jazz improvisation and interaction*. Chicago: University of Chicago Press.

Montero, B. (2010). Does bodily awareness interfere with highly skilled movement? *Inquiry, 53*(2), 105–122.

Montero, B. (2016). *Thought in action. Expertise and the conscious mind*. Oxford: Oxford University Press.

Moore, D. (2003). *The dependent gene*. New York: Henry Holt.

Moore, D. R., & Jeffery, G. (1994). Development of auditory and visual systems in the fetus. In G. D. Thorburn & R. Harding (Eds.), *Textbook of fetal physiology* (pp. 278–286). Oxford: Oxford University Press.

Moors, A. (2013). On the causal role of appraisal in emotion. *Emotion Review, 5*(2), 132–140.

Moran, D. (2017). Intercorporeality and intersubjectivity: A phenomenological exploration of embodiment. In C. Durt, T. Fuchs, & C. Tewes (Eds.), *Embodiment, enaction, and culture: Investigating the constitution of the shared world* (pp. 25–46). Cambridge, MA: MIT Press.

Morley, I. (2013). *The prehistory of music: Human evolution, archeology, and the origins of musicality*. Oxford: Oxford University Press.

Morris, B. B. (1971). Effects of order and trial on Necker cube reversals under free resistive instructions. *Perceptual and Motor Skills, 33*(1), 235–240.

Mukamel, R., Ekstrom, A. D., Kaplan, J., Iacoboni, M., & Fried, I. (2010). Single-neuron responses in humans during execution and observation of actions. *Current Biology, 20*(8), 750–756.

Münte, T. F., Altenmüller, E., & Jäncke, L. (2002). The musician's brain as a model of neuroplasticity. *Nature Reviews Neuroscience, 3*, 473–478.

Nagel, E. (1956). *Logic without metaphysics and other essays in the philosophy of science*. Glencoe, IL: Free Press.

Nagel, T. (1974). What is it like to be a bat? *The Philosophical Review, 83*, 435–450.

Nagy, Z. (2017). *Embodiment of musical creativity: The cognitive and performative causality of musical composition*. London: Routledge.

Nakagawa, Y. (2000). *Education for awakening: An eastern approach to holistic education*. Brandon, MT: Education Renewal.

Nawrot, E. S. (2003). The perception of emotional expression in music: Evidence from infants, children and adults. *Psychology of Music, 31*(1), 75–92.

Nayak, S., Wheeler, B. L., Shiflett, S. C., & Agostinelli, S. (2000). Effect of music therapy on mood and social interaction among individuals with acute traumatic brain injury and stroke. *Rehabilitation Psychology, 45*(3), 274–283.

Necker, L. A. (1832). Observations on some remarkable phenomenon which occurs in viewing a figure of a crystal or geometrical solid. *London and Edinburgh Philosophical Magazine and Journal of Science, 3*, 329–337.

Nettl, B. (1974). Thoughts on improvisation: A comparative approach. *Musical Quarterly, 60*(1), 1–19.

Nettl, B. (1983). *The study of ethnomusicology: Twenty-nine issues and concepts*. Urbana: University of Illinois Press.

Nettl, B. (2000). An ethnomusicologist contemplates universals in musical sound and musical culture. In N. L. Wallin, B. Merker, & S. Brown (Eds.), *The origins of music* (pp. 463–472). Cambridge, MA: MIT Press.

Nettl, B. (2005). *The study of ethnomusicology: Thirty-one issues and concepts*. Urbana: University of Illinois Press.

Nettl, B., & Russell, M. (1998). *In the course of performance: Studies in the world of musical improvisation*. Chicago: University of Chicago Press.

Newen A., de Bruin, L., & Gallagher, S. (2018). *The Oxford handbook of 4e cognition*. New York: Oxford University Press.

Ng, R. (2020). Decolonizing teaching and learning through embodied learning: Toward an integrated approach. In S. Batacharya & Y. R. Wong (Eds.), *Sharing breath: Embodied learning and decolonization* (pp. 33–54). Edmonton: Alberta University Press.

Noddings, N. (1984). *Caring: A feminine approach to ethics and moral education*. Berkeley: University of CA Press.

Noddings, N. (1995). *Philosophy of education*. Boulder, CO: Westview Press.

Noddings, N. (1999). Caring and competence. In G. Griffin (Ed.), *The education of teachers* (pp. 205–220). Chicago: National Society of Education.

Noddings, N. (2012). The caring relation in teaching. *Oxford Review of Education, 38*, 777–786.

Noë, A. (2004). *Action in perception.* Cambridge, MA: MIT Press.

Nollman, J. (1990). *Spiritual ecology: A guide to reconnecting with nature.* New York: Bantam.

Nollman, J. (1999). *The charged border: Where whales and humans meet.* New York: Holt.

Novembre, G., Ticini, L. F., Schütz-Bosbach, S., & Keller, P. E. (2012). Distinguishing self and other in joint action. Evidence from a musical paradigm. *Cerebral Cortex, 22*(12), 2894–2903.

Novembre, G., Ticini, L. F., Schütz-Bosbach, S., & Keller, P. E. (2013). Motor simulation and the coordination of self and other in real-time joint action. *Social Cognitive and Affective Neuroscience, 9*(8), 1062–1068. https://doi.org/10.1093/scan/nst086

Nussbaum, C. O. (2007). *The musical representation: Meaning, ontology, and emotion.* Cambridge, MA: MIT Press.

Olsho, L. W. (1984). Infant frequency discrimination. *Infant Behavior and Development, 7,* 27–35.

O'Neill, S. A. (2009). Revisioning musical understandings through a cultural diversity theory of difference. In L. Bartel (Ed.), *Research to practice: Vol. 4. Exploring social justice: How music education might matter* (pp. 70–89). Waterloo, ON: Canadian Music Educators' Association.

O'Regan, J. K., & Noë, A. (2001a). A sensorimotor approach to vision and visual consciousness. *Behavioral and Brain Sciences, 24*(5), 939–973.

O'Regan, J. K., & Noë, A. (2001b). What it is like to see: A sensorimotor theory of visual experience. *Synthese, 129*(1), 79–103.

Ortony, A., & Turner, T. J. (1990). What's basic about emotions? *Psychological Review, 97*(3), 315–331.

Oudgenoeg-Paz, O., Volman, M. J. M., & Leseman, P. P. M. (2012). Attainment of sitting and walking predicts development of productive vocabulary between ages 16 and 28 months. *Infant Behavior and Development, 35,* 733–736.

Overy, K., & Molnar-Szakacs, I. (2009). Being together in time: Musical experience and the mirror neuron system. *Music Perception, 26*(5), 489–504.

Oyama, S. (1985). *The ontogeny of information.* Cambridge: Cambridge University Press.

Oyama, S., Griffiths, P., & Gray, R. (2001). *Cycles of contingency: Developmental systems and evolution.* Cambridge, MA: MIT Press.

Panksepp, J. (2009). The emotional antecedents to the evolution of music and language. *Musicae Scientiae.* Special Issue 2009–2010, 229–259. https://doi.org/10.1177/1029864909013002111

Panksepp, J., & Bernatzky, G. (2002). Emotional sounds and the brain: The neuroaffective foundations of musical appreciation. *Behavioral Processes, 60*, 133–155.

Pantev, C., Engelien, A., Candia, V., & Elbert, T. (2001). Representational cortex in musicians: Plastic alterations in response to musical practice. *Annals of the NY Academy of Sciences, 930*, 300–314.

Parncutt, R. (2006). Prenatal development. In G. E. McPherson (Ed.), *The child as musician: A handbook of musical development* (pp. 1–31). New York: Oxford University Press.

Parncutt, R. (2007). Systematic musicology and the history and future of western musical scholarship. *Journal of Interdisciplinary Music Studies, 1*(1), 1–32.

Parncutt, R. (2008). Prenatal development and the phylogeny and ontogeny of music. In S. Hallam, I. Cross, & M. Thaut (Eds.), *Oxford handbook of music psychology* (1st ed.). Oxford: Oxford University Press.

Patel, A. D. (2003). Language, music, syntax and the brain. *Nature Neuroscience, 6*(7), 674–681.

Patel, A. D. (2006). Musical rhythm, linguistic rhythm, and human evolution. *Music Perception, 24*(1), 99–104.

Patel, A. D. (2008). *Music, language, and the brain*. Oxford: Oxford University Press.

Patel, A. D. (2010). Music, biological evolution, and the brain. In M. Bailar (Ed.), *Emerging disciplines* (pp. 91–144). Houston, TX: Rice University Press.

Patel, A. D. (2012). Language, music, and the brain: A resource-sharing framework. In P. Rebuschat, M. Rohrmeier, J. Hawkins & I. Cross (Eds.), *Language and music as cognitive systems* (pp. 204–223). Oxford: Oxford University Press.

Patel, A. D. (2018). Music as a transformative technology of the mind: An update. In H. Honing (Ed.), *The origins of musicality* (pp. 113–26). Cambridge, MA: MIT Press.

Patel, A. D., & Iversen, J. R. (2014). The evolutionary neuroscience of musical beat perception: The action simulation for auditory prediction (ASAP) hypothesis. *Frontiers in Systems Neuroscience* 8:57. https://doi.org/10.3389/fnsys.2014.00057

Pearson, L. (2013). Gesture and the sonic event in Karnatak music. *Empirical Musicology Review, 8*(1). https://emusicology.org/article/view/3918/3548

Perani D., Saccuman, M. C., Scifo, P., Spada, D., Andreolli, G., Rovelli R., Baldoli, C., & Koelsch, S. (2010). Functional specializations for music processing in the human newborn brain. *Proceedings of the National Academy of Sciences of the United States of America, 107*, 4758–4763.

Peretz, I. (1993). Audio atonalia for melodies. *Cognitive Neuropsychology, 10*(1), 21–56.

Peretz, I. (2006). The nature of music from a biological perspective. *Cognition, 100*(1), 1–32.

Peretz, I. (2012). Music, language and modularity in action. In P. Rebuschat, M. Rohrmeier, J. Hawkins & I. Cross (Eds.), *Language and music as cognitive systems* (pp. 254–268). Oxford: Oxford University Press.

Peretz, I., & Coltheart, M. (2003). Modularity of music processing. *Nature Neuroscience, 6*(1), 123–144.

Perone, S., & Oakes, L. M. (2006). It clicks when it is rolled and squeaks when it is squeezed: What 10-month-old infants learn about function. *Child Development, 77*, 1608–1622.

Perone, S., Madole, K. L., Ross-Sheehy, S., Carey, M., & Oakes, L. M. (2009). The relation between infants' activity with objects and attention to object appearance. *Developmental Psychology, 44*, 1242–1248.

Perone, S., Madole, K. L., & Oakes, L. M. (2011). Learning how actions function: The role of outcomes in infants' representation of events. *Infant Behavioral Development, 34*(2), 351–362.

Peters, D. (2015). Musical empathy, emotional co-constitution, and the "musical other." *Empirical Musicology Review, 10*(1–2). https://doi.org/10.18061/emr.v10i1-2.4611

Pfeifer, J. H., Iacoboni, M., Mazziota, J. C., & Dapretto, M. (2008). Mirroring others' emotions relates to empathy and interpersonal competence in children. *Neuroimage, 39*, 2076–2085.

Piaget, J. (1964). *Development and learning*. In R. Ripple & V. Rockcastle (Eds.), *Piaget rediscovered*. Ithaca, NY: Cornell University Press.

Pigliucci, M. (2001). *Phenotypic plasticity: Beyond nature and nurture*. Baltimore, MD: Johns Hopkins University Press.

Pinker, S. (1997). *How the mind works*. New York: Norton.

Plutchik, R. (2001). The nature of emotions. *American Scientist, 89*, 334–350.

Podlipniak, P. (2017). The role of the Baldwin effect in the evolution of human musicality. *Frontiers in Neuroscience*, 11:542. https://doi.org/10.3389/fnins.2017.00542

Poggio, T., & Reichardt, W. (1976). Visual control of orientation behavior in the fly (part II). *Quarterly Reviews of Biophysics, 9*(3), 377–438.

Pohjannoro, U. (2020). Embodiment in composition: 4E theoretical considerations and empirical evidence from a case study. *Musicae Scientiae*. https://doi.org/10.1177/1029864920961447

Poldrack, R. A. (2006). Can cognitive processes be inferred from neuroimaging data? *Trends Cognitive Sciences, 10*, 59–63.

Pollack, J. B. (1989). Connectionism: Past, present, and future. *Artificial Intelligence Review, 3*, 3–20.

Polyani, M. (1969). *Knowing and being*. Chicago: University of Chicago Press.

Posner, M., & Rothbart, M. K. (1998). Attention, self-regulation, and consciousness. *Philosophical Transactions of the Royal Society of London. Series B, Biological Sciences, 353*, 1915–1927.

Potter, J. (2000). Post cognitivist psychology. *Theory and Psychology, 10*(1), 31–37.

Powell, K. (2005). The ensemble art of the solo: The social and cultural construction of artistic practice and identity in a Japanese American Taiko ensemble. *Journal of Art and Learning Research, 21*(1), 273–295.

Powell, K., & Lajevic, L. (2011). Emergent places in preservice art teaching: Lived curriculum, relationality, and embodied knowledge. *Studies in Art Education, 53*(1), 35–52.

Prinz, J. J. (2004a). Embodied emotions. In R. C. Solomon (Ed.), *Thinking about feeling* (pp. 44–60). Oxford: Oxford University Press.

Prinz, J. J. (2004b). *Gut reactions: A perceptual theory of emotion*. New York: Oxford University Press.

Rabinowitch, T. C., Cross, I., & Burnard, P. (2012). Long-term musical group interaction has a positive influence on empathy in children. *Psychology of Music, 41*(4), 484–498.

Rabinowitch, T. C., & Knafo-Noam, A. (2015). Synchronous rhythmic interaction enhances children's perceived similarity and Closeness towards each other. *PLoS ONE, 10*(4), e0120878. https://doi.org/10.1371/journal.pone.0120878

Raffman, D. (1993). *Language, music and mind*. Cambridge, MA: MIT Press.

Rahaim, M. (2012). *Musicking bodies: Gesture and voice in Hindustani music*. Middletown, CT: Wesleyan University Press.

Ramachandran, V. S. (2011). *The tell-tale brain: A neuroscientist's quest for what makes us human*. New York: Norton.

Ramachandran, V. S., & Hubbard, E. M. (2001). Synaesthesia: A window into perception, thought and language. *Journal of Consciousness Studies, 8*(12), 3–34.

Ratcliffe, M. (2017). Empathy without simulation. In T. Fuchs, M. Summa, & L. Vanzago (Eds.), *Imagination and social perspectives: Approaches from phenomenology and psychopathology*. London: Routledge.

Ravignani, A., Delgado, T., & Kirby, S. (2016a). Musical evolution in the lab exhibits rhythmic universals. *Nature Human Behaviour*, 1:0007. https://doi.org/10.1038/s41562-016-0007

Ravignani, A., Fitch, W. T., Hanke, F. D., Heinrich, T., Hurgitsch, B., Kotz, S. A., Scharff, C., Stoeger A. C., & Boer, B. (2016b). What pinnipeds have to say about human speech, music, and the evolution of rhythm. *Frontiers in Neuroscience*, 10:274. https://doi.org/10.3389/fnins.2016.00274

Ravignani, A., Honing, H., & Kotz, S. A. (2017). Editorial: The evolution of rhythm cognition: Timing in music and speech. *Frontiers in Human Neuroscience*, 11:303. https://doi.org/10.3389/fnhum.2017.00303

Ravn, S., & Christensen, M. K. (2014). Listening to the body? How phenomenological insights can be used to explore a golfer's experience of the physicality of her body. *Qualitative Research in Sport, Exercise and Health*, 6(4), 462–477.

Ravn, S., & Hansen, H. P. (2013). How to explore dancers' sense experiences? A study of how multi-sited fieldwork and phenomenology can be combined. *Qualitative Research in Sport, Exercise and Health*, 5(2), 196–213.

Rebuschat, P., Rohrmeier, M., Hawkins, J., & Cross, I. (2012). *Language and music as cognitive systems*. Oxford: Oxford University Press.

Reddy, V. (2008). *How infants know minds*. Cambridge, MA: Harvard University Press.

Reddy, V., Markova, G., & Wallot, S. (2013). Anticipatory adjustments to being picked up in infancy. *PLoS ONE*, 8(6), e65289. https://doi.org/10.1371/journal.pone.0065289

Reed, E. S. (1996). *Encountering the world: Toward an ecological psychology*. Oxford: Oxford University Press.

Regelski, T. A. (1998). The Aristotelian bases of praxis for music and music education as praxis. *Philosophy of Music Education Review*, 6(1), 22–59.

Regelski, T. A. (2002). On "methodolatry" and music teaching as critical and reflective praxis. *Philosophy of Music Education Review*, 10(2), 102–123.

Regelski, T. A. (2004). *Teaching general music in grades 4–8: A musicianship approach*. New York: Oxford University Press.

Regelski, T. A. (2012). Ethical dimensions of school-based music education. In W. D. Bowman, & A. L. Frega (Eds.), *The Oxford handbook of philosophy in music education* (pp. 284–304). Oxford: Oxford University Press.

Reichardt, W., & Poggio, T. (1976). Visual control of orientation behavior in the fly (part I). *Quarterly Reviews of Biophysics*, 9(3), 311–375.

Reimer, B. (1970). *A philosophy of music education*. Englewood Cliffs, NJ: Prentice Hall.

Reimer, B. (1989). *A philosophy of music education* (2nd ed.). Englewood Cliffs, NJ: Prentice Hall.

Reimer, B. (2003). *A philosophy of music education: Advancing the vision*. Upper Saddle River, NJ: Prentice Hall.

Reybrouck, M. (2001). Biological roots of musical epistemology: Functional cycles, Umwelt, and enactive listening. *Semiotica, 134*(1–4), 599–633.

Reybrouck, M. (2004). Music cognition, semiotics and the experience of time: Ontosemantical and epistemological claims. *Journal of New Music Research, 33*(4), 411–428.

Reybrouck, M. (2005a). A biosemiotic and ecological approach to music cognition: Event perception between auditory listening and cognitive economy. *Axiomathes. An International Journal in Ontology and Cognitive Systems, 15*(2), 229–266.

Reybrouck, M. (2005b). Body, mind and music: Musical semantics between experiential cognition and cognitive economy. *Trans: Transcultural Music Review 9.* https://doi.org/10.1080/07494460600647451

Reybrouck, M. (2006). Music cognition and the bodily approach: Musical instruments as tools for musical semantics. *Contemporary Music Review, 25*(1/2), 59–68.

Reybrouck, M. (2010). Music cognition and real-time listening: Denotation, cue abstraction, route description and cognitive maps. *Musicae Scientiae, 14*, 187–202.

Reybrouck, M. (2012). Musical sense-making and the concept of affordance: An ecosemiotic and experiential approach. *Biosemiotics, 5*, 391–409.

Reybrouck, M. (2014). Music teaching and learning: Reflecting on the musical experience. In T. De Baets & T. Buchborn (Eds.), *The reflective music teacher* (pp. 57–70). Innsbruck: Helbling.

Reybrouck, M. (2015a). Deixis in musical narrative: Musical sense-making between discrete particulars and synoptic overview. *Chinese Semiotic Studies, 11*, 79–90.

Reybrouck, M. (2015b). Music as environment: An ecological and biosemiotic approach. *Behavioral Sciences, 5*, 1–26.

Reybrouck, M. (2016). Musical information beyond measurement and computation: Interaction, symbol processing and the dynamic approach. In P. Kostagiolas, K. Martzoukou & C. Lavranos (Eds.), *Trends in music information seeking, behavior, and retrieval for creativity* (pp. 100–120). Hershey, PA: IGI Global.

Reybrouck, M. (2017a). Music knowledge construction. Enactive, ecological, and biosemiotic claims. In M. Lesaffre, P-J. Maes, & M. Leman (Eds.), *The Routledge companion to embodied music interaction* (pp. 58–65). New York: Routledge.

Reybrouck, M. (2017b). Perceptual immediacy in music listening Multimodality and the "in time/outside of time" dichotomy. *Versus-Quaderni Di Studi Semiotci, 46*(1), 89–104.

Reybrouck, M. (2020). *Musical sense-making: Enaction, experience and computation.* London: Routledge.

References

Reybrouck, M., & Eerola, T. (2017). Music and its inductive power: A psychobiological and evolutionary approach to musical emotions. *Frontiers in Psychology*, 8:494. https://doi.org/10.3389/fpsyg.2017.00494

Rhodes, M. (1961). An analysis of creativity. *The Phi Delta Kappan*, 42(7), 305–310.

Richerson, P. J., & Boyd, R. (2005). *Not by genes alone: How culture transformed human evolution*. Chicago: University of Chicago Press.

Ridley, M. (2003). *Nature via nurture*. New York: Harper Collins.

Ridout, R., & Habron, J. (2020). Three flute players' lived experiences of Dalcroze eurhythmics in preparing contemporary music for performance. *Frontiers in Education*, 5:18. https://doi.org/10.3389/feduc.2020.00018

Rizzolatti, G., Fadiga, L., Fogassi, L., & Gallese, V. (2002). From mirror neurons to imitation: Facts and speculations. In A. N. Meltzoff & W. Prinz (Eds.), *The imitative mind: Development, evolution, and brain bases* (pp. 247–266). Cambridge: Cambridge University Press.

Rizzolatti, G., & Sinigaglia, C. (2008). *Mirrors in the brain. How our minds share actions and emotions*. Oxford: Oxford University Press.

Roberts, T. (2015). Extending emotional consciousness. *Journal of Consciousness Studies*, 22(3–4), 108–128.

Robinson, D. (2020). *Hungry listening: Resonant theory for indigenous sound studies*. Minneapolis: University of Minnesota Press.

Robson, S. J., & Kuhlmeier, V. A. (2016). Infants' understanding of object-directed action: An interdisciplinary synthesis. *Frontiers in Psychology*, 7:111. https://doi.org/10.3389/fpsyg.2016.00111

Rochat, P. (1989). Object manipulation and exploration in 2- to 5-month-old infant. *Developmental Psychology*, 25(6), 871–884.

Rodriguez, C. X. (2004). Popular music in music education: Toward a new conception of musicality. In C. X. Rodriguez (Ed.), *Bridging the gap: Popular music and music education* (pp. 13–28). Carlos Reston, VA: MENC.

Roholt, T. C. (2014). *Groove: A phenomenology of rhythmic nuance*. New York: Bloomsbury.

Rosenberg, M. E. (2010). Jazz and emergence (part one) from calculus to Cage, and from Charlie Parker to Ornette Coleman: Complexity and the aesthetics and politics of emergent form in jazz. *Inflexions: A Journal of Research Creation*, 4. http://www.inflexions.org/n4_rosenberghtml.html

Rothenberg, D. (2005). *Why birds sing*. New York: Allen Lane.

Rothenberg, D. (2014). *Bug music: How insects gave us rhythm and noise*. New York: Picador.

Rothenberg, D., & Ulvaeus, M. (Eds.) (2009). *The book of music and nature: An anthology of sounds, words, thoughts*. Middletown, CT: Wesleyan University Press.

Rowlands, M. (1999). *The body in mind: Understanding cognitive processes*. Cambridge: Cambridge University Press.

Rowlands, M. (2010). *The new science of the mind: From extended mind to embodied phenomenology*. Cambridge, MA: MIT Press.

Ruff, H. A. (1984). Infants' manipulative exploration of objects: Effects of age and object characteristics. *Developmental Psychology, 20*, 9–20.

Ruff, H. A. (1986). Components of attention during infants' manipulative exploration. *Child Development, 57*, 105–114.

Ruff, H. A., McCarton, C., Kurtzberg, D., & Vaughan, H. G. (1984). Jr- Preterm infants' manipulative exploration of objects. *Child Development, 55*(4), 1166–1173.

Runco, M. (2014). *Creativity: Theories and themes: Research, development, and practice* (2nd ed.). Amsterdam: Academic Press.

Runeson, S., & Frykholm, G. (1983). Kinematic specification of dynamics as an informational basis for person-and-action perception: Expectation, gender recognition, and deceptive intention. *Journal of Experimental Psychology: General, 112*(4), 585–615.

Russell, J. A. (1994). Is there universal recognition of emotion from facial expression? A review of the cross-cultural studies. *Psychological Bulletin, 115*, 102–141.

Russell, J. A. (2003). Core affect and the psychological construction of emotion. *Psychological Review, 110*(1), 145–172.

Ryan, K., & Schiavio, A. (2019). Extended musicking, extended mind, extended agency. Notes on the third wave. *New Ideas in Psychology, 55*, 8–17. https://doi.org/10.1016/j.newideapsych.2019.03.001

Sacks, O. (2007). *Musicophilia: Tales of music and the brain*. New York: Knopf.

Salice, A., Høffding, S., & Gallagher, S. (2017). Putting plural self-awareness into practice: The phenomenology of expert musicianship. *Topoi, 2017*, 1–13.

Sander, D., Grandjean, D., & Scherer, K. R. (2005). A systems approach to appraisal mechanisms in emotion. *Neural Networks: The Official Journal of the International Neural Network Society, 18*(4), 317–352.

Savage, P., Loui, P., Tarr, B., Schachner, A., Glowacki, L., Mithen, S., & Fitch, W. (2020). Music as a coevolved system for social bonding. *Behavioral and Brain Sciences*, 1–36. https://doi:10.1017/S0140525X20000333

References

Sawyer, R. K. (2003). *Group creativity: Music, theater, collaboration*. Mahwah, NJ: Erlbaum.

Sawyer, R. K. (2006). Group creativity: Musical performance and collaboration. *Psychology of Music, 34*, 148–165.

Sawyer, R. K. (2007). Improvisation and teaching. *Critical Studies in Improvisation, 2*(2), 1–4. https://doi.org/10.21083/csieci.v3i2.380

Sawyer, R. K., & DeZutter, S. (2009). Distributed creativity: How collective creations emerge from collaboration. *Psychology of Aesthetics, Creativity, and the Arts, 3*(2), 81–92.

Schafer, R. M. (1994). *The soundscape: Our sonic environment and the tuning of the world*. Rochester, VT: Destiny Books.

Scheler, M. (1954). *The nature of sympathy*. (P. Heath, Trans.). London: Routledge and Kegan Paul.

Scherer, K. R. (2001). Appraisal considered as a process of multi-level sequential checking. In K. R. Scherer, A. Schorr, & T. Johnstone (Eds.), *Appraisal processes in emotion: Theory, methods, research* (pp. 92–120). New York: Oxford University Press.

Scherer, K. R. (2005). What are emotions? And how can they be measured? *Social Science Information, 44*(4), 695–729.

Scherer, K. R. (2007). Component models of emotion can inform the quest for emotional competence. In G. Matthews, M. Zeidner, & R. D. Roberts (Eds.), *The science of emotional intelligence: Knowns and unknowns* (pp. 101–126). New York: Oxford University Press.

Scherer, K. R., & Coutinho, E. (2013). How music creates emotion: A multifactorial process approach. In T. Cochrane, B. Fantini, & K. R. Scherer (Eds.), *The emotional power of music, multidisciplinary perspectives on musical arousal, expression, and social control* (pp. 121–145). Oxford: Oxford University Press.

Scherer, K. R., & Zentner, M. R. (2001). Emotional effects of music: Production rules. In P. N. Juslin & J. A. Sloboda (Eds.), *Music and emotion: Theory and research* (pp. 361–392). Oxford: Oxford University Press.

Schiavio, A., & Altenmüller, E. (2015). Exploring music-based rehabilitation for Parkinsonism through embodied cognitive science. *Frontiers in Neurology*, 6:217. https://doi.org/10.3389/fneur.2015.00217

Schiavio, A., & Benedek, M. (2020). Dimensions of musical creativity. *Frontiers in Neuroscience*, 14:578932. https://doi.org/10.3389/fnins.2020.578932

Schiavio, A., Biasutti, M., van der Schyff, D., & Parncutt, R. (2020). A matter of presence: A qualitative study on teaching individual and collective music classes. *Musicae Scientiae, 24*(3), 356–376.

Schiavio, A., & Cummins, F. (2015). An inter(en)active approach to musical agency and learning. In R. Timmers, N. Dibben, Z. Eitan, R. Granot, T. Metcalfe, A. Schiavio & V. Williamson (Eds.), *Proceedings of the International Conference on the Multimodal Experience of Music 2015*.

Schiavio, A., & De Jaegher, H. (2017). Participatory sense-making in joint musical practices. In M. Lesaffre, M. Leman & P. J. Maes (Eds.), *The Routledge companion to embodied music interaction* (pp. 31–39). New York: Routledge.

Schiavio, A., & Høffding, S. (2015). Playing together without communicating? A pre-reflective and enactive account of joint musical performance. *Musicae Scientiae, 19*(4), 366–388.

Schiavio, A., & Menin, D. (2011). *Mirroring teleomusical acts* [Poster presentation]. IV Neuromusic Conference, University of Edinburgh, Edinburgh, Scotland.

Schiavio, A., & Timmers, R. (2016). Motor and audiovisual learning consolidate auditory memory of tonally ambiguous melodies. *Music Perception, 34*(1), 21–32.

Schiavio, A., & van der Schyff, D. (2016). Beyond musical qualia. Reflecting on the concept of experience. *Psychomusicology: Music, Mind, and Brain, 26*(4), 366–378.

Schiavio, A., & van der Schyff, D. (2018). 4E music pedagogy and the principles of self-organization. *Behavioral Sciences, 8*(8), 72. https://doi.org/10.3390/bs8080072

Schiavio, A., van der Schyff, D., Céspedes-Guevara, J., & Reybrouck, M. (2017). Enacting musical emotions. sense-making, dynamic systems, and the embodied mind. *Phenomenology and the Cognitive Sciences*, 1–25.

Schiavio, A., van der Schyff, D., Gande, A., & Kruse-Weber, S. (2019). Negotiating individuality and collectivity in community music. A qualitative case study. *Psychology of Music, 47*(5), 706–721.

Schiavio, A., van der Schyff, D., Kruse-Weber, S., & Timmers, R. (2017). When the sound becomes the goal. 4E cognition and teleomusicality in early infancy. *Frontiers in Psychology, 8*:1585. https://doi.org/10.3389/fpsyg.2017.01585

Schlaug, G. (2001). The brain of musicians. A model for functional and structural adaptation. *Annals of the NY Academy of Sciences, 930*, 81–99.

Schneck, D. J., & Berger, D. S. (2006). *The music effect: Music physiology and clinical applications*. London: Jessica Kingsley Publishers.

Schöner, G. (2008). Dynamical systems approaches to cognition. In R. Sun (Ed.), *The Cambridge handbook of computational cognitive modeling* (pp. 101–126). Cambridge: Cambridge University Press.

Schöner, G., & Kelso, J. A. S. (1988a). Dynamic pattern generation in behavioral and neural systems. *Science, 239*, 1513–1520.

Schöner, G., & Kelso, J. A. S. (1988b). A synergetic theory of environmentally-specified and learned patterns of movement coordination. II. Component oscillator dynamics. *Biological Cybernetics, 58*, 81–89.

Schubert, E. (2017). Musical identity and individual differences in empathy. In R. Macdonald, D. J. Hargreaves, & D. Miell (Eds.), *Handbook of musical identities* (pp. 322–344). Oxford: Oxford University Press.

Schuldberg, D. (1999). Chaos theory and creativity. In M. Runco & S. Pritzker (Eds.), *The encyclopedia of creativity* (Vol. 1, pp. 259–272). London: Academic Press.

Schwadron, A. A. (1966). Information theory and esthetic perception. *Music Educators Journal, 53*(1), 116–117.

Searle, J. (1967). *Speech acts*. London: Cambridge University Press.

Searle, J. (1990). Is the brain's mind a computer program? *Scientific American, 262*(1), 26–31.

Seashore, C. (1938). *Psychology of music*. New York: McGraw-Hill.

Seddon, F. A., & Biasutti, M. (2009). A comparison of modes of communication between members of a string quartet and a jazz sextet. *Psychology of Music, 37*(4), 395–415.

Service, V. (1984). Maternal styles and communicative development. In A. Lock & E. Fisher (Eds.), *Language development* (pp. 132–140). London: Croom Elm.

Sessions, R. (1941). The composer and his message. In A. Centeno (Ed.), *The intent of the artist* (pp. 101–134). Princeton, NJ: Princeton University Press.

Shapiro, L. A. (2011). *Embodied cognition*. New York: Routledge.

Sheets-Johnstone, M. (1999). *The primacy of movement*. Amsterdam: John Benjamins.

Sheets-Johnstone, M. (2010). Thinking in movement. Further analyses and validations. In J. Stewart, O. Gapenne, & E. A. Di Paolo (Eds.), *Enaction: Toward a new paradigm for cognitive science* (pp. 165–182). Cambridge, MA: MIT Press.

Sheets-Johnstone, M. (2012). Fundamental and inherently interrelated aspects of animation. In A. Foolen, U. Lüdtke, T. Racine, & J. Zlatev (Eds.), *Moving ourselves, moving others: Motion and emotion in intersubjectivity, consciousness and language* (pp. 27–56). Amsterdam: J. Benjamins.

Shevock, D. (2018). *Eco-literate music pedagogy*. New York: Routledge.

Shusterman, R. (2008). *Body consciousness: A philosophy of mindfulness and somaesthetics*. Cambridge: Cambridge University Press.

Silverman, M. (2012). Virtue ethics, care ethics, and "The good life of teaching." *Action, Criticism, and Theory for Music Education, 11*(2), 96–122.

Silverman, M. (2014). Empathy in music and music education. In W. F. Thompson (Ed.), *Music in the social and behavioral sciences: An encyclopedia*. Los Angeles, CA: SAGE Publications.

Silverman, M. (2020). Sense-making, meaningfulness, and instrumental music education. *Frontiers in Psychology*, 11:837. https://doi.org/10.3389/fpsyg.2020.00837

Singer, T., Seymour, B., O'Doherty, J., Kaube, H., Dolan, R. J., & Frith, C. D. (2004). Empathy for pain involves the affective but not sensory components of pain. *Science*, *303*(5661), 1157–1162.

Skånland, M. S. (2013). Everyday music listening and affect regulation: The role of MP3 players. *International Journal of Qualitative Studies on Health and Well-Being*, 8, 20595. https://doi.org/10.3402/qhw.v8i0.20595

Skinner, M. M., Stephens, N. B., Tsegai, Z. J., Foote, A. C., Nguyen, N. H., Gross, T., Pahr, D. H., Hublin, J. J., & Kivell, T. L. (2015). Human-like hand use in Australopithecus africanus. *Science*, *347*, 395–399.

Slaby, J. (2014). Emotions and the extended mind. In C. von Scheve & M. Salmela (Eds.), *Collective emotions: Perspectives from psychology, and sociology* (pp. 32–46). Oxford: Oxford University Press.

Slaby, J., & Gallagher, S. (2015). Critical neuroscience and socially extended minds. *Theory, Culture and Society*, *32*(1), 33–59.

Sloboda, J. A. (1985). *The musical mind: The cognitive psychology of music*. Oxford: The Clarendon Press.

Sloboda, J. A. (Ed.) (1988). *Generative processes in music*. Oxford: Oxford University Press.

Sloboda, J. A. (2000). Musical performance and emotion: Issues and developments. In S. W. Yi (Ed.), *Music, mind and science* (pp. 220–238). Seoul: Western Music Research Institute.

Small, C. (1998). *Musicking: The meaning of performing and listening*. Middletown, CT: Wesleyan University Press.

Smith, A. (1790 [1759]). *Theory of moral sentiments* (6th ed.). London: A. Millar.

Smith, C. (1998). A sense of the possible: Miles Davis and the semiotics of improvisation. In B. Nettl & M. Russell (Eds.), *In the course of performance: Studies in the world of musical improvisation* (pp. 261–89). Chicago: University of Chicago Press.

Smith, G. D., & Silverman, M. (Eds.) (2020). *Eudaimonia: Perspectives for music learning*. London: Routledge.

Smith, L., & Thelen, E. (2003). Development as a dynamic system. *Trends in Cognitive Science*, *7*(8), 343–344.

Smith, S. L., Gerhardt, K. J., Griffiths, S. K., Huang, X., & Abrams, R. M. (2003). Intelligibility of sentences recorded from the uterus of a pregnant ewe and from the fetal inner ear. *Audiology and Neuro Otology, 8,* 347–353.

Smolensky, P. (1988). On the proper treatment of connectionism. *Behavioral Brain Sciences, 11,* 1–74.

Smolensky, P. (1990). Tensor product variable binding and the representation of symbolic structures in connectionist systems. *Artificial Intelligence, 46,* 159–216.

Sober, E. (1984). *The nature of selection: Evolutionary theory in philosophical focus.* Cambridge, MA: MIT Press.

Sober, E. (Ed.) (1993). *Conceptual issues in evolutionary biology: An anthology.* Cambridge, MA: MIT Press.

Sokolowski, R. (2000). *Introduction to phenomenology.* Cambridge: Cambridge University Press.

Soley, G., & Hannon E. (2010). Infants prefer the musical meter of their own culture: A cross-cultural comparison. *Developmental Psychology, 46,* 286–292 10.1037/a0017555.

Solis, G., & Nettl, B. (2009). *Musical improvisation: Art, education, and society.* Urbana: University of Illinois Press.

Solomon, R. C. (1976). *The passions: Emotions and the meaning of life.* New York: Doubleday.

Solomon, R. C. (1977). Husserl's concept of the noema. In F. Elliston & P. McCormick (Eds.), *Husserl: Expositions and appraisals* (pp. 168–181). Notre Dame, IN: University of Notre Dame Press.

Song, Y. Y., Zeng, R. S., Xu, J. F., Li, J., Shen, X., & Yihdego, W. G. (2010). Interplant communication of tomato plants through underground common mycorrhizal networks. *PLoS ONE, 5*(10), e13324. https://doi.org/10.1371/journal.pone.0013324

Sparshott, F. (1982). *The theory of the arts.* Princeton, NJ: Princeton University Press.

Spencer, J. P., Clearfield, M., Corbetta, D., Ulrich, B., Buchanan, P., & Schöner, G. (2006). Moving toward a grand theory of development: In memory of Esther Thelen. *Child Development, 77,* 1521–1538.

Sperber, D. (1996). *Explaining culture.* Oxford: Blackwell.

Sperber, D., & Hirschfield, L. A. (2004). The cognitive foundations of cultural stability and diversity. *Trends in Cognitive Science, 8*(1), 40–47.

Standley, J. (1995). Music as a therapeutic intervention in medical and dental treatment: Research and applications. In T. Wigram, B. Saperston, & R. West (Eds.), *The art and science of music therapy: A handbook* (pp. 3–22). Amsterdam: Harwood Academic Publishers.

Stein, E. (1989[1917]). *On the problem of empathy* (3rd ed., W. Stein, Trans.). Washington: ICS Publications. Originally published as *Zum Problem der Einfühlung*, Halle: Buchdruckerei des Waisenhauses.

Steinbeis, N., & Koelsch, S. (2008). Shared neuronal resources between music and language indicate semantic processing of musical tension-resolution patterns. *Cerebral Cortex, 18*(1), 1169–1178.

Steinbeis, N., & Koelsch, S., & Sloboda, J. (2006). The role of harmonic expectancy violations in musical emotions: Evidence from subjective, psychological, and neural responses. *Journal of Cognitive Neuroscience, 18*(1), 1380–1393.

Stephen, D. G., & Dixon, J. A. (2009). The self-organization of insight: Entropy and power laws in problem solving. *Journal of Problem Solving, 2*, 72–101.

Stephen, D. G., Dixon, J. A., & Isenhower, R. W. (2009). Dynamics of representational change: Entropy, action, and cognition. *Journal of Experimental Psychology: Human Perception and Performance, 35*, 1811–1832.

Sterelny, K. (2012). *The evolved apprentice: How evolution made humans unique*. Cambridge, MA: MIT Press.

Sterelny, K. (2014). Constructing the cooperative niche. In G. Barker, E. Desjardins, & T. Pearce (Eds.), *Entangled life: History, philosophy and theory of the life sciences* (Vol. 4, pp. 261–279). New York: Springer.

Stern, D. (1985). *The interpersonal world of the infant: A view from psychoanalysis and developmental psychology*. Amsterdam: Vrije Universiteit Amsterdam.

Sternberg, R. J. (2005). Intelligence, competence, and expertise. In A. J. Elliot & C. S. Dweck (Eds.), *Handbook of competence and motivation*. New York: Guilford Press.

Stewart, J., Gapenne, O., & Di Paolo, E. A. (Eds.) (2010). *Enaction: Toward a new paradigm for cognitive science*. Cambridge, MA: MIT Press.

Still, A. & Costall, A. (Eds.) (1991). *Against cognitivism: Alternative foundations for cognitive psychology*. London: Harvester-Wheatsheaf.

Stoffregen, T. A. (2003). Affordances as properties of the animal–environment system. *Ecological Psychology, 15*(2), 115–134.

Stokes, M. (2010). *The republic of love: Cultural intimacy in Turkish popular music*. Chicago: University of Chicago Press.

Streek, J. (1980). Speech acts in interaction: A critique of Searle. *Discourse Processes, 3*(2), 133–153.

Strogatz, S. (1994). *Nonlinear dynamics and chaos: With applications to physics, biology, chemistry, and engineering*. New York: Perseus Books.

Strogatz, S. (2001). Exploring complex networks. *Nature, 410*(6825), 268–276.

References

Strunk, O. (1950). *Source readings in music history from classical antiquity through the romantic era.* New York: Norton.

Stuart, S. (2012) Enkinaesthesia: The essential sensuous background for co-agency. In Z. Radman (Ed.), *The background: Knowing without thinking: Mind, action, cognition and the phenomenon of the background* (pp. 167–86). New York: Palgrave Macmillan.

Sudnow, D. (1978). *Ways of the hand: The organization of improvised conduct.* Cambridge, MA: Harvard University Press.

Sudnow, D. (2001). *Ways of the hand: A rewritten account.* Cambridge, MA: MIT Press.

Sur, M., & Leamey, C. A. (2001). Development and plasticity of cortical areas and networks. *Nature Reviews Neuroscience, 2*, 251–262.

Sutton, J. (2006). Distributed cognition: Domains and dimensions. *Pragmatics and Cognition, 14*(2), 235–247.

Sutton, J. (2010). Exograms and interdisciplinarity: History, the extended mind, and the civilizing process. In R. Menary (Ed.), *The extended mind* (pp. 189–225). Cambridge, MA: MIT Press.

Sutton, J., Harris, C. B., Keil, P. G., & Barnier, A. J. (2010). The psychology of memory, extended cognition, and socially distributed remembering. *Phenomenology and the Cognitive Sciences, 9*(4), 521–560.

Sutton, J., McIlwain, D., Chrisensen, W., & Geeves, A. (2011). Applying intelligence to the reflexes: Embodied skills and habits between Dreyfus and Descartes. *Journal of the British Society for Phenomenology, 42*(1), 78–103.

Swanwick, K. (1979). *A basis for music education.* London: Routledge.

Tamplin, J., & Baker, F. A. (2017). Therapeutic singing protocols for addressing acquired and degenerative speech disorders in adults. *Music Therapy Perspectives, 35*(2), 113–123.

Taruskin, R. (1995). *Text and act: Essays on music and performance.* New York: Oxford University Press.

Tettamanti, M., & Weniger, D. (2006). Broca's area: A supramodal hierarchical processor? *Cortex, 42*, 491–494.

Thelen, E. (1989). Self-organization in developmental processes: Can systems approaches work? In M. Gunnar & E. Thelen (Eds.), *Systems and development: The Minnesota symposia on child psychology* (Vol. 22, pp. 17–171). Hillsdale, NJ: Erlbaum.

Thelen, E. (1994). Three-month-old infants can learn task-specific patterns of interlimb coordination. *Psychological Science, 5*, 280–285.

Thelen, E. (2000). Grounded in the world: Developmental origins of the embodied mind. *Infancy, 1*(1), 3–28.

Thelen, E., Schoner, G., Scheier, C., & Smith, L. B. (2001). The dynamics of embodiment: A field theory of infant preservative reaching. *Behavioral and Brain Sciences, 24*, 1–86.

Thelen, E., & Smith, L. B. (1994). *A dynamic systems approach to the development of cognition and action*. Cambridge, MA: MIT Press.

Thompson, E. (2004). Life and mind: From autopoiesis to neurophenomenology. A tribute to Francisco Varela. *Phenomenology and the Cognitive Sciences, 3*, 381–398.

Thompson, E. (2007). *Mind in life: Biology, phenomenology, and the sciences of mind*. Cambridge, MA: Harvard University Press.

Thompson, E., & Stapleton, M. (2009). Making sense of sense-making: Reflections on enactive and extended mind theories. *Topoi, 28*, 23–30.

Thompson, W. I. (1998). *Coming into being: Artifacts and texts in the evolution of consciousness*. New York: Palgrave Macmillan Trade.

Titchener, E. B. (1909). *Lectures on the elementary psychology of feeling and attention*. New York: Macmillan.

Toiviainen, P. (2000). Symbolic AI versus connectionism in music research. In E. R. Miranda (Ed.), *Readings in music and artificial intelligence* (pp. 47–68). London: Routledge.

Tolbert, E. (2001). Music and meaning: An evolutionary story. *Psychology of Music, 29*(1), 84–94.

Tomaino, C. M. (2009). Clinical applications of music therapy in neurologic rehabilitation. In R. Haas & V. Brandes (Eds.), *Music that works: Contributions of biology, neurophysiology, psychology, sociology, medicine and musicology* (pp. 211–220). Vienna: Springer.

Tomasello, M. (1999). *The cultural origins of human cognition*. Cambridge, MA: Harvard University Press.

Tomasello, M. (2014). *A natural history of human thinking*. Cambridge, MA: Harvard University Press,

Tomasello, M., Carpenter, M., Call, J., Behne, T., & Moll, H. (2005). Understanding and sharing intentions: The origins of cultural cognition. *Behavioral and Brain Sciences, 28*(5), 675–691.

Tomkins, S. S. (1962). *Affect, imagery, and consciousness* (Vol. 1). New York: Springer.

Tomkins, S. S. (1963). *Affect, imagery, and consciousness* (Vol. 2). New York: Springer.

Tomlinson, G. (2015). *A million years of music: The emergence of human modernity*. Brooklyn, NY: Zone Books.

Tooby, J., & Cosmides, L. (1989). Evolutionary psychology and the evolution of culture, part I. *Ethology and Sociobiology, 10*(1), 29–49.

Tooby, J., & Cosmides, L. (1992). The psychological foundation of culture. In J. H. Barkow, L. Cosmides, & J. Tooby (Eds.), *The adapted mind* (pp. 19–136). Oxford: Oxford University Press.

Torrance, S., & Froese, T. (2011). An inter-enactive approach to agency: Participatory sense-making, dynamics, and sociality. *Humana.Mente, 15*, 21–53.

Torrance, S., & Schumann, F. (2019). The spur of the moment: What jazz improvisation tells cognitive science. *AI & Society, 34*(2), 251–268.

Trainor L. J. (2015). The origins of music in auditory scene analysis and the roles of evolution and culture in musical creation. *Philosophical Transactions of the Royal Society of London. Series B, Biological Sciences, 370*(1664), 20140089. https://doi.org/10.1098/rstb.2014.0089

Trainor L. J., & Trehub, S. E. (1992). A comparison of infants' and adults' sensitivity to Western musical structure. *Journal of Experimental Psychology: Human Perception and Performance, 18*(2), 394–402.

Trehub, S. E. (2000). Human processing predispositions and musical universals. In N. L. Wallin, B. Merker, & S. Brown (Eds.), *The origins of music* (pp. 427–448). Cambridge, MA: MIT Press.

Trehub, S. E. (2003a). Absolute and relative pitch processing in tone learning tasks. *Developmental Science, 6*, 46–47.

Trehub, S. E. (2003b). In the beginning, there was music. *Bulletin of Psychology and the Arts, 4*, 42–44.

Trehub, S. E. (2003c). Musical predispositions in infancy: An update. In R. Zatorre & I. Peretz (Eds.), *The cognitive neuroscience of music* (pp. 3–20). Oxford: Oxford University Press.

Trehub, S. E. (2003d). The developmental origins of musicality. *Nature Neuroscience, 6*, 669–673.

Trehub, S. E. (2003e). Toward a developmental psychology of music. *Annals of the New York Academy of Sciences, 999*, 402–413.

Trehub, S. E. (2009). Music lessons from infants. In S. Hallam, I. Cross, & M. Thaut (Eds.), *Oxford handbook of music psychology* (pp. 229–234). Oxford: Oxford University Press.

Trehub, S. E., Bull, D., & Thorpe, L. A. (1984). Infants' perception of melodies: The role of melodic contour. *Child Development, 55*, 821–830.

Trehub, S. E., & Hannon, E. E. (2006). Infant music perception: Domain-general or domain-specific mechanisms? *Cognition, 100*, 73–99. https://doi.org/10.1016/j.cognition.2005.11.006

Trehub, S. E., & Hannon, E. E. (2009). Conventional rhythms enhance infants' and adults' perception of musical patterns. *Cortex: A Journal Devoted to the Study of the Nervous System and Behavior, 45*(1), 1100118.

Trehub, S. E., & Nakata, T. (2001–2002). Emotion music in infancy. *Musicae Scientiae, 6*, 37–61.

Trevarthen, C. (1979). Communication and cooperation in early infancy: A description of primary intersubjectivity. In M. Bullowa (Ed.), *Before speech*. Cambridge: Cambridge University Press.

Trevarthen, C. (1998). The concept and foundations of infant intersubjectivity. In S. Bråten (Ed.), *Intersubjective communication and emotion in early ontogeny* (pp. 15–46). Cambridge: Cambridge University Press.

Trevarthen, C. (1999). Musicality and the intrinsic motive pulse: Evidence from human psychobiology and infant communication. *Musicae Scientiae,* special issue, 157–213.

Trevarthen, C. (2002). Origins of musical identity: Evidence from infancy for musical social awareness. In R. A. R. MacDonald, D. J. Hargreaves, & D. Miell (Eds.), *Musical identities* (pp. 21–38). Oxford: Oxford University Press.

Trevarthen, C. (2017). Play with infants: The impulse for human story-telling. In T. Bruce, P. Hakkarainen, & M. Bredikyte (Eds.), *The Routledge international handbook of play in early childhood* (pp. 198–215). Abingdon: Taylor & Francis/Routledge.

Trevarthen, C. & Hubley, P. (1978). Secondary intersubjectivity: Confidence, confiding and acts of meaning in the first year. In A. Lock (ed.), *Action, gesture and symbol: The emergence of language* (pp. 183–229). London: Academic Press.

Trost, W., Ethofer, T., Zentner, M., & Vuilleumier, P. (2013). Mapping aesthetic musical emotions in the brain. *Cerebral Cortex, 22*, 2769–2783.

Tunçgenç, B., & Cohen, E. (2016). Movement synchrony forges social bonds across group divides. *Frontiers in Psychology,* 7:782. https://doi.org/10.3389/fpsyg.2016.00782

Turino, T. (2008). *Music as social life: The politics of participation*. Chicago: University of Chicago Press.

Turvey, M. (1992). Affordances and prospective control: An outline of the ontology. *Ecological Psychology, 4*(3), 173–187.

Tuuri, K., & Koskela, O. (2020). Understanding human-technology relations within technologization and appification of musicality. *Frontiers in Psychology,* 11:416. https://doi.org/10.3389/fpsyg.2020.00416

Tye, M. (1995). *Ten problems of consciousness*. Cambridge, MA: MIT Press.

Tye, M. (2000). Knowing what it is like: The ability hypothesis and the knowledge argument. In G. Preyer & F. Siebert (Eds.), *Reality and human supervenience*. Lanham, MD: Rowman and Littefield.

Tye, M. (2002). *Consciousness, color, and content*. Cambridge, MA: MIT Press.

Tye, M. (2014). Transparency, qualia realism, and representationalism. *Philosophical Studies, 170*, 39–57.

Urban, P. (2014). Toward an expansion of an enactive ethics with the help of care ethics. *Frontiers in Psychology*, 5:1354. https://doi.org/10.3389/fpsyg.2014.01354

Uttal, W. (2001). *The new phrenology*. Cambridge, MA: MIT Press.

Van den Tol, A. J. M., & Edwards, J. (2013). Exploring a rationale for choosing to listen to sad music when feeling sad. *Psychology of Music, 41*(4), 440–465.

van der Meer, A. L. H., van der Weel, F. R., & Lee, D. N. (1995). The functional significance of arm movements in neonates. *Science, 267*, 693–695.

van der Schyff, D. (2013). Emotion, embodied mind, and the therapeutic aspects of musical experience in everyday life. *Approaches: Music Therapy & Special Music Education, 5*(1), 20–58.

van der Schyff, D. (2015). Music as a manifestation of life: exploring enactivism and the 'eastern perspective' for music education. *Frontiers in Psychology*, 6:345. https://doi.org/10.3389/fpsyg.2015.00345

van der Schyff, D. (2016). From Necker cubes to polyrhythms: Fostering a phenomenological attitude in music education. *Phenomenology and Practice, 10*(1), 4–24.

van der Schyff, D. (2019). Improvisation, enaction and self-assessments. In D. Elliott, G. McPherson, & M. Silverman (Eds.), *The Oxford Handbook of philosophical and qualitative perspectives on assessment in music education* (pp. 319–346). New York: Oxford University Press.

van der Schyff, D., & Krueger, J. (2019). Musical empathy, from simulation to 4E interaction. In A. F. Corrêa (Ed.), *Music, speech, and mind* (pp. 73–108). Curitiba, Brazil: Brazilian Association of Cognition and Musical Arts.

van der Schyff, D., & Schiavio, A. (2017a). Evolutionary musicology meets embodied cognition: Biocultural coevolution and the enactive origins of human musicality. *Frontiers in Neuroscience*, 11:519. https://doi.org/10.3389/fnins.2017.00519

van der Schyff, D., & Schiavio, A. (2017b). The future of musical emotions. *Frontiers in Psychology* 8:988. https://doi.org/10.3389/fpsyg.2017.00988

van der Schyff, D., Schiavio, A., & Elliott, D. J. (2016). Critical ontology for an enactive music pedagogy. *Action, Criticism & Theory for Music Education, 15*(4), 81–121.

van der Schyff, D., Schiavio, A., Walton, A., Velardo, V., & Chemero, A. (2018). Musical creativity and the embodied mind: Exploring the possibilities of 4E cognition and dynamical systems theory. *Music & Science, 1.* https://doi.org/10.1177/2059204318792319

van Duijn, M., Keijzer, F., & Franken, D. (2006). Principles of minimal cognition: Casting cognition as sensorimotor coordination. *Adaptive Behavior, 14*(2), 157–170.

van Gelder, T. (1990). Compositionality: A connectionist variation on a classical theme. *Cognitive Science, 14,* 355–384.

van Gulick, R. (1982). Mental representation: A functionalist view. *Pacific Philosophical Quarterly, 63*(1), 3–20.

Van Orden, G. C., Pennington, B. F., & Stone, G. O. (2001). What do double dissociations prove? Modularity yields a degenerating research program. *Cognitive Science, 25*(1), 111–117.

Varela, F. (1979). *Principles of biological autonomy.* New York: Elsevier North Holland.

Varela, F. (1988). Structural coupling and the origin of meaning in a simple cellular automata. In E. Secarez, F. Celada, N.A. Mitchinson, & T. Tada (Eds.), *The semiotics of cellular communications in the immune system* (pp. 151–161). New York: Springer-Verlag.

Varela, F., Lachaux, J. P., Rodriguez, E., & Martinerie, J. (2001). The brainweb: Phase synchronization and large-scale integration. *Nature Reviews Neuroscience, 2*(4), 229–239.

Varela, F., Thompson, E., & Rosch, E. (1991). *The embodied mind: Cognitive science and human experience.* Cambridge, MA: MIT Press.

Velardo, V. (2016). *Towards a music systems theory: Theoretical and computational modelling of creative music agents* [Unpublished doctoral dissertation]. University of Huddersfield.

von Hofsten, C. (1982). Eye-hand coordination in the newborn. *Developmental Psychology, 18*(3), 450–461.

Vuoskoski, J. K., Clarke, E. F., & DeNora, T. (2017). Music listening evokes implicit affiliation. *Psychology of Music, 45*(4), 584–599.

Wallas, G. (1926). *Art of thought.* New York: Harcourt-Brace.

Wallin, N. L., Merker, B., & Brown, S. (Eds.) (2000). *The origins of music.* Cambridge, MA: MIT Press.

Wallmark, Z., Deblieck, C. & Iacoboni, M. (2018). Neurophysiological effects of trait empathy in music listening. *Frontiers in Behavioral Neuroscience, 12:*66. https://doi.org/10.3389/fnbeh.2018.00066

References

Walton, A., Richardson, M. J., & Chemero, A. (2014). Self-organization and semiosis in jazz improvisation. *International Journal of Signs and Semiotic Systems*, *3*(2), 12–25.

Walton, A., Richardson, M. J., Langland-Hassan, P., & Chemero, A. (2015). Improvisation and the self-organization of multiple musical bodies. *Frontiers in Psychology*, 6:313. https://doi.org/10.3389/fpsyg.2015.00313

Walton, A., Richardson, M. J., Langland-Hassan, P., Chemero, A., & Washburn, A. (2017). Creating time: Affording social collaboration in music improvisation. *Topics in Cognitive Science*, *10*(1), 95–119.

Ward, L. M. (2003). Synchronous neural oscillations and cognitive processes. *Trends in Cognitive Science*, *12*, 553–559.

Ward, T. B. (1995). What's old about new ideas. In S. M. Smith, T. B. Ward, & R. A. Finke (Eds.), *The creative cognition approach* (pp. 157–178). Cambridge, MA: MIT Press.

Weber, A. (2001). Cognition as expression: The autopoietic foundations of an aesthetic theory of nature. *Sign Systems Studies*, *29*(1), 153–168.

Weber A., & Varela, F. J. (2002). Life after Kant: Natural purposes and the autopoietic foundations of biological individuality. *Phenomenology and the Cognitive Sciences*, *1*(2), 97–125.

Wentink, C. (2017). *Exploring lived experiences of ensemble performers with dalcroze eurhythmics: An interpretative phenomenological analysis* [Unpublished doctoral dissertation]. North-West University.

Werner, K. (2020). Cognitive confinement, embodied sense-making, and the (de)colonization of knowledge. *Philosophical Papers*, *49*, 339–364.

Werner, L. A. (2002). Infant auditory capabilities. *Current Opinion in Otolaryngology & Head and Neck Surgery*, *10*, 398–402.

Wertheimer, M. (1938). Laws of organization in perceptual forms. In W. Ellis (Ed.), *A source book of Gestalt psychology* (pp. 71–88). London: Routledge and Kegan.

Wexler, P. (2000). *The mystical society: Revitalization in culture, theory, and education*. Boulder, CO: Westview.

Wheeler, M. (2005). *Reconstructing the cognitive world: The next step*. Cambridge, MA: MIT Press.

Wheeler, M. (2010). In defense of extended functionalism. In R. Menary (Ed.), *The extended mind* (pp. 245–270). Cambridge, MA: MIT Press.

Wicker, B., Keysers, C., Plailly, J., Royet, J. P., Gallese, V., & Rizzolatti, G. (2003). Both of us disgusted in my insula: The common neural basis of seeing and feeling disgust. *Neuron*, *40*, 655–664.

Wiggins, G. (2012). Computer models of (music) cognition. In P. Rebuschat, M. Rohrmeier, J. Hawkins & I. Cross (Eds.), *Language and music as cognitive systems* (pp. 169–188). Oxford: Oxford University Press.

Willis, P. E. (1978). *Profane culture*. London: Routledge.

Wilson, M., & Cook, P. F. (2016). Rhythmic entrainment: Why humans want to, fireflies can't help it, pet birds try, and sea lions have to be bribed. *Psychonomic Bulletin & Review*, 23, 1647–1659. https://doi.org/10.3758/s13423-016-1013-x

Wittgenstein, L. (1980). *Remarks on the philosophy of psychology* (Vol. 1). (G. H. von Wright & H. Nyman, Eds.; C. G. Luckhardt & M. A. E. Aue, Trans.). Oxford: Blackwell.

Woodruff-Smith, D. (2007). *Husserl*. New York: Routledge.

Woodward, A. L. (1998). Infants selectively encode the goal object of an actor's reach. *Cognition*, 69, 1–34.

Woodward, A. L., & Gerson, S. A. (2014). Mirroring and the development of action understanding. *Philosophical Transactions of the Royal Society of London B, Biological Sciences*, 369, 20130181.

Wrangham, R. (2009). *Catching fire: How cooking made us human*. London: Profile.

Wynn, M. (2004). Musical affects and the life of faith. *Faith and Philosophy: Journal of the Society of Christian Philosophers*, 21(1), 25–44.

Wynn, T. G. (1996). The evolution of tools and symbolic behavior. In A. Lock & C. R. Peters (Eds.), *Handbook of human of human symbolic evolution* (pp. 263–287). Oxford: Clarendon Press.

Wynn, T. G. (2002). Archaeology and cognitive evolution. *Behavioral Brain Sciences*, 25, 389–438.

Yu, L., & Tomonaga, M. (2015). Interactional synchrony in chimpanzees: Examination through a finger-tapping experiment. *Scientific Reports*, 5:10218. https://doi.org/10.1038/srep10218

Zabelina, D. L., Robinson, M. D., Council, J. R., & Bresin, K. (2012). Patterning and non-patterning in creative cognition: insights from performance in a random number generation task. *Psychology of Aesthetics, Creativity, and the Arts*, 6, 137–145.

Zahavi, D. (2011). Empathy and direct social perception: A phenomenological proposal. *Review of Philosophy and Psychology*, 2(3), 541–558.

Zahavi, D. (2014). *Self and other: Exploring subjectivity, empathy, and shame*. Corby: Oxford University Press.

References

Zahavi, D., & Michael, J. (2018). Beyond mirroring: 4E perspectives on empathy. In A. Newen, L. de Bruin, & S. Gallagher (Eds.), *The Oxford handbook of 4e cognition*. Oxford: Oxford University Press.

Zatorre, R., Chen, J., & Penhune, V. (2007). When the brain plays music: Auditory–motor interactions in music perception and production. *Nature Reviews Neuroscience*, 8, 547–558.

Zatorre, R. J., & Salimpoor, V. N. (2013). Music perception and pleasure. *Proceedings of the National Academy of Sciences Jun 2013*, 110 (Suppl 2), 10430–10437.

Zbikowski, L. M. (2005). *Conceptualizing music: Cognitive structure, theory, and analysis*. New York: Oxford University Press.

Zbikowski, L. M. (2017). *Foundations of musical grammar*. New York: Oxford University Press.

Zentner, M. (2012). A language for musical qualia. *Empirical Musicology Review*, 7, 80–83.

Zentner, M., Grandjean, D., & Scherer, K. R. (2008). Emotions evoked by the sound of music: Characterization, classification, and measurement. *Emotion*, 8(4), 494–521.

Zubrow, E., Cross, I., & Cowan, F. (2001). Musical behaviour and the archaeology of the mind. *Archaeologica Polona*, 39, 111–126.

Index

Absorption (musical), 74–75, 233
Acousmatic music, 51
Action loops, 142–143, 149
Adaptation. *See* Evolution
Adaptivity, 31–34, 40, 49, 105, 154, 158, 160, 172
Aesthetics, 9–11, 17, 94, 96, 98–101, 141, 206, 211, 223–224, 232. *See also* Emotion
   aesthetic view (music education), 190, 192, 194–196, 199
   expressionism, 92
   formalism, 10, 91–92, 191–194
Affectivity, 36, 89, 104, 197, 207, 224
Affect programs. *See* Emotion
Affordances, 26–28, 106–107, 125, 137, 140–141, 149–150, 221
   development and growth, 46–47, 104, 161–162, 176, 175–185
   enactment of, 27, 34–36, 83–84
   musical instruments, 47, 79
   perception of, 39
   shared, interpersonal, 46–48, 79, 118, 199
   sonic, 155, 164, 167, 181
Agency, 75, 78, 81, 195, 198, 206–207, 221
   enactive conception of, 34–37, 216
Aizawa, K., 44, 213, 215
Aleatoric composition, 181
Amazeen, E., 174
Ambrosini, E., 230

Amusia, 15, 134
Ancient Greece, 5–6
Anthropomorphism, 149, 151, 216
Anticipation, 58–59, 94, 221
   musical, 94, 101, 105, 151. *See also* Emotion
Aphasia, 15, 126
Apollo, 5–6
Appraisal. *See* Emotion
Archeology, 2, 4–5, 129, 132, 135, 152–153
   bone flutes, 4, 135, 145
   hand axes, 141
Arendt, H., 200, 233
Aristotle, 7–8, 190–197, 213–215, 233
Aristoxenus, 8, 196
Arousal, 95, 160, 223. *See also* Emotion
Attractors, 41, 143, 173, 174–177, 180–181, 232. *See also* Dynamical systems theory
Audience, 46, 107, 110, 121, 183, 204, 223
Auditory system, 56, 156, 230
Autonomic nervous system, 96
Autonomy, 21, 23, 40, 104, 162, 186, 234
   enactive conception of, 28, 31–33, 120–121, 182
   relational, 34–37, 48, 79, 83, 116, 121, 198–199, 206–208, 211
Autopoiesis, 21, 23, 28–35, 40, 84, 104, 126, 186, 198, 207–209, 215–216

Baboons, 148–149
Bach, J. S., 9, 92, 183
Bacteria, 31, 216
Bailey, D., 219
Baroque music, 8–9, 180
Barrett, L., 3, 30, 38, 96, 132, 148–149, 223
Basic emotion theory (BET). See Emotion
Basins of attraction, 41, 103, 105, 125, 143, 174, 176, 179–181, 232. See also Dynamical systems theory
Bateson, G., 234
Beauty, 10, 91, 98, 192
Bechtel, W., 12
Beer, R., 41
Beethoven, L. v., 180
Berliner, P., 219
Bernstein, L., 215
Besson, M., 229
Bidirectionality, 27, 34, 70, 84, 159, 200, 221
Biocultural approach, 23, 86, 132, 136–137, 140, 144, 146–148, 151–154, 189. See also Evolution: coevolution (biocultural)
Bistable images, 71, 73
Boden, M., 170, 232
Borgo, D., 20, 61, 178, 181
Bowman, W., 21, 192, 195, 235
Brain plasticity, 16, 85, 103, 126, 132, 146–148, 153, 229
BRECVEMA. See Emotion
Brooks, R., 26, 216
Buddhist psychology, 235
Bundle hypothesis, 64–66, 73
Byard, J., 63

Cadence, 12, 51, 60
Cage, J., 181, 231
Canonical neurons, 38–39, 160
Carman, T., 219
Cells, 9, 28–30, 34, 139, 216, 228

Celys lyre, 5
Ceremony, 1, 6, 107, 123
Céspedes-Guevara, J., 222–224
Chalmers, D., 47, 51, 147, 214
Cheesecake (music as), 133
Chemero, A., 19, 25–28, 42–43, 220, 232
Chemotaxis, 216
Cheney, D., 148–149
Childhood, 23, 155, 203, 224
Chomsky, N., 14, 214–215
Chords, 13, 15, 51, 59, 64–65, 164, 203
Churchland, P.M., 27, 54, 218
Circularity, 40, 70, 84, 104, 151, 172, 217
Clarke, E. F., 20–21, 64, 89, 120, 122, 172, 225
Classical music, 8–9, 157, 178–179, 183, 190–191, 194–195, 202
  common practice, 8
  Western canon, 11, 92, 187
Co-arising (organism-environment relationship), 27, 32, 36, 40, 48–49, 121, 151, 235
Cochlea, 156, 230
Coevolution (biocultural). See Evolution
Cognitive enclosure, 208
Cognitive modules. See Modular approach to mind
Collaboration, 76, 177, 182, 191, 204–205
Colombetti, G., 97, 101, 103–104, 215
Colonialism, 191, 187, 206, 208
Coltrane, J., 181
Combinatorial cognition, 144–147, 151
Commodification of music, 11, 180, 191, 206, 214
Communication, 34, 46, 104, 111, 118, 135, 137, 152, 158, 189, 234
  musical, 81, 88, 92, 110, 123, 203–204, 229
  nonverbal (bodily, emotional-empathic), 1, 5, 43, 61–63, 81, 84,

Index

107, 116, 144–145, 161–162, 200, 219–220, 230. *See also* Interaction; Empathy
plants, 213
verbal, 62
Communicative musicality, 135
Comparative studies with non-human animals, 18, 133, 136, 143, 152. *See also* Evolution
Component process model (CPM). *See* Emotion
Composition (musical), 2, 9, 15, 164–165, 170, 191, 205, 214–215, 231–232
Computational approach to mind, 8, 19, 26–28, 34, 64, 147–149, 214, 220, 227, 229. *See also* Modular approach to mind
consciousness, 21, 50–61
music cognition, 12–13, 15, 79–81, 131
Connectionism, 16, 20, 27–28, 147–148, 171, 215
Consciousness, 21–22, 83, 120, 157, 193, 210. *See also* Dualism; Phenomenology
access consciousness, 53
computational view, 21, 50–61
executive functions, 54
false consciousness, 267
hard problem, 51, 54, 57, 217
homunculus, 53, 57, 217
inner theatre, 57
intentionality, modes of experience, 69–70, 74–75
mind-mind problem, 54–55
musical consciousness as situated embodiment, 61–67, 153
sentience, 54, 92, 193
Constancy hypothesis, 64–66, 73
Constraint, 34, 41–42, 45, 103–104, 121, 126, 151, 176
in creativity, 173, 178–179, 181

in evolutionary processes, 136, 138–140, 142–143, 145, 149
Cooke, D., 92
Cooperation, 107, 123, 199, 234
Coordination, 107, 121, 124, 141, 226
bodily, 78, 82, 136, 152–153, 164, 172–175, 185
development in infants, 158–159
interpersonal (social, cultural), 42–43, 132, 136–137, 143, 147, 152–153, 199, 234, 232
Coplan, A., 109–111, 224
Creativity, 23, 165, 169–173, 177–187, 189, 191, 195, 211, 231–233
Cross, I., 11, 137, 195
Cross-modal experience, 38, 80, 83, 98–102, 113, 120, 157, 159, 221, 223, 230
metaphorical mind, 22, 54–55, 59–63, 71, 82–89, 102, 114, 122–124, 182, 196
musical learning, 182, 184–185, 189, 204
Cultural epicycles, 140–141, 143, 145, 151–152. *See also* Evolution: coevolution (biocultural)
Cummins, F., 145, 216
Cybernetics, 37

Dalcroze Eurythmics, 203–204
Damasio, A., 13, 83, 96, 104
Darwin, C., 131–133, 226–228, 230
Davidson, J., 16
Davis, R., 62
De Jaegher, H., 37, 216
Delalande, F., 164–165
Dennett, D., 21, 52–53, 56–60, 63–65, 80–81, 105, 147, 219, 223
DeNora, T., 17, 127–128, 194, 215
Descartes, R., 9, 13, 53
Deutsch, D., 13, 131
Developmental systems theory, 138, 139, 227

Dewey, J., 26, 86, 222, 233
Differential equations, 40, 174. *See also* Dynamical systems theory
Di Paolo, E., 32–34, 44, 47, 202, 216
Direct description, 59–61
Direct social perception. *See* Empathy
Distributed cognition, 43, 82, 233
Divje Babe flute (Neanderthal flute), 4
Dodecaphony, 170, 180
Donald, M., 149
Dreyfus, H., 52–53, 184, 234
Drumming, drummers, 1, 45, 61, 76–82, 111–112, 122, 162, 168, 178, 229.
  See also Rhythm
  great ape drumming, 136
  Ghanaian drumming, 22, 51, 71, 76–77, 79, 82, 120, 158, 172, 189
Dualism, 9, 13, 203, 214
Dynamical attending, 82, 143, 226.
  See also Dynamical systems theory
Dynamical systems theory (DST), 21, 28, 149, 152, 231
  affective science, 102–106
  biocultural coevolution, 142–143.
    See also Cultural epicycles
  development, learning, and creativity, 173–180
  key concepts, 40–42. *See also* Attractors; Repellors; Basins of attraction; Phase portrait; Phase transitions

4E approach, 3, 43–50, 82, 152–154, 177, 180, 186, 198–199, 201–205, 211–212
Ecological psychology, 20, 21, 25, 26, 28, 43, 138
Education (music), 23, 49, 121, 129, 154, 179, 187–191, 210, 213, 225, 234. *See also* Pedagogy
  curriculum, 193, 200, 202, 235
  music education as aesthetic education (MEAE), 192–194

praxial music education, 11, 190, 198–200, 202–203, 205–206
Eerola, T., 96, 105, 107, 223–225
Ekstasis, 234
Embeddedness, 3, 21–22, 43–49, 119–125, 152–158, 177, 180–182, 194, 198
Embodied music cognition, 16–19, 22, 25, 108, 113, 124, 225
Embodiment, 3, 27, 43, 46–47, 61, 121, 203, 215
Emergent properties, 34, 36, 41, 63, 137, 142, 146, 151, 233
emotions, 91, 101–105, 125
Emotion, 63, 83, 89–109, 133, 155, 173, 189, 193–195, 211, 218, 225, 233.
  See also Mood
  activation spreading, 87
  aesthetic emotions, 95, 99–101, 223
  affect programs, 22, 91, 95–96, 101, 223
  anticipation and expectation, 15, 59, 94, 101, 151, 222–223
  appraisal, 91, 93–95, 98–99, 101, 105
  associative processing, 94
  basic emotion theory (BET), 95–98, 223
  BRECVEMA, 98, 100–101, 223
  component process model (CPM), 98–101, 223
  constructionism, 223–224
  contagion, 95, 98–99, 109, 114, 223
  emotion and empathy, 109, 112–128, 143, 161
  epistemic access to affective states, 108
  gestures and utterances, 37, 62
  induction, 95, 97
  misattribution theory, 93
  music and regulation of, 1, 127
Empathy, 49, 62–63, 89, 99–100, 107–129, 159, 173, 187, 193, 197, 211, 224–225
  direct social perception, 116–119, 122, 124

Index

embodied simulation theory, 39, 110–119, 122, 124–125, 234, 225
empathic space, 119, 123, 124–125, 127, 142, 153, 199, 205. *See also* Scaffolding (musical)
theory-theory, 109, 110, 113
Enculturation, 94, 157
Enlightenment (aesthetics), 9, 121, 190, 196
Ensembles (musical), 46, 51, 79, 120, 177, 179, 181, 191, 199, 203
Entrainment, 41, 47, 119, 121, 128, 140, 122
  infants, 156
  rhythmic, 95, 98, 101, 122, 127, 138, 150, 178, 223, 226
  social, 138, 142–144, 147, 150, 152
Entropy, 17, 41, 103, 143, 175, 178, 180. *See also* Dynamical systems theory
Epistasis, 228
Ethics, 7, 197, 202, 235
  ethics of care, 199–200, 209
Eudaimonia, 197–198
Evolution, 26, 133, 135–151, 230
  adaptation, 14, 22, 104, 106, 136, 137–139, 146, 226–228
  adaptationism, 131–132, 138–139, 226
  coevolution (biocultural), 86, 140–141, 145–146, 150, 153–154, 228. *See also* Biocultural approach
  evolutionary musicology, 22, 129, 131–133, 137, 143
  evolution of emotions, 93, 95, 101
  exaptation, 133, 227–228
  optimization (survival of the fittest), 131, 138–139, 226–228
  phenotypes, 132–133, 139, 145, 151, 154, 189, 227–228
  selective pressures, 131, 137–138, 140, 145, 227–228
  viability, 139

Extended mind, 38–39, 44, 47–49, 82, 88, 104, 119–128, 152–153, 201, 212
Externalism, 47–48

Facial expression, 96, 103, 107, 115, 117, 121, 219, 225–226
Feedback, 17, 24, 120, 122, 137, 140, 153, 182, 185, 204
Fetus, 156, 230
Fibers (theory of), 9
Finger wagging, 173–176
Fitch, T., 143–144, 230
Flourishing, 7, 18, 126, 173, 187, 189, 198, 210–211
Flow, 184, 233
Fodor, J., 148, 214, 229
Folk psychology, 95, 105
Frequency (perception in infants), 156
Fuchs, T., 217
Functionalism, 53, 57

Gallagher, S., 39, 82–83, 89, 115, 118, 184, 213
Gallese, V., 38, 111–113, 217, 225
Generative theory of tonal music, 14
Genetics, 131, 134, 141, 146, 154, 158–159, 161, 227–229. *See also* Evolution
  epigenetics, 146, 228–229
  DNA, 139
  genome, 138–139, 144, 151–152
Gestalt, 52, 66–67, 73, 85, 218, 184, 222
Gesture-calls, 144
Gibson, E. J., 157
Gibson, J. J., 25–27, 34, 230
Globalization, 205–206
Goal-directed action, 42, 78, 80, 100, 122, 155, 175, 177, 181, 186
  infants, 158–168
  perception of, 38–39, 111–112, 113, 115–117, 217, 230. *See also* Mirror neurons
Grammar, 12, 14, 214, 219, 229
Guitar, 51, 64–65, 89, 181, 219

Haken-Kelso-Bunz model (HKB), 41, 173–174
Hanslick, E., 10, 91, 194. *See also* Aesthetics
Harmony, 12, 64–65, 94, 157, 165, 170, 178, 181, 203, 230
Heidegger, M., 206, 213, 233, 235
Helmholtz, H., 214
Hendrix, J., 89, 181, 215
Hermes, 5–6
Hindustani music, 62
Høffding, S., 225, 233
Homeostasis, 48, 81
Homeric Hymns, 5
Hormones, 117, 228, 230
Housefly, 215–216
Huron, D., 59, 94, 101, 222
Husserl, E., 44, 69, 70–71, 115, 220–221
Hutto, D., 115, 151, 217

Identity, 1, 17, 30, 32, 83, 110, 123, 189, 199, 210, 216
Ihde, D., 70, 72–75
Imagination, 39, 69, 87, 109–110, 186, 197
Imberty, M., 163
Imitation, 76, 109, 217
Improvisation, 154, 172, 197, 202–205, 208, 219, 233
  musical, 20, 42, 61–62, 81, 135, 164–165, 178, 180–181, 187–188, 211
  nature of mind, 43, 48–49
Indigenous cultures, 191, 208, 210
Industrial attitude, 206, 207
Ineffability, 56, 58, 60, 61, 64, 64–65
Information-processing. *See* Computational approach to mind; Modular approach to mind
Instability, 41, 81, 119, 143, 173–179, 181, 185–186. *See also* Dynamical systems theory
Institutions, 195, 208–209

Intelligence, 38, 53, 148, 213, 216, 229
Intentionality, 70, 74, 82, 87, 136–137, 157, 220–221, 69–70, 74–75
Interaction, 28, 45, 40–41, 67, 122, 146–150, 190, 199, 233–234
  between body, brain, environment, 23, 51, 153, 186
  creativity, 171–176, 178, 183
  in infant development, 157–163.
  See also Teleomusicality
  interactional asymmetry, 32–35, 120
  musical, 17, 20, 42, 76, 106, 151, 178, 203–204, 219, 225–226
  organism-environment, 19, 22, 30, 75, 83, 139
  robots, 38
  with social and material environment, 32, 35–37, 39, 44–49, 80, 102–105, 115–125, 129, 132, 141, 184–187.
  See also Extended mind; Interaction theory
  with caregivers, 37, 100, 107, 110, 116–117, 119, 127, 135, 155, 161–162, 164–169, 200
Interaction theory, 114–119. *See also* Empathy
Internalist conception of mind, 13, 16, 19, 21, 47–48, 63, 91
Interoception, 83
Intersubjectivity, 34, 76, 79, 115–117, 120, 199, 217, 221, 234

Jackendoff, R., 14, 54–55, 58, 151, 220
James, W., 26, 86, 222, 233
Javanese music, 97
Jazz, 61–62, 120, 180, 190, 200, 208, 219, 232, 235
Johnson, M., 26, 63, 86–88, 196
Joint action, 101, 117, 145, 154, 177, 211
Judgment, 12, 57–58, 69–71, 98
Julliard String Quartet, 121
Juslin, P., 93–97, 100, 222

Index

Kant, I., 10, 99, 101, 194, 233. *See also* Aesthetics
Karmiloff-Smith, A., 134
Kelso, S., 26, 173–174
Killin, A., 137–138, 141
Kincheloe, J., 207–209
Kinematics, 112, 160, 163–164, 166, 172
Kirchhoff, M., 47
Knapping, 145
Koelsch, S., 14–15, 100
Köhler, W., 66, 85
Krueger, J., 20, 122–125, 224

Lakoff, G., 26, 88
Langer, S., 192–194
Language, 13–16, 51–52, 58, 60, 63, 86–87, 133–138, 144–148, 150–151, 219–220
  music-as-language, 14, 87, 195
Large, E., 232
Ledoux, J., 83, 96
Leman, M., 87, 113, 221
Lerdahl, F., 14, 151
Lewontin, R., 32, 138–140
Life-mind continuity, 19, 28, 30, 33, 43, 44, 51, 71, 189, 210–211
Life philosophy, 207
Limbic system, 229
Linguistic access, 51, 54, 60, 67
Loneliness, 75
Lullabies, 135, 156

Malafouris, L., 153–154, 229
Material engagement theory, 153–154
Mathews, F., 210
Maturana, H., 28, 30
McClary, S., 194
Melody, 1, 8, 62, 66, 124, 156–157, 165, 168, 185, 229–230
Membrane (cellular), 29–30
Memory, 39, 48, 61, 83, 94, 97, 98, 185–186, 206, 223
Menary, R., 47

Mendel, G., 228
Mentalese, 12–13, 52, 148–149. *See also* Computational approach to mind
Merleau-Ponty, M., 26, 50, 73, 86
Metabolism, 29–30, 44, 83, 117, 125, 142, 183, 207, 216, 228
Metaphorical mind. *See* Cross-modal experience
Metaplasticity, 153
Meyer, L., 92, 192, 222
Mimesis, 80, 95, 107, 123, 138, 140–144, 147, 149–150, 152
Mindful awareness, 235
Mirror neurons, 17, 38–39, 84, 111–114, 120, 160, 215, 217
Mithen, S., 135
Modular approach to mind, 12, 14, 16, 132, 134, 146–147, 214, 227, 229. *See also* Computational approach to mind
Mood, 1, 63, 83, 95–96, 108, 122–123, 126, 193. *See also* Emotion
Morphology, 141, 149
Motivation, 84, 120, 160, 162, 164, 186, 197, 218, 228, 231
Mozart, W. A., 183
Multistability, 66–67, 71, 73, 76
Muses, 5–6
Musicing, 18, 198
Musicking, 18, 127, 136, 145, 189, 191, 200, 202, 210–211, 235
Music of the spheres, 6
Musicology, 9, 22, 60, 92, 131–133, 137, 143, 154, 194
  critical musicology, 11
  ethnomusicology, 18, 20, 62, 195, 135, 202
  evolutionary musicology, 22, 129, 131–133, 137, 143
  interdisciplinary musicology, 17–20, 129, 212
Music therapy, 2, 17, 23, 126–128, 205, 212

Musilanguage, 135. *See also* Evolution
Musique concrète, 51
Mutation, 226
Myin, E., 217

Narrative, 59, 61–62, 75, 102, 114, 118
Necker cube, 71–73, 75–77, 158, 220. *See also* Bistable images; Phenomenology; Multistability
Negotiation, 17, 118, 179, 199
Neoliberalism, 206
Neonates, 127–128
Nettl, B., 195
Neuroendocrine effects, 126, 134
Neuroscience, 38, 81, 86, 100, 103, 111, 113, 126, 134, 229. *See also* Brain plasticity; Canonical neurons; Mirror neurons
 encephalography, 88
 critical neuroscience, 89
 functional magnetic resonance imaging (fMRI), 88, 111
 magnetoencephalography (MEG), 88
 transcranial magnetic stimulation (TMS), 88
Newborns, 231
Niche, 20, 35–36, 47–49, 83–84, 137–140, 145, 150, 204, 215, 228
Noddings, N., 200
Noë, A., 56
Noise, 1, 89, 165, 225
Nonconceptual content, 59
Normativity, 30, 34, 208, 225
Notation, 8, 76–78, 80–81, 203
Nuances (musical), 21, 59–61, 112, 163, 167

Offloading, 121–122, 177–178. *See also* Extended mind
Ontogenesis, 36, 134, 137, 150, 153, 165–166, 196, 211, 228

Ontology, 7, 10–11, 28, 30, 32, 53, 153–154, 213, 220, 234
 critical ontology, 205, 207–209
Operational closure, 30, 83, 207
Orchestra, 46
Orpheus, 6
Oscillation, 82, 142, 147, 149, 174, 226. *See also* Dynamical systems theory
Oyama, S., 139, 142–143

Paleolithic, 4, 141–143, 145, 150, 152–153, 228. *See also* Evolution
Panksepp, J., 94, 146, 229
Parity principle, 47
Parncutt, R., 100, 230
Patel, A., 126–127, 134, 137, 228
Pedagogy, 11, 187, 190, 198, 200, 202–204, 206–207, 210, 231. *See also* Education (music)
Peirce, C. S., 137, 222
Pélog scale, 97
Perceptually guided action, 28, 31–32, 49, 169, 216
Peretz, I., 15, 132, 134
Performance (musical), 16, 18, 46, 61–62, 88–89, 106, 110, 121–122, 177–179, 184, 190–191
Peripersonal space, 39, 160
Personality, 225, 228
Personal listening device, 122
Persona theory, 93, 107–108, 114. *See also* Emotion; Empathy
Personhood, 191, 198
Perturbation, 29–30, 41–42, 46, 119, 142, 172–173, 175, 177
Phase portrait, 41, 103, 173–174, 176, 179. *See also* Dynamical systems theory
Phase transitions, 41, 143, 174. *See also* Dynamical systems theory
Phenomenal information properties (pips), 58, 65, 105, 223

Phenomenology, 22, 26, 45, 49–50, 91, 115, 126, 201, 204. *See also* Intentionality
empathy, 115
epoché, 69
metaphorical mind, cross-modal experience, 79–89
multistable images, 71–73. *See also* Necker cube
musical learning, 75–81, 176, 182, 201, 204
natural attitude, 69, 157
noema and *noesis*, 69–70, 83–85, 220
passive synthesis, 83–84, 86, 157, 182, 234
sedimentation, 69–72, 76, 79, 157–158, 172, 208, 221
structure of experience, 70–71
Phronēsis, 197–199, 201, 233–234. *See also* Praxis
Phusis, 213, 233–234
Phylogenetic, 133, 149, 196
Physiology, 85–86, 95–98, 106, 119, 223, 125, 227, 230
Piaget, J., 165
Piano, 51, 62, 164, 167, 180, 183–185, 232
Pinker, S., 14, 113, 133
Plato, 6–9, 13, 196, 213
Poiesis, 197–198, 234. *See also* Praxis
Polyphony, 8
Polytonality, 203
Praxis, 189–191, 193, 195, 197–201, 203–205, 207–211, 233–234
Precarity, 32, 35, 205, 216
Prediction, 44, 58, 94, 101, 105, 151, 230
Predictive processing, 17
Prehuman ancestors, 4, 18, 23, 129, 132–135, 141, 144, 151–155, 227, 230. *See also* Evolution
Prelinguistic, 5, 84, 86, 100, 107
Primary intersubjectivity, 116–117

Primates, 133, 144
Primordial affectivity, 36, 104
Proposition (cognition), 52, 57–60, 63, 80–81, 87, 94, 149–150, 220, 229–230
Proprioception, 79, 83, 117, 121
Prostheses, 45, 154
Protolinguistic communication, 145
Protomusicality, 129, 135, 145, 149, 155–156, 161–162, 164, 166–167, 229–230
Psyche, 6–7
Psychoacoustics, 214
Psychophysical experience, 75, 102, 115–116, 119, 158
Pulse, 47, 66, 76–78, 82, 95. *See also* Rhythm; Entrainment
Pythagoras, 6, 8, 196

Qualia, 55–58, 80, 218–220. *See also* Consciousness
Qualitative research, 88, 187, 204, 234
Quantitative research, 88–89

Radical embodied cognitive science, 19, 25, 28
Radical enactivism, 151, 215, 217
Raffman, D., 21, 54, 59–61
Ramachandran, V. S., 22, 84–86
Ratcliffe, M., 118
Ravignani, A., 136, 143–145
Reddy, V., 116
Regelski, T., 194, 197, 200–201
Reification, 191, 208, 214
Reimer, B., 192–193
Relationality, 33, 35, 48, 83, 117, 199
Renaissance, 9
Repellors, 41, 173. *See also* Dynamical systems theory
Representation. *See* Computational approach to mind
Resource-sharing framework (music and language), 15

Respiration, 29, 124, 156
Reybrouck, M., 20, 226
Rhythm, 94–95, 101, 134, 181, 221, 228–229. *See also* Entrainment; Drumming
  Adowa rhythms, 77
  evolution of music, 141, 143–145
  hemiola, 77–78
  infants, 156–157, 161, 167–168
  polyrhythm, 66–67, 71, 76–82, 120, 158, 172, 174, 182, 189
Ritual, 1, 107, 123, 128, 135, 145, 194
Rizzolatti, G., 111
Robotics, 26, 38, 53, 216
Roholt, T., 60, 65–66
Romantic music, 8, 180
Rosch, E., 21, 28, 49, 55, 131, 172, 215–216, 220, 228

Salience, 36, 104, 148–149, 159
Satisfaction, 15, 59, 101, 222
Sawyer, R. K., 191, 206, 233
Saxophone, 181
Scaffolding (musical), 22, 119–128, 153, 178, 199, 212, 226. *See also* Empathy: empathic space
Schenkerian analysis, 14, 214
Scherer, K., 97–101, 223
Schoenberg, A., 170, 180, 232
Schön, D., 229
Schubert, F., 128
Schumann, C., 56
Schumann, F., 211
Seashore, C., 214
Secondary intersubjectivity, 116–117
Second-wave approach (extended mind), 47, 211
Selfhood, 121, 198, 209
Self-organization, 28–30, 34, 40–42, 78–82, 121–122, 126, 210, 233–234
  creativity and learning, 173–175, 181, 183, 186–189
  emotion, 101–102, 104–105
  evolution, 140–143, 150, 153
  musical development, 156, 158–159, 161, 164, 169, 203
  music education, 205–207
  neural, 118, 132
  praxis, 197, 199
Semiotics, 137
Sense-making, 19–23, 30–40, 67, 81–88, 110, 149, 154–169, 173, 186–191, 199, 208–211
  participatory, 37, 42–43, 48, 63, 79–82, 101–106, 119–126, 152, 196, 204–206
Sensorimotor activity, 33, 34, 94
Sensorimotor autonomy, 31
Sensorimotor coupling, 83–84
Sensorimotor development, 168
Sensorimotor enactivism, 215
Sensorimotor experience, 87
Sensorimotor interactions, 35, 106, 116
Sensorimotor loops, 46
Sensorimotor neurons, 38–39, 112
Sensorimotor understanding, 113, 165
Serialism, 180, 232
Sessions, J., 222
Seyfarth, R., 148–149
Significance, 28–29, 33–35, 67, 186, 189, 194, 216
Silverman, M., 20–21, 197–199, 205, 213
Situatedness, 75, 106, 195
Skånland, M., 108
Skilled coping, 74, 184, 233
Sloboda, J., 97
Small, C., 18
Smith, L., 160–161
Social cognition, 22, 34, 37, 100, 110, 111, 113–117, 119, 233. *See also* Empathy; Interaction
Socialization, 42, 107, 126, 135, 149, 205, 224
Socrates, 213

Index

Solitary experience of music, 34, 107–108, 114, 122, 155, 162, 166, 225
Sonata form, 9
Songwriting, 127
Soothing infants, 136
Speech, 15, 59, 86, 118, 126–127, 134–136, 140, 219, 224
  prosodic speech, 135, 161
Sperber, D., 134
Stability, 17, 40–41, 81, 103, 142–143, 158, 161, 173–178. *See also* Instability; Entropy; Dynamical systems theory
  infant behavior, 119, 128
Stapleton, M., 31–32, 36, 48, 216
Star Spangled Banner (Hendrix), 89, 215
Stein, E., 115
Stern, D., 86
Stewart, J., 29
Stochastic music, 181
String quartet, 120–121, 181, 233
Structural coupling, 34, 36, 47, 83, 103, 105, 125
Subjectivity, 70, 88, 120, 135, 195
Sudnow, D., 185, 232
Sutton, J., 47
Synchronization, 41, 95, 101, 159, 123–124, 142, 167, 226, 232. *See also* Entrainment
Synergistics, 35, 117, 120–122, 142, 158–159, 169, 176, 184, 210
Synesthesia, 84, 220
Syntax (cognition), 12, 14, 52–53, 146, 147–149, 215, 229

Tactile experience, 17, 46, 55, 76, 78–79, 87, 121, 152
Taruskin, R., 195
Tchaikovsky, P. I., 231
Techné, 7, 197–198. *See also* Praxis
Technology, 45–46, 134, 146, 190, 205, 211, 235
Teleodynamic systems, 42

Teleomusicality, 155–157, 159, 161, 163–167, 182, 230
  teleomusical acts, 161–168
Texture, 59, 62, 86–87, 102, 153, 220. *See also* Cross-modal experience
Thelen, E., 159–161, 176
Theoria, 197–198. *See also* Praxis
Theory-theory. *See* Empathy
Thermodynamics, 30, 48, 83, 175
Third-wave approach (extended mind), 47, 211–212
Thompson, E., 21, 28, 30–33, 36, 48–49, 55, 120, 131, 172, 182, 215–220, 227
Thompson, W. F., 223
Timbre, 56, 59, 64–65, 144–145, 165
Timmers, R., 185
Titchener, E., 26, 224
Tomaino, C., 126–127
Tomasello, M., 107, 136, 145, 225
Tomkins, S., 95, 97
Tomlinson, G., 131–132, 137–145, 147–151, 153, 227–228
Tonal music, 14–15, 51, 157, 193, 214
Toolmaking, 141, 143–144
Torrance, S., 208, 211
Trainor, L., 157
Trehub, S., 157, 230
Trevarthen, C., 116–117, 135, 234
Turino, T., 195

Umwelt, 33, 36
Utterance, 12, 37, 93, 107, 145, 155, 230

Valence, 33, 83, 87, 94, 125, 186, 216, 224
Varela, F., 21, 28, 30, 49, 55, 131, 172, 215–216, 220, 228
Viability, 30–31, 33, 36–37, 44, 126
  in evolution, 139
Vibration, 6, 9, 65
Vitality affect contours, 86–87
Vivaldi, A., 180
Vocalists, 62

Vocalization, 136, 156, 225, 230
Vocal learning, 62, 134, 144, 230
Vuoskoski, J., 226

Wallas, G., 170–171
Walton, A., 178–179
Williams, J., 170, 231
Wittgenstein, L., 225
Worldview, 179–181, 186–187, 206, 232. *See also* Creativity
Worship, 1, 18
Wundt, W., 26

Xenakis, I., 181
Xylophone, 172

Zahavi, D., 69, 115, 118
Zbikowski, L., 222
Zeus, 5